Semiempirical Methods of
Electronic Structure Calculation

Part B: Applications

MODERN THEORETICAL CHEMISTRY

Editors: **William H. Miller**, *University of California, Berkeley*
Henry F. Schaefer III, *University of California, Berkeley*
Bruce J. Berne, *Columbia University, New York*
Gerald A. Segal, *University of Southern California, Los Angeles*

Semiempirical Methods of

Electronic Structure Calculation

Part B: Applications

Edited by
Gerald A. Segal
University of Southern California, Los Angeles

PLENUM PRESS · NEW YORK AND LONDON

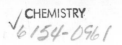
Library of Congress Cataloging in Publication Data

Main entry under title:

Semiempirical methods of electronic structure calculation.

(Modern theoretical chemistry; v. 7-8)
Includes bibliographical references and indexes.
CONTENTS: pt. A. Techniques. pt. B. Applications.
1. Molecular orbitals. 2. Quantum chemistry. I. Segal, Gerald A. II. Series.
QD461.S396   541'.28   76-48060
ISBN 0-306-33508-5 (v. 8)

© 1977 Plenum Press, New York
A Division of Plenum Publishing Corporation
227 West 17th Street, New York, N.Y. 10011

Printed in the United States of America

Contributors

C. J. Ballhausen, Department of Physical Chemistry, H. C. Ørsted Institutet, Københavns Universitet, Copenhagen, Denmark

David L. Beveridge, Chemistry Department, Hunter College of the City University of New York, New York City, New York

R. L. Ellis, Department of Chemistry, University of Illinois, Urbana, Illinois

Marie C. Flanigan, Department of Chemistry, State University of New York at Buffalo, Buffalo, New York

H. H. Jaffé, Department of Chemistry, University of Cincinnati, Cincinnati, Ohio

Andrew Komornicki, Department of Chemistry, State University of New York at Buffalo, Buffalo, New York

James W. McIver, Jr., Department of Chemistry, State University of New York at Buffalo, Buffalo, New York

Richard P. Messmer, General Electric Corporate Research and Development, Schenectady, New York

Josef Michl, Department of Chemistry, University of Utah, Salt Lake City, Utah

Donald G. Truhlar, Department of Chemistry and Chemical Physics Program, University of Minnesota, Minneapolis, Minnesota

Preface

If one reflects upon the range of chemical problems accessible to the current quantum theoretical methods for calculations on the electronic structure of molecules, one is immediately struck by the rather narrow limits imposed by economic and numerical feasibility.

Most of the systems with which experimental photochemists actually work are beyond the grasp of *ab initio* methods due to the presence of a few reasonably large aromatic ring systems. Potential energy surfaces for all but the smallest molecules are extremely expensive to produce, even over a restricted group of the possible degrees of freedom, and molecules containing the higher elements of the periodic table remain virtually untouched due to the large numbers of electrons involved. Almost the entire class of molecules of real biological interest is simply out of the question. In general, the theoretician is reduced to model systems of variable appositeness in most of these fields.

The fundamental problem, from a basic computational point of view, is that large molecules require large numbers of basis functions, whether Slater-type orbitals or Gaussian functions suitably contracted, to provide even a modestly accurate description of the molecular electronic environment. This leads to the necessity of dealing with very large matrices and numbers of integrals within the Hartree–Fock approximation and quickly becomes both numerically difficult and uneconomic. Ignoring the expense of integral computation, the largest matrices that can be effectively and accurately diagonalized on a digital computer are of the order of a few hundred if all roots are desired. If thermochemistry or electronic excitation energies are desired, some treatment of the correlation problem is required. In even the most rudimentary of treatments this increases as at least the square of the number of basis functions, so that large numbers of orbitals rapidly make these calculations impossible as well.

Volumes 7 and 8 of this series deal with semiempirical methods for quantum chemical calculations. Historically, the most popular semiempirical methods have been: (1) Hückel theory, which is fundamentally a tool for exploring the implications of the connectedness or topology of a molecule in regard to the form of the equations to be solved as discussed by Trinajstić in Volume 7, and (2) the manifold of zero-differential-overlap techniques. The rationale usually offered for the use of semiempirical techniques during the period when digital computers have commonly been available is that by avoiding the computation of large numbers of integrals, larger molecules can be efficiently dealt with.

This has come to be only partially true. While the range of applicability of these methods is larger than *ab initio* methods, STO-3G level calculations have come close to competitiveness. At the same time, these semiempirical approaches exhibit the same limits with respect to the matrices to be diagonalized that *ab initio* methods do, so that many photochemical problems, for instance, must still be dealt with in the π-electron Pariser–Parr–Pople approximation. In addition, attempts to treat elements beyond Cl in the periodic table by ZDO techniques have been distinctly unencouraging.

The definition of a semiempirical technique has been cast considerably more generally in these volumes to include methods which recognize the problems just mentioned and which attempt to meet them in a variety of ways, sometimes without the actual use of experimentally determined parameters.

Noteworthy among these are pseudopotentials which enable one to treat large numbers of electrons which are unaffected in the molecular environment implicitly rather than explicitly. The Xα approach, also included, has, among others, the advantage that it evades the basis set problem by dividing the molecule into coarsely grained areas, solving the problem within these regions, then joining these solutions to form the final global solution to the problem. Lastly, the consistent force field method joins classical potentials to quantum mechanical techniques in such a way that one can even explore the conformational surfaces of really complex systems such as hydrated enzymes.

Volume 7 attempts to survey the leading semiempirical techniques in current use. Most of the theories described are readily available to all scientists via computer programs obtainable from the Quantum Chemistry Program Exchange at Indiana University. Each of these methods has its respective advantages and limits, but some are so constructed that for certain properties and limited classes of molecules they provide a level of numerical accuracy far exceeding that of the basic *ab initio* approaches from which they are derived. While these theories have frequently been disparaged as "numerical fits," they are extremely useful and sometimes have a sound theoretical basis as discussed by Freed in Volume 7. All, however, also have areas in which they fail to be at all dependable. Such a situation is inescapable when one is dealing with approximate solutions since full dependability in all applications would constitute a more or less exact solution to our problem.

The nonspecialist scientist who has a problem and wants to carry out a theoretical calculation is faced by a somewhat bewildering array of possible approaches, some of which may be really useful and even semiquantitatively accurate for what he or she wants to learn from the calculation, and some of which may be almost totally useless in the application envisaged. The literature is full of such overextensions of semiempirical approaches.

Volume 8 therefore attempts to consider some distinct subfields of chemistry and contiguous areas usually classed as physics, such as solids, and to summarize the advantages and disadvantages of each of the common approaches when applied to the problems of each area. This volume is therefore an attempt at the synthesis of the methods described in Volume 7. It necessarily touches upon only a few of the possible areas of theoretical interest, but will, it is hoped, give the reader a feel for what each approach can and cannot tell one.

It seems clear to this editor that the present thrust of research in this area of theoretical chemistry is, and should be, bifurcate. On the one hand, further work is going on in extending the range of accuracy of the present approaches. While it is not too hard to be skeptical of schemes such as the various MINDO approximations which claim semiquantitative accuracy, they are frequently useful. On the other hand, theories which avoid the basis set size problem are sorely needed, for it is this difficulty which is presently limiting quantum chemistry in both its *ab initio* and semiempirical forms.

Gerald A. Segal

Contents

Chapter 1. Ground-State Potential Surfaces and
Thermochemistry

*Marie C. Flanigan, Andrew Komornicki, and
James W. McIver, Jr.*

Chapter 2.   Electronic Excited States of Organic Molecules

R. L. Ellis and H. H. Jaffé

Chapter 3. Photochemistry
Josef Michl

Chapter 4. Approximate Methods for the Electronic Structures of Inorganic Complexes
C. J. Ballhausen

Chapter 5. Approximate Molecular Orbital Theory of Nuclear and Electron Magnetic Resonance Parameters
David L. Beveridge

Chapter 6.   The Molecular Cluster Approach to Some Solid-State Problems

Richard P. Messmer

Chapter 7.   Electron Scattering

Donald G. Truhlar

Contents of Part A

Ground-State Potential Surfaces and Thermochemistry

Marie C. Flanigan, Andrew Komornicki,
and James W. McIver, Jr.

1. Introduction

It has been known since the early days of quantum mechanics that with existing, well-understood, physical and mathematical concepts, it is in principle possible to accurately predict and account for all "chemical" behavior. Realization of this exciting and challenging goal requires only the development of an adequate computational technology and there have been significant advances toward this end in the last ten years. One of these has been the evolution of approximate, semiempirical molecular orbital (MO) methods. These methods can be used to study properties of molecules large enough to be interesting to the organic chemist and often to the biochemist. Although it is perhaps unduly optimistic to say that these methods have brought us to the state in which the computer effectively competes with existing laboratory measurement techniques,[1] they do have a number of unique and valuable features. In addition to their computational economy and broad applicability, the interpretation of computed results can be couched in the simple and familiar vocabulary of LCAO molecular orbital theory.

Other chapters in this volume document the more formal and mathematical aspects of these methods, together with a number of their applications. This chapter is mainly concerned with the application of semiempirical molecular orbital theory to structural, thermodynamic, and kinetic properties of

Marie C. Flanigan, Andrew Komornicki, and James W. McIver, Jr. • Department of Chemistry, State University of New York at Buffalo, Buffalo, New York

molecules, i.e., properties derivable from the ground-state potential energy surface. Rather than attempt a broad review of existing molecular orbital methods and their computational results, the aim of this chapter will be to address the question of how one goes about calculating macroscopic properties from the potential surface, given any particular semiempirical MO theory. The word "semiempirical" is stressed because it is not clear that the techniques developed for these methods will necessarily be the best ones to use with an *ab initio* method. Selected results of calculations with a few of the more popular methods will be presented in order to give the reader a rough estimate of their overall quality. Comparisons of these results with experiment and those obtained by *ab initio* theories will be made. It is worth noting that the development and parametrization of these methods is currently a rather active research area and results of the more recently developed methods appear to show a gradual improvement in quality. A critical assessment of the current state of this technology would be possible only after extensive applications of these methods have been made.

2. Macroscopic Properties from Molecular Calculations

2.1. A Scheme for Thermodynamic Parameters

Thermochemistry and kinetics, as they are often practiced in the chemistry laboratory, are concerned with the measurement of such macroscopic quantities as rate and equilibrium constants, enthalpies, entropies, and heat capacities and their dependence on temperature and pressure. The practitioner of molecular quantum mechanics, on the other hand, is primarily concerned with Hamiltonians and wave functions of molecules. Both the experimentalist and the theoretician might report that molecule A is 10 kcal more stable than its isomer, molecule B, yet both could easily be talking about two entirely different quantities; the former may have reported the free energy (or perhaps the enthalpy) of a chemical reaction obtained under certain experimental conditions, whereas the latter may have quoted the energy difference between two minima of a potential energy surface. Moreover, a translation of potential energy differences into the macroscopic language may change the result considerably, turning apparent perfect agreement between theory and experiment into a 100 kcal discrepancy. What is needed is a clearly defined scheme for carrying out such translations in order that meaningful comparisons between theory and experiment can be made.

One such scheme, valid for ideal gases, is outlined in Fig. 1. In this figure, each level represents an intermediate computational result, and the arrows connecting the levels represent a simplifying assumption or a method of attaining the result. The validity of the assumptions must be examined for each

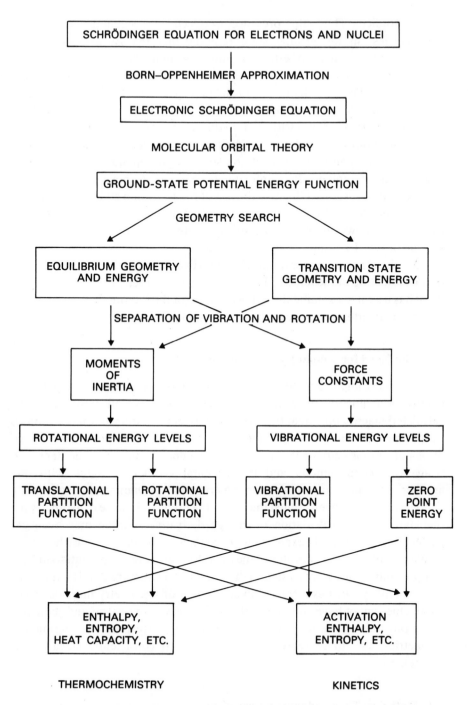

Fig. 1. Outline of an approximate computational scheme for obtaining thermodynamic and kinetic quantities of ideal gases from molecular properties.

system studied. Thus, the presence of electronic degeneracies (or near degeneracies) in the ground state requires that the Born–Oppenheimer approximation itself be questioned. Similarly, molecules with free internal rotations need special techniques to evaluate the rotational–vibrational partition function. The right-hand side of the scheme of Fig. 1 involves the use of transition state theory to evaluate macroscopic kinetic properties. Even if the approximations associated with steps in this scheme are valid, it is by no means clear that transition state theory itself represents an accurate method of calculating rate constants. This of course will also depend on the particular system being examined. It is unfortunate that one cannot yet determine in advance which systems will allow the use of the transition state hypothesis. In view of the current level of accuracy of molecular quantum mechanical calculations, however, it appears to be a safe assumption that the overall scheme of Fig. 1 is valid.

Much of the remainder of this chapter will be concerned with the methods of implementing the scheme of Fig. 1 using semiempirical MO theory. At this point it is useful to examine the details of a few of the intermediate steps of this scheme in an attempt to gauge their relative importance.

2.2. The Need for Geometry Calculations

The most important step is the calculation of the ground-state potential energy as a function of molecular geometry. The overall accuracy of the calculated macroscopic quantities is limited by the accuracy of this step. Since "experimental" molecular geometries lie at or very near potential minima, an important measure of the accuracy of the energy surface is the degree to which calculated minima and experimental geometries agree with one another. One may, of course, use experimental geometries for calculating energies, and it has been argued that this is the preferred choice of geometry for the Hartree–Fock calculation of energy derivatives.[2] One may even get perfect agreement with experimental energies in doing so. This, however, would be a misleading indicator of the overall accuracy of the potential surface. If, for example, one is interested in the energy of a structure that is not known experimentally, such as a transition state, then this geometry must be calculated. One cannot have confidence that the energy of this calculated geometry is accurate unless the energy minima are known to correspond to experimental geometries. For these reasons it seems best to consistently calculate all geometries.

2.3. Statistical Thermodynamic Formalism

The formalism relating properties of the potential energy surface to thermodynamic quantities can be found in many textbooks on statistical thermodynamics. What is described below is adapted from Herzberg.[3]

For an ideal gas all thermodynamic variables can be obtained from U, the potential energy of the calculated equilibrium geometry; E_{zp}, the zero-point vibrational correction; Q, the molecular partition function; and the dependence of Q on the temperature T. Macroscopic kinetic properties depend on analogous quantities (Section 2.4), except that the transition state geometry is used in place of an equilibrium geometry. For one mole of an ideal gas,

$$E = U + E_{zp} + RT^2 \frac{d \ln Q}{dT} \tag{1}$$

$$H = E + RT$$

$$C_p = \frac{dH}{dT} \tag{3}$$

$$S = R(1 - \ln N_0) + RT \frac{d \ln Q}{dT} + R \ln Q \tag{4}$$

$$G = H - TS \tag{5}$$

where E, H, C_p, S, and G are, respectively, the internal energy, enthalpy, heat capacity at constant pressure, entropy, and Gibbs free energy of the system. N_0 is Avogadro's number and R is the gas constant. In quantum mechanical calculations, it is convenient to take the zero of energy to be the potential energy of all the nuclei and electrons separated by infinity.

2.4. Activation Parameters[4]

The calculation of the specific rate constant $k_r(T)$ of an elementary (unimolecular, bimolecular, termolecular) gas phase reaction using transition state theory is relatively straightforward,

$$k_r(T) = \kappa \frac{kT}{h} K^\dagger \tag{6}$$

where $K^\dagger$ is the "equilibrium constant" for reactants A, B, etc. in equilibrium with the transition state,

$$K^\dagger = \frac{Q^\dagger / V}{(Q_A / V)(Q_B / V)} \exp \frac{-\Delta E_0^a}{RT} \tag{7}$$

This form of the equilibrium constant implies that concentration units (molecules cm^{-3}) are being used. However, Eq. (6) does not depend on units provided that k_r and $K^\dagger$ are expressed in the same units. The transmission coefficient κ cannot be calculated from the properties of reactant and transition state alone and is usually set equal to unity. The partition functions Q_A, Q_B, etc., are the partition functions of the reactants, $Q^\dagger$ is the partition function of the transition state (Section 2.6), and V is the volume. The term ΔE_0^a is given as

$$\Delta E_0^a = U^\dagger + E_{zp}^\dagger - (U_A + E_{zp}^A + U_B + E_{zp}^B + \cdots) \tag{8}$$

where $U^\dagger$ and U_A, U_B, etc. are the potential energies of the transition state and reactants, respectively, and $E_{zp}^\dagger$ and E_{zp}^A, E_{zp}^B, etc. are the corresponding zero-point vibrational corrections. Transition state theory is considerably weaker than equilibrium state theory (Section 2.3), primarily because of uncertainties in the equilibrium assumption and the transmission coefficients.

Experimental uncertainties are also greater in the kinetics of elementary reactions than in equilibrium studies, mainly because of problems in sorting out rate constants of elementary steps in mechanisms which are often complicated, temperature dependent, and speculative. For most elementary reactions, the activation energy,

$$E_a = -R \frac{d \ln k_r}{d(1/T)} \tag{9}$$

is fairly constant over a reasonable range of temperatures. The experimental E_a of Eq. (9) and the ΔE_0^a of Eq. (7) are not equal (except at absolute zero), although their difference may be small, and neither quantity is equal to the potential energy difference between the transition state and reactants. The experimental quantities most often reported are $\Delta H^\dagger$ and $\Delta S^\dagger$, the enthalpy and entropy of activation. These are obtained by imposing a pseudothermodynamic interpretation on the "equilibrium" constant $K^\dagger$,

$$\Delta G^\dagger = \Delta H^\dagger - T \Delta S^\dagger = -RT \ln K^\dagger \tag{10}$$

If pressure units are used for k_r and $K^\dagger$ (standard state of 1 atm),

$$\Delta H^\dagger = E_a - RT \tag{11}$$

while for concentration units (assuming ideal behavior),

$$\Delta H^\dagger = E_a - nRT \tag{12}$$

where $n = 1, 2, 3$ for uni-, bi-, and termolecular reactions, respectively. At low to moderate temperatures, the difference between $\Delta H^\dagger$ and E_a can be neglected. For ideal gases, $\Delta H^\dagger$ is independent of the standard state. The relation between $\Delta S_p^\dagger$ and $\Delta S_c^\dagger$, where the subscripts refer to pressure and concentration standard states, respectively, is

$$\Delta S_c^\dagger = \Delta S_p^\dagger + (n-1)R \ln pV \tag{13}$$

Again, n is 1, 2, or 3 for uni-, bi-, and termolecular reactions. Here p is the standard pressure (1 atm) and V is the molar volume,

$$V = 22,412 T/298 \text{ cm}^3 \tag{14}$$

Thus for bimolecular gas reactions at 300°K, $\Delta S_c^\dagger - \Delta S_p^\dagger = 20.1$ eu.

The enthalpies and entropies can be obtained from Eqs. (2) and (4) and using the appropriate standard state in the partition function [e.g., setting $p = 1$ in Eq. (22) below].

2.5. The Zero-Point Vibrational Correction

At low to moderate temperatures, the most important contribution to the energy quantities is U. The second most important contribution is E_{zp}, the zero-point energy correction,

$$E_{zp} = N_0 \sum_i \tfrac{1}{2} h\nu_i \tag{15}$$

where the ν_i are the fundamental harmonic vibrational frequencies.

The index i runs over all vibrational degrees of freedom. For a structure with M atoms there are $3M - 6$ vibrations for nonlinear equilibrium geometries, $3M - 5$ for linear equilibrium geometries, $3M - 7$ for nonlinear transition states, and $3M - 6$ for linear transition states.

Some care must be used in treating this quantity. In an extreme case, one might be interested in comparing experimental and calculated heats of atomization at $0°K$ (for simplicity),

$$\Delta H_{0°K}^{\text{atomization}} = \Delta U^{\text{atomization}} + E_{zp} \tag{16}$$

Here $\Delta U^{\text{atomization}}$ is the depth of the potential well of the molecule relative to isolated atoms and E_{zp} is the zero-point correction for the molecule. Thus the simple equating of potential energy to enthalpy in atomization reactions will be in error by E_{zp}. Experimental values of E_{zp} for a number of hydrocarbons are listed in Table 1. As can be seen from this table, molecules the size of cyclohexane can give an error on the order of 100 kcal. Relatively large errors will also be encountered for formation reactions and hydrogenation reactions. On the other hand, reactions involving only small hydrocarbons, such as the theoretically important bond separation reaction,[5]

$$C_3H_8 + CH_4 \rightarrow 2C_2H_6$$

or simple isomerizations, will generally give a fairly small error (on the order of 1 or 2 kcal) at absolute zero if zero-point corrections are ignored. This is because these molecules tend to obey simple stoichiometric additivity in their zero-point energies. The estimated values in Table 1 were obtained by setting the zero-point energies of hydrogen and carbon equal to 7 and 2 kcal, respectively. If di- and triatomic molecules are exempted, this additivity phenomena will hold for nonhydrocarbons as well by also setting equal to 2 kcal the zero-point energies of N, O, and F. As seen in Table 1, though, the additivity begins to deteriorate for the larger molecules. Note that even a small deviation from additivity, such as the 3 kcal value for methane, can become significant in reactions in which methane is multiplied by a large stoichiometric coefficient. Whether or not ignoring the zero-point corrections is justifiable, one can frequently account for them in stable molecules by simply using Eq. (15) and experimental values of the vibrational frequencies. No data are available for transition states, on the other hand, and zero-point corrections

Table 1. A Comparison of Experimental and
Estimated Zero-Point Energies

	E_{zp}, kcal	
Molecule	Exp.[a]	Est.
H	0	7
C	0	2
H_2	6.3	14
CH_4	27.0	30
C_2H_2	16.2	18
C_2H_4	30.9	32
C_2H_6	45.3	46
$H_3C—C≡CH$	33.8	34
$H_2C=C=CH_2$	33.4	34
△	34.2	34
C_3H_6	48.5	48
△	49.4	48
C_3H_8	62.5	62
$H_2C=CHCH_2CH_3$	66.1	64
☐	67.3	64
n-C_4H_{10}	80.6	78
i-C_4H_{10}	80.2	78
⬠	86.0	80
n-C_5H_{12}	98.5	94
neo-C_5H_{12}	96.8	94
◎	61.1	54
∿	104.5	96
n-C_6H_{14}	116.2	110

[a] Calculated as $\sum_i \frac{1}{2}h\nu_i$, where the ν_i are experimental vibrational frequencies. Taken from References 5 and 42.

must also be made for these species. An error of several kilocalories can occur if the zero-point corrections for reactants are used for the transition state. Here it seems most appropriate to obtain them from the calculated vibrational frequencies of the transition states.

2.6. The Partition Function

Very few enthalpies are accurately known at 0°K. In order to correct either the experimental enthalpy to absolute zero or the calculated result to the experimental temperature, one must have the molecular partition function over a wide temperature range. Since this correction can amount to several kilocalories, it cannot normally be neglected.

Entropies are also important thermochemical and kinetic quantities. From Eq. (4) it is seen that terms involving the partition function make the only significant contribution to the entropy.

Given the potential energy surface, it is in principle possible to calculate the partition function Q exactly. Since such a calculation is extremely tedious, it is necessary to make the several simplifying assumptions implied in the scheme of Fig. 1. In the ideal gas approximation (zero intermolecular forces) the translational and internal degrees of freedom can be separated,

$$Q = Q_{tr}Q_{int} \tag{17}$$

Using the particle-in-a-box model, the translational partition function is given as

$$Q_{tr} = V\left(\frac{mkT}{2\pi h^2}\right)^{3/2} \tag{18}$$

where V is the volume and m is the absolute mass of the gas molecule. Since the thermodynamic quantities depend only on the logarithm of Q, they can be separated into additive contributions from the internal and translational degrees of freedom, e.g.,

$$E = U + E_{zp} + RT^2 \frac{d \ln Q_{int}}{dT} + E_{tr} \tag{19}$$

$$S = R(1 - \ln N_0) + RT\frac{d \ln Q_{int}}{dT} + R \ln Q_{int} + S_{tr} \tag{20}$$

where, using the ideal gas law, $PV = RT$,

$$E_{tr} = \tfrac{3}{2}RT \tag{21}$$

$$S_{tr} = \tfrac{5}{2}R \ln T + \tfrac{3}{2}R \ln m + R \ln\left[\left(\frac{2\pi}{N_0}\right)^{5/2}\frac{k^{5/2}}{h^3}\right]$$

$$+ \tfrac{3}{2}R - R \ln P + \tfrac{5}{2}R \ln N_0 \tag{22}$$

In the above two expressions only the second term in S_{tr} contains any reference to a molecular property. The partition function for the internal degrees of freedom depends on the nuclear spin, electronic, rotational, and vibrational states of the system. Since the energy gaps between the ground and excited electronic states are generally large (although this might be questioned for some transition states), the contribution from electronic states can be ignored. Except at very low temperatures, the contribution from nuclear spin can also be ignored. Finally, if the interaction between rotational and vibrational degrees of freedom is ignored, Q_{int} can be simplified to

$$Q_{int} = Q_r Q_{vib} \tag{23}$$

If the temperature is not too high and if there are no free rotations in the molecule, the harmonic oscillator approximation can be used to evaluate Q_{vib},

$$Q_{vib} = \prod_i [1 - \exp(h\nu_i/kT)]^{-1} \tag{24}$$

where the ν_i are the vibrational frequencies and the index i runs over all vibrational degrees of freedom. If the temperature is again not too high (or if the moments of inertia are sufficiently large), classical equipartition can be used to evaluate the rotational partition function, which for linear molecules is

$$Q_r = 8\pi^2 I_A kT/\sigma h^2 \tag{25}$$

and for nonlinear molecules

$$Q_r = \frac{\pi^{7/2}}{\sigma} \left(\frac{8kT}{h^2}\right)^{3/2} (I_A I_B I_C)^{1/2} \tag{26}$$

Here σ is the symmetry number and I_A, I_B, and I_C are the principal moments of inertia.

When calculating the enthalpy and entropy of a transition state two modifications in the formalism are required. The first involves omitting the imaginary frequency vibration in Eqs. (15) and (24) (see Section 4.3). The second involves the symmetry number σ which appears in the formulas for the rotational partition function, Eqs. (25) and (26). Although the situation is not completely understood,[6] it has been argued that since the use of symmetry numbers can give incorrect results for rate expressions, they should all be set equal to unity and a statistical factor, the number of ways in which the reaction can occur, should be used to correct the final rate expression.[7] Particular care must be taken, when calculating entropies of activation, to include the same correction made in the reported experimental activation entropy, or to specify which method, symmetry numbers or statistical factors, has been used.

Except in the cases noted, the above approximations to the molecular partition function will give an accuracy sufficient for the calculation of the thermodynamic properties at normal laboratory temperatures. Alternatively, the experimental ν_i and moments of inertia, when known, can be used to estimate experimental enthalpies of reaction at 0°K.

3. Semiempirical Molecular Orbital Theory for Closed Shells

3.1. The Nature of Semiempirical Theory

Semiempirical molecular orbital theory, at its current stage of development, occupies a unique position in the canon of theoretical chemistry. It is neither fish nor fowl. On the one hand it models the *ab initio* single-determinant, minimal Slater basis set theory, which is a purely mathematical approximation to a tested and accepted fundamental quantity—the ground-state solution to the electronic Schrödinger equation. On the other hand, it has the trappings of pure empiricism—the fitting of a function to a discrete set of data points by varying adjustable parameters. The adjustable parameters are the weakness of the theory in the sense that their presence allows the violation of fundamental principles of quantum mechanics. For example, it is possible by adjusting parameters to obtain total energies lower than the exact ground-state energy, which by the variational principle is impossible for wave functions satisfying the Pauli exclusion principle. Yet, paradoxically, the parameters strengthen the theory in the sense that they allow sufficient flexibility for the theory to potentially transcend in accuracy the results obtained by its *ab initio* counterpart.

3.2. Parametrization

A detailed description of the formalism of several of the currently used semiempirical MO methods can be found in Chapter 2 of Vol. 7 of this treatise and will not be repeated here. It is of interest, however, to mention the various ways in which the parameters in the theory are introduced and determined.

There are basically two types of parameters in the "neglect of differential overlap" methods: those that are determined by the variational principle and those that are determined by reference to atomic or molecular properties. The first type includes only the linear expansion coefficients in the LCAO approximation of the molecular orbitals. The second type are introduced into the approximations of the atomic integrals over the basis functions (parameters are also introduced into the core repulsion terms in the MINDO/2,[8] MINDO/3,[9] and PNDDO[10] methods). In contrast to *ab initio* theory, the latter parameters cannot be determined variationally since they must be formally regarded as being implicitly contained in both the wave function and the Hamiltonian (see Chapter 7 of Vol. 7 of this treatise). Many of the atomic integrals are zero due to the neglect-of-differential-overlap approximation. Of those that remain, the one-center integrals are related to *experimental* atomic properties such as ionization potentials, electron affinities, and the energies of the atoms in various states.

There are two schools of thought concerning the disposition of the remaining parameters. One approach, exemplified by the CNDO[11] and INDO[12] methods, argues that the theory represents an approximation to the minimal Slater basis *ab initio* method. Accordingly, the parameters were adjusted to fit the charge densities and bond orders obtained by *ab initio* calculations of small molecules (after an appropriate orthogonalization of the basis set). In this type of approach, the theory maintains some of its *a priori* character in the sense that no appeal is made to experimental molecular (as opposed to atomic) properties. One expects, however, that the accuracy of the computed properties will be limited by the accuracy of the parent *ab initio* method, which in many cases is not very good.

This limitation does not necessarily exist in the second approach, which adjusts the parameters in such a manner that the theory directly reproduces experimental properties. In other words, this form of the theory can give results which exceed the accuracy of *ab initio* methods. In the MINDO family of methods the parameters are adjusted to give the best fit in a least squares sense to experimental enthalpies of atomization at 298°K, and hence avoids the problem of correcting for the zero-point energy or temperature effects discussed in Section 2 of this chapter. This advantage comes at a price, however, since one is forced to adopt the position that the parameters somehow incorporate macroscopic dynamical effects into a static molecular calculation. The calculated energy as a function of molecular geometry must be regarded as an "enthalpy at 298°K" function. From Eqs. (1), (2), and (15) it is seen that this function depends, among other things, on vibrational frequencies and hence on molecular force constants. Its use for calculating force constants, for example, is thus questionable. In some applications of MINDO/2 this problem is cavalierly ignored; the calculated energy is treated as potential energy in spite of the parametrization method.[13]

3.3. Strengths and Limitations for Potential Surface Calculations

There are two fundamental advantages that semiempirical methods enjoy over *ab initio* theory. The first has already been mentioned, namely that the presence of empirically adjustable parameters allows in principle the accuracy of computed energies to exceed those of single-determinant *ab initio* methods, methods which cannot exceed the Hartree–Fock limit. Whether or not this can be or has been achieved in practice is subject to some dispute.[14–16]

There is no question about the second advantage: the computational speed of semiempirical relative to *ab initio* methods. The computer time t required for an SCF molecular orbital calculation varies with the number of atomic orbitals N (a measure of the size of a molecule) roughly as

$$t = \alpha + \beta N + \gamma N^2 + \delta N^3 + \epsilon N^4 + \cdots \tag{27}$$

The coefficients depend on the individual molecule, its geometry, the starting orbitals for the SCF procedure, etc., so comparisons are only general. The point is that an *ab initio* molecular orbital calculation involves steps (calculating and processing two-electron integrals) on the order of N^4, whereas a semiempirical MO method has only an N^3 step (matrix multiplication and diagonalization) as the rate-determining step. As the size of the molecule increases, the ratio of *ab initio* to semiempirical computer times increases linearly. Base time comparisons for a few molecules run on the same computer (CDC 6400) indicate that a semiempirical method, MINDO/2, is about 30 times faster than a minimal-basis-set *ab initio* method, STO-3G, for C_3 molecules such as propene, 60 times faster for C_4 molecules, and over 120 times faster for C_7 molecules.[17] One soon reaches a molecule for which it is totally impractical to carry out an *ab initio* calculation.

The above comparisons are for a single set of bond lengths and angles in each case. When one is exploring a potential surface in a search for equilibrium geometries and transition states one utilizes either explicitly or implicitly the first derivatives of the energy with respect to the nuclear positions. The set of all such derivatives, the potential energy gradient, can be obtained by finite difference techniques involving as few as $3M-6$ (M is the number of atoms) evaluations of the potential energy function. Alternatively, they may be obtained accurately by a first-order perturbation method. In semiempirical MO theory, all of the first derivatives can be calculated analytically in less than 10% of the time required to evaluate the energy itself.[18] In other words, the additional expense of calculating the gradient is negligible. For *ab initio* methods, on the other hand, it is estimated that the gradient requires more than four times as much computer time as that needed to calculate the energy itself. Thus, when geometry searches are a requirement of the calculation, as they are in studying thermochemistry and kinetics, the computational advantage of semiempirical MO theory over that of the corresponding *ab initio* theory is enormous.

If the goal of a computation is understanding the nature of a physical process or measurement, rather than the simple prediction of an experimental number, then *accurate ab initio* calculations are preferred. The parameters of the semiempirical theory tend to mask the underlying physical principles responsible for the result.

A limitation shared by all single-determinant SCF methods manifests itself when one is computing reaction paths on an energy surface, as would occur in the search for a transition state, for example. In reactions which are forbidden by the Woodward–Hoffmann rules[19] a crossing of highest occupied and lowest vacant orbitals occurs along a path connecting reactant and product. This crossing manifests itself as a cusp in the energy surface, a cusp which does not exist on the exact energy surface. Pople[20] has shown that such cusps are the result of the limited flexibility of the single-determinant wave function and can be removed either by employing configuration interaction or by allowing the orbitals to become complex. Gregory and Paddon-Row[21] have further pointed out that limited configuration interaction may not be sufficient for removing the crossing of the lowest energy surfaces and that either extensive configuration-interaction or a multiconfiguration SCF treatment would be required. Since configuration interaction will be more important in the vicinity of a cusp than at an equilibrium geometry, one expects that the energy of transition states located at or near a cusp could be in serious error. (Recall that semiempirical methods are parametrized for equilibrium geometries of molecules.)

4. Exploring Potential Energy Surfaces

4.1. The Size and Shape of Potential Surfaces

4.1.1. Computing Molecular Properties

Most molecular properties subjected to quantum mechanical calculations can be evaluated by ancient and well-known techniques, such as calculating the expectation value of an operator or using perturbation theory in one form or another. Examples include dipole and multipole moments, electric and magnetic susceptibilities, high-resolution magnetic resonance parameters, etc. These are essentially electronic properties and are evaluated at a fixed molecular geometry. In contrast, the properties discussed in this chapter depend on the potential energy evaluated as a function of molecular geometry, i.e., they depend on the shape of the potential energy surface. In particular, these properties are derived from the equilibrium and transition state geometries and their surroundings. These geometries are stationary points on the potential energy surface and require special search techniques for their location. The development of these techniques has been a fairly recent undertaking, since they are designed for implementation on a digital computer.

4.1.2. The Dimension Problem

If someone were asked to calculate the equilibrium bond length of a diatomic molecule, he might calculate the potential energy U at a number of values, e.g., ten, of the interatomic distance R, plot U vs. R, look at the plot, and immediately pick out the minimum. For a (hypothetical) linear triatomic molecule ABC, which has two isomeric forms A=B—C and A—B=C, he might calculate U at ten values of R_{AB} and R_{BC}, giving a two-dimensional grid containing $10^2 = 100$ points. These energies could then be plotted as a contour map, such as shown in Fig. 2. After a few moments of study he could pick out the positions of the two minima, the barrier separating them, and the relative energies of these structures. Figure 2 might also represent the nonlinear triatomic evaluated at one value of the bond angle. For ten such values, the energy grid now containing 10^3 points, the corresponding maps could be stacked and after some considerable study the positions and energies of the minima and barriers could be determined. As the number of degrees of

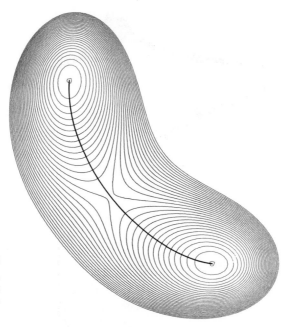

Fig. 2. A hypothetical potential energy surface showing reactant, transition state, product, and a steepest descent reaction path connecting these three points.

freedom increases, two difficulties with this process emerge: The number of required grid points increases exponentially and the global view of the energy surface is lost. A large number of very fast future computers operating in parallel might handle the gridding problem for small molecules, but this would be useless unless there is a major breakthrough in automated pattern recognition theory. The only alternative, it seems, is to use chemical intuition to estimate the geometry of a molecule and use a search method to refine the structure. Having lost the global picture, however, one can never be absolutely certain that one has located the structure of interest, e.g., the most stable conformer of a large molecule. For transition state geometries, chemical intuition is less reliable and search methods for locating a suitable initial estimate of the geometry must be used.

4.1.3. Reaction Paths

We define a reaction path as a continuous line in configuration space connecting two potential energy minima, reactant and product. These paths are used in conjunction with transition state calculations in two ways: to locate a geometry suitable for an initial guess of the transition state geometry and to ensure that no intervening barriers of high energy separate a calculated transition state and reactant or product.

It is tempting to visualize a reaction path in much the same way one would picture a mountain stream flowing from a col down a ravine and into a valley as illustrated in Fig. 2. It is also tempting to ascribe physical significance to such

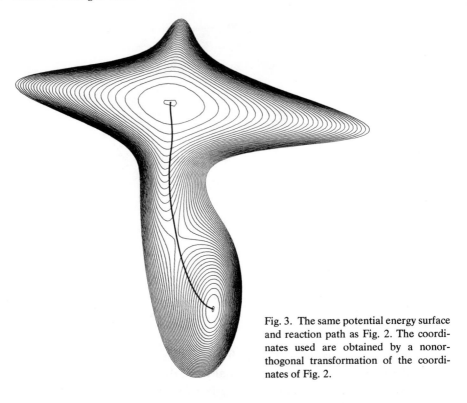

Fig. 3. The same potential energy surface and reaction path as Fig. 2. The coordinates used are obtained by a nonorthogonal transformation of the coordinates of Fig. 2.

paths. The analogy between a mountain range and a potential energy surface can be misleading since the former is a physical entity while the latter is a mathematical construction which, except for the stationary points, depends upon the coordinate system used to describe it.[22] Although a particular set of internal coordinates may be a "natural choice" for a particular system, no physical property will depend on this choice. Coordinate transformations are generally nonorthogonal. Examples are the transformation of Cartesian to internal coordinates (bond lengths and angles) or from one set of internal coordinates to another. Figure 3 illustrates the effect of such a transformation of the coordinates used in plotting Fig. 2. The "valleys" dramatically change their shape and direction.

The reaction path illustrated in Fig. 2 is a steepest descent path, a commonly used definition of reaction path. It follows the direction of the energy gradient $\mathbf{g}$ and is everywhere perpendicular to a contour line. The same path plotted in Fig. 3, obtained by a nonorthogonal coordinate transformation of Fig. 2, is no longer steepest descent. This is because angles are not conserved under a nonorthogonal transformation. The following algebraic argument further illustrates this point. Let the column vectors $\mathbf{x}$ and $\mathbf{x}_0$ represent two adjacent points on a steepest descent path in the $\mathbf{x}$ coordinate system,

$$\mathbf{x} - \mathbf{x}_0 = \epsilon \mathbf{g}_x \tag{28}$$

where $\mathbf{g}_x$ is the gradient in the $\mathbf{x}$ coordinate system and ϵ is a small scalar. Let $\mathbf{y}$ and $\mathbf{y}_0$ be the coordinates of these same two points in a new coordinate system $\mathbf{y}$ and let $\mathbf{S}$ be the nonsingular transformation relating the $\mathbf{x}$ and $\mathbf{y}$ coordinates,

$$\mathbf{y} = \mathbf{Sx} \tag{29}$$

The potential energy of these two points will be unaffected by the coordinate transformation, i.e., $U(\mathbf{x}) = U(\mathbf{y})$, so that expanding the energy about $\mathbf{x}_0$ gives

$$U(\mathbf{x}) - U(\mathbf{x}_0) = \mathbf{g}_x^+ \cdot (\mathbf{x} - \mathbf{x}_0) + \cdots$$

$$= \mathbf{g}_y^+ \cdot (\mathbf{y} - \mathbf{y}_0) + \cdots$$

$$= \mathbf{g}_y^+ \cdot \mathbf{S} \cdot (\mathbf{x} - \mathbf{x}_0) + \cdots$$

$$= (\mathbf{S}^+ \cdot \mathbf{g}_y)^+ \cdot (\mathbf{x} - \mathbf{x}_0) + \cdots \tag{30}$$

from which it follows that $\mathbf{g}_x = \mathbf{S}^+ \cdot \mathbf{g}_y$. Inserting this result and Eq. (29) into Eq. (28) gives

$$\mathbf{y} - \mathbf{y}_0 = \epsilon (\mathbf{SS}^+) \mathbf{g}_y \tag{31}$$

In other words, the two points will lie on a steepest descent path in the $\mathbf{y}$ coordinate system if and only if $\mathbf{S}$ is an orthogonal transformation.

In addition to suffering from a lack of uniqueness, some definitions of reaction paths can give rise to strange results. For example, the path obtained by varying one geometrical variable in small increments from its reactant to its product value while requiring that the energy be a minimum with respect to the other variables can give a discontinuous path. The same procedure applied in reverse (beginning at the product value) can result in a different discontinuous path, giving rise to an apparent "chemical hysteresis."[23] The occurrence of this effect in connection with the Woodward–Hoffmann rules has been clearly described by Gregory and Paddon-Row.[21] It has also been illustrated that this effect can result from the procedure itself and does not necessarily depend on any special features of the potential energy function.[24]

It is possible to define a unique, coordinate-independent "reaction path" and assign to it a physical interpretation, but only if one takes into account the dynamics of the system. One such path is the classical mechanical trajectory of lowest possible energy which passes through the transition state.[22] Although potentially useful in studies of dynamical effects of reactions, this reaction path does not "terminate" at reactants or products in the usual sense, but will indefinitely oscillate in the reactant and product regions unless the kinetic energy is somehow removed.

4.2. Geometry Optimization

4.2.1. Criteria for Selecting a Method

Assuming one has estimated an initial set of coordinates, the problem becomes one of refining the geometry until the optimum is reached. There are a large variety of methods for optimizing functions of many variables.[25] In selecting a method, one must answer such questions as, "Is the function constrained?" "Are the first derivatives easily available?" "The second?" "Can the function be written as a sum of squares?" etc.

4.2.2. Equilibrium Geometries

The class of optimization techniques known as variable metric methods appear to be most suitable for locating equilibrium geometries using semiempirical molecular orbital theory.[18,26] Geometries of molecules the size of LSD (49 atoms) have now been calculated with these methods.[27]

Basically, variable metric methods generate improved geometries $\mathbf{X}^{n+1}$ from old geometries $\mathbf{X}^n$ by using the recursion

$$\mathbf{X}^{n+1} = \mathbf{X}^n - \alpha_n \mathbf{A}^n \mathbf{g}^n \tag{32}$$

where $\mathbf{g}^n$ is the energy gradient. The various methods differ by the way in which the scalar α_n and the symmetric matrix $\mathbf{A}^n$ are determined. For example, the steepest descent method is recovered by fixing $\mathbf{A}^n$ to be the identity matrix and varying α_n to minimize the energy at each step. The process is repeated by replacing $\mathbf{X}^n$ by the $\mathbf{X}^{n+1}$ so determined. The Newton–Raphson method is recovered by setting $\alpha_n = 1$ and $\mathbf{A}^n$ equal to the inverse of the matrix of second partial derivatives at each step. The true variable metric methods lie intermediate between the extremes represented by steepest descent and Newton–Raphson. In the Murtagh–Sargent method,[28] for example, $\mathbf{A}^n$ is initially set equal to the identity matrix $\mathbf{A}^0 = \mathbf{1}$ and is updated by the recursion

$$\mathbf{A}^n = \mathbf{A}^{n-1} + \mathbf{Z}^n \mathbf{Z}^{n+}/C^n \tag{33}$$

where the column vector $\mathbf{Z}^n$ is given as

$$\mathbf{Z}^n = -\mathbf{A}^{n-1}[\mathbf{g}^n - (1 - \alpha_{n-1})\mathbf{g}^{n-1}] \tag{34}$$

and C^n as

$$C^n = (\mathbf{g}^n - \mathbf{g}^{n-1})^+ \mathbf{Z}^n \tag{35}$$

The α_n is set equal to one at each step unless the calculated energy increases, in which case α_n is halved and the step repeated. The power of methods such as Murtagh–Sargent lies in the facts that, unlike steepest descent, they are quadratically convergent (they will minimize a quadratic function exactly in no more than m steps, where m is the dimension), they avoid a one-dimensional search, and, unlike Newton–Raphson, they do not require second derivatives. An additional feature is that when Cartesian coordinates (or symmetry-adapted internal coordinates) are used, the method will conserve molecular symmetry, effectively reducing the number of variables to the number of totally symmetric degrees of freedom.[18] This is useful in calculating geometries of certain symmetric transition states.

4.2.3. Transition State Geometries

Transition state geometries in general cannot be determined by energy minimization techniques. This is because they are not energy minima, but are saddle points on the potential energy surface. They are similar to minima, however, in that they are also stationary points, points of zero potential energy gradient. Thus, given a suitable initial estimate of the geometry, minimization of the norm of the energy gradient

$$\sigma(\mathbf{x}) = \sum_i g_i^2(\mathbf{x}) \tag{36}$$

can lead to a transition state provided that σ is zero at the minimum.[24] Further calculations are necessary to characterize the resulting structure as a transition state (see below).

Since σ is a sum of squares, the very powerful least squares methods can be used. In particular, a generalized least squares method due to Powell[29] has been found to be ideally suited to this task. In fact this method, which can also be used for equilibrium geometries, has been found to be comparable in speed to the Murtagh–Sargent method for geometry determination.[24]

A difficulty with transition state calculations is that a reasonable initial geometry is required for the σ-minimization procedure to converge on the desired structure. Since no experimental structural data are known for transition states, and since chemical intuition can be treacherous for estimating them, particularly regarding their symmetry,[30,31] additional search methods are needed to provide the initial estimate. These methods involve the generation of reaction paths. One of the simplest and least expensive is the "linear internal coordinate path" method.[32] This involves defining a set of internal coordinates $\mathbf{X}$ common to both the reactant and product. As the parameter λ varies from zero to one, the coordinates vary linearly from their reactant $\mathbf{X}^R$ to product $\mathbf{X}^P$ values,

$$\mathbf{X}(\lambda) = \mathbf{X}^R + \lambda(\mathbf{X}^P - \mathbf{X}^R) \tag{37}$$

The energy and σ along this path are calculated and examined for maxima and minima, respectively. The position of a secondary minimum in σ (it is zero at reactant and product) is used as the initial geometry in the least squares minimization. If no secondary minimum is observed, the maximum energy point along this path can be displaced in the steepest descent direction. The position of this lower energy displaced maximum X^m together with reactant and product (or any two points along the linear path that bracket the maximum) can be used to construct a new quadratic internal coordinate path,

$$X(\lambda) = (1 - 2\lambda)[(1 - \lambda)X^R - \lambda X^P] + 4\lambda(1 - \lambda)X^m \tag{38}$$

The above procedure can be repeated until the desired secondary minimum in σ is observed.

A recently described technique due to Halgren et al.[33] avoids the use of internal coordinates and does not require the energy gradient, so that it holds promise for use with *ab initio* methods. This method also uses linear and quadratic paths, except that the complete set of $M(M-1)/2$ interatomic distances is used in place of $3M - 6$ internal coordinates. Since for systems containing more than four atoms, these interatomic distances will lie outside the space of $3M - 6$ internal coordinates, the path involving the $3N$ Cartesian coordinates that is closest to the linear interatomic distance path (in a least squares sense) is used. The distances from reactant and from product to the maximum along this path are defined and the energy of the maximum is minimized keeping the ratio of these distances fixed. A new path passing through the new maximum is defined and the process is repeated.

4.3. Force Constants

Given the $3N \times 3N$ force constant matrix $\mathbf{F}$,

$$F_{ij} = \frac{\partial^2 U}{\partial x_i \, \partial x_j} \tag{39}$$

where the x_i are Cartesian coordinates of an N-atom molecule, the quantities $4\pi^2 \nu_i^2$ are obtained as the eigenvalues of

$$\mathbf{F}' = \mathbf{M}^{-1/2} \mathbf{F} \mathbf{M}^{-1/2} \tag{40}$$

where M is a diagonal matrix containing the triplets of the atomic masses. Six of the frequencies ν_i are zero, corresponding to overall translations and rotations of the rigid molecule (five are zero for linear molecules). The remaining $3N - 6$ (or $3N - 5$) ν_i are the fundamental vibration frequencies used to calculate the zero-point energy correction, Eq. (15), and the vibrational contribution to the partition function, Eq. (24).

The eigenvalues and eigenvectors of $\mathbf{F}$ are used to identify transition states. In particular, a theorem due to Murrell and Laidler[34] states that $\mathbf{F}$ must have exactly one negative eigenvalue. The corresponding vibrational frequency ν_i must therefore be imaginary. The corresponding eigenvector is called the transition vector. The symmetry species of the transition vector must satisfy certain theorems which act as selection rules for the permitted symmetry of the transition state.[22,35] Thus, in order to qualify as a transition state, the computed structure, in addition to satisfying the Murrell–Laidler requirement, must have a transition vector of the proper symmetry.

As was the case with the methods described above for geometry optimization, the calculation of the force constants takes advantage of the easy availability of the potential energy gradient $\mathbf{g}$. In this case an entire column of $\mathbf{F}$ is obtained by finite differences,[24]

$$\mathbf{F}_j = \frac{\partial \mathbf{g}}{\partial x_j} \approx \frac{1}{2h}[\mathbf{g}(x_1, x_2, \ldots, x_j + h, \ldots, x_{3N}) - \mathbf{g}(x_1, x_2, \ldots, x_j - h, \ldots, x_{3N})]$$

(41)

where h is a small scalar on the order of 0.01 Bohr. This is identical to the method first proposed by Pulay,[36] except for the use of Cartesian rather than internal coordinates. Errors introduced in the finite difference approximation can be detected by examining the asymmetry of $\mathbf{F}$ ($F_{ij} - F_{ji}$) and the number of decimal places to which the translational and rotational eigenvalues are zero.

5. Selected Results and Comparisons

5.1. Introduction

The main objective of this section is the comparison of results calculated by three of the more popular semiempirical molecular orbital methods with those obtained by *ab initio* methods and experiment. One wishes to see how well these methods fare in practice in order to assess their range of applicability and their overall utility. The inclusion of the *ab initio* results is partly motivated by the recent lively debate on the relative merits of semiempirical and *ab initio* methods.[14–16,37,38] The semiempirical methods INDO, MINDO/2, and MINDO/3 are the only ones discussed, not because they are necessarily the best of the semiempirical methods, but only because they have been the most extensively applied. Other methods, both existing and yet to be developed, may very well prove to be superior to the best of these three.

Section 5.2 presents and discusses calculated results of selected equilibrium geometries of molecules containing hydrogen, carbon, nitrogen, and oxygen. Section 5.3 examines calculated relative stabilities of molecules. To facilitate comparisons, the results are organized as enthalpies of reactions.

The final portion of Section 5.3 examines some of the difficulties encountered when attempting to compare and interpret the quality of the results for reaction enthalpies obtained by various methods. This is followed by a short section comparing a few calculated activation energies with experiment.

Finally, calculated vibrational frequencies of four molecules are compared with experiment. In the latter two sections, only the results of two semiempirical methods are given.

5.2. Molecular Geometries

Some comments were made in Section 2.2 regarding the use of calculated as opposed to experimental geometries, and reasons were given why calculated geometries are advantageous, the most important being that they provide a check on the potential function being used. With this in mind, a comparison of the calculated geometries obtained using INDO, MINDO/2, MINDO/3, and the *ab initio* method STO-3G[5] is made with experiment. Tables 2 and 3 present the calculated and experimental geometries of various hydrocarbons and nonhydrocarbons, respectively. Although not all geometrical parameters are reported, all geometries in Tables 2 and 3 were completely optimized.

From Table 2 it is seen that all methods except MINDO/2 reproduce the C—H bond lengths within a few hundredths of an angstrom and all methods, including MINDO/2, reproduce the important trends, such as the progressive shortening of this bond going from ethane to ethylene to acetylene. MINDO/2 tends to overestimate the C—H bond length by about 0.1 Å. STO-3G consistently gives the best result for this bond, closely followed by MINDO/3.

All methods similarly account for the important trends in the C—C bond lengths, again with STO-3G giving the best agreement with experiment, closely followed by MINDO/3. The C—C single bond lengths tend to be underestimated by the semiempirical methods by a few hundredths of an angstrom.

MINDO/3 and STO-3G give results of comparable quality for the H—C—H bond angles, the error being at most about 3° (except in norbornane), with the calculations tending to underestimate this angle. INDO and MINDO/2 tend to underestimate it by 5–6°. All methods generally account for the experimental trends.

The skeletal C—C—C bond angles are also well accounted for by the *ab initio* STO-3G method. The situation is somewhat less satisfactory for the semiempirical methods. Although the sample is small and the difference not great, INDO and MINDO/2 consistently give better results than MINDO/3 for this angle.

Table 3 shows that MINDO/2 inadequately accounts for the geometries of nonhydrocarbons. The X—H bond lengths (X = N, O) are again consistently overestimated. More serious, however, is the MINDO/2 prediction of linear water and planar ammonia. This is also reflected in the bond angles for the alcohols, ethers, and amines. Of the remaining methods, STO-3G gives the best overall results except for water, where MINDO/3 excels. In general, INDO tends to give somewhat better overall results than MINDO/3 for these bond lengths, the most serious errors of the latter being the excessive shortening of the C=O and C—N bond lengths, and the large errors in the H—N—O and H—O—N bond angles.

The discussion of the calculated geometries in these tables is by no means intended to be a statistical analysis but rather a broad survey employing only some, hopefully representative compounds. Several points can be noted about the results. One is that semiempirical methods tend to reproduce multiple bond lengths more accurately than single bond lengths (recall C—C, C—O, N—O, and C—N). In contrast, *ab initio* fares better with single bonds than with multiple bonds. A weakness in MINDO/2 which appears to have been improved in MINDO/3 is the overestimation of the C—H bond length (now overestimated by only about 0.02 Å in MINDO/3). As noted, MINDO/2 also greatly overestimates bond angles in which the central atom is either O or N. Two problems in MINDO/3 pointed out by Bingham *et al.*[39] are a tendency to overestimate the flatness of rings and a tendency to underestimate H—C—H bond angles in methylenes. In general, the *ab initio* STO-3G method appears to be superior to all the semiempirical methods in geometry predictions. However, semiempirical methods do reasonably well in reproducing trends and the overall magnitudes of the errors are not large enough to undermine the utility of these methods in structure prediction. This, coupled with the fact that semiempirical methods are far more economically feasible than *ab initio* methods, will ensure their continued use and development. However, the crucial test of the methods studied above lies in how well they reproduce experimental energies. This will be explored in the next section.

5.3. Energies of Equilibrium States

5.3.1. Chemical Accuracy

This section will examine how well semiempirical and *ab initio* methods reproduce experimental relative energies. However, before discussing detailed results, it may be helpful to consider what is actually being attempted when a quantum mechanical energy calculation is made. The total molecular potential energy (the energy required to separate all of the electrons and nuclei from one another by infinity) is on the order of hundreds of thousands of kilocalories for the molecules considered here. This is the energy obtained in a quantum mechanical calculation. On the other hand, chemical energies, that is, relative energies obtained from rate and equilibrium constant measurements, are measured in terms of kilocalories or at most tens of kilocalories. These energies reflect the differences in stability of molecules. Computing these relative energies to an accuracy of 1 kcal requires taking the difference between two very large numbers, each of which must therefore be accurate to within about

Table 2. Hydrocarbon Geometries[a,b]

Molecule	Coordinate	Exp.	INDO[43]	MINDO/2[44]	MINDO/3[17]	STO-3G[45]
1. Ethane	C–C	1.53[46]	1.46	1.49	1.48	1.54
	C–H	1.10	1.12	1.21	1.11	1.11
	H–C–H	107.8	106.6	105.2	105.5	108.3
2. Ethylene	C–C	1.33[47]	1.31	1.31	1.31	1.31
	C–H	1.08	1.11	1.19	1.10	1.08
	H–C–H	116.6	111.4	111.4	117.7	115.4
3. Acetylene	C–C	1.20[48]	1.20	1.19	1.20	1.17
	C–H	1.06	1.10	1.15	1.07	1.07
4. Allene	C–C	1.31[49]	1.31[44]	1.29	1.31	1.29[50]
	C–H	1.09	1.11	1.19	1.10	1.08
	H–C–H	118.2	112.5	113.5	117.8	116.2
5. Propane	C–C–C	112.4[51]	111.9[44]	116.3	119.6	112.4[50]
6. Propene	C–C=C	124.3[52]	129.3[44]	126.4	128.9	125.1[50]
7. *Trans*-2-Butene	C–C=C	123.0[53]	—	127.0	131.9	124.5
8. *Cis*-2-Butene	C–C=C	127.8[53]	—	127.2	134.3	128.0
9. *Trans*-1,3-butadiene	C–C=C	123.1[54]		129.6	131.0	124.2[55]
	C–C		1.48[44]			
10. Cyclopropane	C–C	1.51[56]		1.48	1.49	1.50[50]
	C–H	1.09	1.12	1.20	1.11	1.08
	H–C–H	115.1	109.2	106.1	107.8	113.8

11. Cyclopropene

Coordinate					
C_1—C_2	1.30[57]	1.32[44]	1.30	1.32	1.28[50]
C_2—C_3	1.52	1.47	1.47	1.48	1.49
C_3—H_3	1.09	1.13	1.21	1.11	1.09
C_1—H_1	1.07	1.10	1.18	1.09	1.08
H—C_3—H	114.7	106.7	107.7	106.4	112.5

12. Benzene

Coordinate					
C—C	1.40[58]	1.39	1.38	1.41	1.39
C—H	1.08	1.12	1.20	1.11	1.08

13. Norbornane

Coordinate					
C_1—C_2	1.56[59]	—	1.52	1.54	—
C_2—C_3	1.55	—	1.52	1.55	—
C_1—C_7	1.56	—	1.52	1.55	—
C—H (av)	1.13	—	1.22	1.12	—
C_1—C_2—C_3	104.2	—	103.1	103.5	—
C_3—C_4—C_5	103.5	—	109.9	110.8	—
H—C_7—H	108.0	—	99.0	102.6	—
H—C_2—H	108.0	—	99.6	102.6	—
C_1—C_7—C_4	96.0	—	93.4	93.4	—
C_2—C_1—C_4—C_5	113.1	—	114.5	115.7	—

[a] Bond distances in Å, bond angles in degrees, all geometries optimized. Unless otherwise noted, reference cited at column heading refers to entire column.
[b] Geometries optimized with respect to all coordinates, but only those coordinates of interest for the comparison are listed.

Table 3. Nonhydrocarbon Geometries[a,b]

Molecule	Coordinate	Exp.	INDO[43]	MINDO/2[44]	MINDO/3[17]	STO-3G[45]
1. CH_4	C—H	1.09[60]	1.12	1.20	1.10	1.08
2. NH_3	N—H	1.01[61]	1.07	1.10	1.03	1.03
	H—N—H	106.7	106.4	120.0	104.2	104.2
3. OH_2	O—H	0.96[62]	1.03	1.10	0.95	0.99
	H—O—H	104.5	104.7	180.0	103.9	100.0
4. CO_2	C—O	1.16[63]	1.23	1.18	1.18	1.19
5. H_2CO	C—O	1.20[64]	1.25	1.22	1.18	1.22
	H—C—H	116.5	115.0	108.2	106.8	114.5
6. CH_3CHO	C—O	1.22[65]	1.26[66]	1.21	1.19	—
	H—C—H	108.3	108.1	104.9	105.6	—
	C—C—O	123.9	127.0	125.9	129.2	—
7. H_3COH	C—O	1.43[67]	1.37	1.32	1.34	1.43[68]
	H—C—H	109.5	108.2	102.4	104.9	108.1
	H—C—O	109.5	110.7	115.8	107.9	107.7
8. $CH_3—O—CH_3$	C—O	1.41[69]	—	1.33	1.34	—
	H—C—H	108.7	—	102.6	105.2	—
	C—O—C	111.7	—	171.2	125.1	—
9. HNO	N—O	1.21[70]	1.19	1.16	—	1.23[68]
	H—N—O	108.6	112.2	128.7	—	107.6
10. H_2NOH	N—O	1.46[71]	1.28	1.27	—	1.42[68]
	N—O—H	103.0	111.8	136.8	—	105.0

11. CH_3NH_2	C—N	$1.47^{(72)}$	1.40	1.41	1.40	$1.49^{(68)}$
	H—C—N	112.9	111.8	111.3	118.6	113.7
12. HCN	C—N	$1.15^{(73)}$	1.18	1.17	1.15	1.15
13. CH_3CN	C—N	$1.16^{(73)}$	—	1.18	1.16	—
	H—C—C	109.5	—	111.9	112.7	—
14. CH_3NC	C—N	$1.42^{(73)}$	—	1.40	1.40	—
	N—C	1.17	—	1.20	1.16	—
	H—C—N	109.1	—	110.6	113.3	—
15. $CH_3N_2CH_3$	C—N	$1.47^{(74)}$	—	1.41	1.41	—
	H—C—N	109.4	—	107.7	121.8	—
16. H_2O_2	O—O	$1.48^{(75)}$	1.04	1.27	1.38	1.40
	H—O—O	94.8	108.8	140.9	103.0	101.1
	H—O—O—H	111.6	83.5	0.0	84.6	125.0
17. O_3	O—O	$1.28^{(76)}$	1.17	1.20	1.25	1.29
	O—O—O	116.8	121.0	147.9	126.1	116.2
18. H_2N_2	N—N	$1.45^{(77)}$	1.33	1.19	1.31	$1.46^{(68)}$
	H—N—N	112.0	110.6	119.1	—	105.4
	H—N—N—H	90.0	69.0	91.0	0.0	91.5
19. $CH_3N_2CH_3$	N—N	$1.25^{(74)}$	—	1.09	1.20	—
	C—N—N	111.9	—	126.3	122.6	—

[a] Bond distances in Å, bond angles in degrees, all geometries optimized. Unless otherwise noted, reference cited at column heading refers to entire column.
[b] Geometries optimized with respect to all coordinates, but only those coordinates of interest for the comparison are listed.

0.0001% for small molecules. It is only because of a not completely understood cancellation of errors that even ballpark estimates of relative stabilities of molecules can be made. Thus, in focusing attention on the shortcomings of various methods in energy prediction, one is justified in asking why they do so well.

Toward the goal of accurate energy computations, an important observation was made by Snyder and Basch.[40] They noted that the energies of complete hydrogenation of a molecule could be calculated within 30 kcal using modest basis set Hartree–Fock methods. This suggests that the correlation errors (on the order of hundreds of kilocalories) for such reactions will largely cancel. It also implies that the experimentally known energies of the small molecules that result from complete hydrogenation can be used in conjunction with these Hartree–Fock calculations to predict the stabilities of large molecules. Hehre *et al.*[5] confirmed this suggestion with additional Hartree–Fock calculations. They further noted that accurate Hartree–Fock calculations for large molecules are not computationally feasible and proposed dividing the complete hydrogenation reaction into two steps, a bond separation reaction followed by the hydrogenation of the products of the bond separation step. The suggestion here is that the bond separation reaction, which contains the large molecule, can be adequately treated by small basis set calculations, and the accurate calculations need only be carried out for the hydrogenation step, which contains molecules having no more than two heavy atoms (i.e., atoms other than hydrogen). The calculated results, some of which are included here, are consistent with this hypothesis.[5]

There are a number of features of this scheme which deserve some attention. First, it should be noted that the total molecular energy is relegated to a position of relative unimportance. Only calculated energy differences are considered. Moreover, implementation of the scheme involves using two different methods for calculating the energy of the same molecule. Second, the scheme focuses only on reactions involving stable closed-shell molecules. This is because of the limitations of closed-shell single-determinant molecular orbital theory. The energies of atomization reactions and many formation reactions are thus inaccessible by this procedure (see discussion in Section 5.3.6 for further elaboration). Third, the success of the procedure depends strongly on the cancellation of various types of errors. In addition to correlation errors, the errors incurred in a small basis set calculation must cancel for the bond separation reactions. Finally, an attractive feature of procedures such as these is that the quality of a given method can be assessed for various types of well-defined and easily interpretable processes, e.g., the separation of bonds in a large molecule.

The tables presented and discussed below include bond separation and hydrogenation reactions as well as isomerizations and a number of other types of artificial reactions which are designed to illustrate the relative energies of molecules in various types of formal bonding situations. The final part of this subsection discusses the difficulties in interpreting the errors in the calculated energies and attempts to identify the strengths and weaknesses of the various methods.

5.3.2. Isomerization Energies

Table 4 contains relative energies of isomers of a number of hydrocarbons. In addition to the three semiempirical methods, results of three *ab initio* methods appear: the minimal basis STO-3G, the extended bases 4-31G[5] and 6-31G*[41] (an extended basis set with polarization functions). The latter method is assumed to lie close to the Hartree–Fock limit. The experimental numbers are obtained from enthalpies of formation at 25°C. As implied in Section 2.5, the neglect of zero-point corrections for isomerizations should not introduce errors that would seriously affect the comparisons. The underlined results are those that give closest agreement with experiment in each case. The geometries used were all optimized for the semiempirical results and the STO-3G results, although only partially for some of the latter, as indicated by the table footnotes. The 4-31G and 6-31G* results were obtained using the STO-3G geometries.

INDO appears to consistently give the worst results. Comparing the INDO results for allene and propyne and butadiene and 2-butyne, it would appear that excessive stability is given to the triple bond and/or the single bond relative to double bonds. This is also reflected in the MINDO/3 and STO-3G results, although somewhat less severely than in INDO. From the data given, MINDO/2, 4-31G, and 6-31G* appear to properly account for this effect.

INDO also gives excessive stability to ring compounds relative to their open-chain analogs. This tendency is reflected to a less severe extent by MINDO/2, the results being between 10 and 30 kcal different from experiment. Except for cyclopropene, the effect is still present in MINDO/3, but to a much lesser extent. In accounting for ring versus open-chain stability, STO-3G lies near MINDO/3 in quality with the ring compounds favored. The effect is not present at all in the 6-31G* results, which give excellent agreement with experiment.

5.3.3. Bond Separation Reactions

Table 5 lists the energies of bond separation reactions calculated by three semiempirical methods, INDO, MINDO/2, and MINDO/3, and two *ab initio* methods representing minimal and extended basis set Hartree–Fock calculations. Also listed are the experimental enthalpies at 298°K and the estimated "experimental" potential energies (the enthalpies at absolute zero minus the zero-point vibrational energy). For ease of interpretation, the underlined result

Table 4. Isomerization Energies[a]

| | Semiempirical | | | Ab initio | | |
Molecule	Exp.[b]	INDO[66]	MINDO/2[44]	MINDO/3[17]	STO-3G[c]	4-31G[c]	6-31G*[c]
C$_3$H$_4$							
Propyne	0.0	0.0	0.0	0.0	0.0	0.0	0.0
Allene	1.6	19.4	2.0	6.9	17.1[d]	0.8[d]	1.7[e]
Cyclopropene	21.9	−42.4	−0.1	24.2	30.0[d]	36.4[d]	25.4[e]
C$_3$H$_6$							
Propene	0.0	0.0	0.0	0.0	0.0	0.0	0.0
Cyclopropane	7.8	−94.0	−10.5	2.4	−3.7[d]	13.2[d]	7.8[e]
C$_4$H$_6$							
Butadiene-1,3 (C$_{2h}$)	0.0	0.0	0.0	0.0	0.0	0.0	0.0
Butadiene-1,3 (C$_{2v}$)	—	0.2	2.2	0.7	4.4	4.3	5.7
2-Butyne	8.6	−40.5	—	−19.8	−10.2	6.5	6.7
Cyclobutene	11.2	−129.3	−20.3	1.2	−12.5	19.9	12.4
Methylene cyclopropane	21.7	—	—	2.0	5.8	25.3	20.2
Bicyclobutane	25.6	−167.7	15.0	17.8	11.6	47.2	30.4
1-Methyl cyclopropene	32.1	−75.5	—	10.1	17.8	43.3	31.4

[a]Energies in kcal/mole. Value underlined is best result among all methods. Unless otherwise noted, reference cited at column heading is reference for given column. All geometrics completely optimized unless otherwise indicated.
[b]Obtained from heats of formation at 298°K.
[c]Standard model bond lengths and angles used, with the exception of all C—C—C angles, which are optimized. 1,3-Butadiene is optimized at STO-3G level. See Ref. 55.
[d]Energies calculated using STO-3G optimized geometries. See Ref. 50.
[e]Results obtained using STO-3G optimized geometries. See Ref. 14.

Table 5. Bond Separation Energies[a]

Reaction	Exp. 0°K[b]	Exp. 298°K	Semiempirical INDO[66]	MINDO/2[44]	MINDO/3[17]	Ab initio STO-3G[d]	Extended basis[h]
1. $CH_3CH_2CH_3 + CH_4 \rightarrow 2CH_3CH_3$	1.5	2.3	−5.3	−3.4	−6.9	0.6	1.0[i]
2. $CH_3CH{=}CH_2 + CH_4 \rightarrow CH_3CH_3 + H_2C{=}CH_2$	5.0	5.3	3.1	4.3	−0.5	4.4	3.9[i]
3. $CH_3C{\equiv}CH + CH_4 \rightarrow CH_3CH_3 + HC{\equiv}CH$	7.2	7.6	15.4	9.5	9.2	8.4	8.1[i]
4. $CH_2{=}C{=}CH_2 + CH_4 \rightarrow 2H_2C{=}CH_2$	−4.1	−2.7	10.9	1.4	2.8	0.5	−4.4[i]
5. $CH_3CH_2CH_2 + 3CH_4 \rightarrow 3CH_3CH_3$	−23.5	−19.6	−116.9	−15.9	−49.3	−45.0	−27.0[i]
6. $C_6H_6 + 6CH_4 \rightarrow 3CH_3CH_3 + 3H_2C{=}CH_2$	61.1	64.5	70.9	43.6	7.2	69.9[e]	63.6[k]
7. $CH{=}CHCH_2 + 3CH_4 \rightarrow 2CH_3CH_3 + H_2C{=}CH_2$	−45.2	−40.4	−141.5	−27.1	−61.0	−65.1	−57.6[i]
8. $CH_2{=}CHCH{=}CH_2 + 2CH_4 \rightarrow 2H_2C{=}CH_2 + CH_3CH_3$	—	14.3	6.5	13.9[c]	−0.7[(9)]	12.6	11.9[i]
9. $CH_3C{\equiv}CCH_3 + 2CH_4 \rightarrow HC{\equiv}CH + 2CH_3CH_3$	—	15.0	32.1	—	18.7[(9)]	13.7[f]	16.6[i]
10. $CH_2CH{=}CHCHCH_2 + 4CH_4 \rightarrow H_2C{=}CH_2 + 3CH_3CH_3$	—	−14.0	−78.3	6.7[c]	−48.1[(9)]	−28.1[f]	−27.0[i]
11. $H_2C{=}C{=}O + CH_4 \rightarrow H_2C{=}CH_2 + H_2C{=}O$	16.5	15.8	17.3	28.4	20.6	15.6[g]	12.8
12. $O{=}C{=}O + CH_4 \rightarrow 2H_2C{=}O$	57.9	60.0	54.1	61.2	50.8	52.5[g]	52.2
13. $CH_3OCH_3 + H_2O \rightarrow 2CH_3OH$	5.3	5.6	−41.3	−2.8	−5.0	2.3[g]	3.3
14. $NH_2CH{=}O + CH_4 \rightarrow CH_3NH_2 + H_2C{=}O$	29.8	30.9	23.3	28.2	27.8	12.5[g]	32.4
15. $CH_3CH{=}O + CH_4 \rightarrow CH_3CH_3 + H_2C{=}O$	8.5	11.4	6.5	13.9	4.6	7.1[g]	11.3

[a] Energies are given in kcal/mole. Value underlined is best result among all methods for a given reaction. Unless otherwise noted, reference cited at column heading is reference for given column. Also, geometries are completely optimized unless otherwise indicated.
[b] Heats of reaction at 0°K corrected for zero-point vibration. See Ref. 5.
[c] Results using partially optimized geometries. See Refs. 8 and 78.
[d] Results using optimized STO-3G geometries. See Ref. 50 and, for C₄ hydrocarbons, Ref. 55.
[e] As footnote d except standard geometry used for benzene. See Ref. 5.
[f] As footnote d except C₄ hydrocarbons only partially optimized.
[g] Results using experimentally determined equilibrium geometries. See Ref. 5.
[h] 4-31G results using experimentally determined equilibrium geometries. See Ref. 5.
[i] 6-31G results using optimized STO-3G geometries. See Ref. 50.
[j] 6-31G* results using optimized STO-3G geometries. See Ref. 14.
[k] As footnote j except standard geometries used for benzene. See Ref. 14.
[l] 4-31G results using STO-3G optimized geometries, except for C₄ hydrocarbons, which are partially optimized. See Refs. 50 and 55.

is that which best agrees with experiment. Due to the different approaches in parametrizing the semiempirical methods, the MINDO/2 and MINDO/3 results should be compared to the enthalpy at 298°K. The INDO and *ab initio* results, on the other hand, should be compared to the corrected energy at 0°K.

In contrast to the isomerizations, bond separation reactions preserve the number of bonds of each formal type. Although these reactions are largely artificial, the energies of bond separation reactions are a measure of the stabilities of adjacent bonds relative to the same bonds separated.[5] Positive bond separation energies indicate stabilization, such as occurs in resonance or hyperconjugation when bonds are juxtaposed. Negative bond separation energies occur in strained systems such as small rings and antiaromatic molecules.

Examination of Table 5 reveals that all methods slightly underestimate the stabilization of adjacent carbon–carbon single bonds (reaction 1) and all methods except MINDO/2 overestimate the strain energy of the small rings (reactions 5, 7, and 10). MINDO/2 tends to underestimate this ring strain slightly in cyclopropane and more seriously in cyclopropene. It is only slightly surpassed in accuracy by the extended basis calculation.

The stability of adjacent C—C single and double bonds in open-chain compounds relative to these same bonds separated is illustrated in reactions 2 and 8. All methods tend to slightly underestimate this stability, although the errors are not great. MINDO/2 in particular gives excellent agreement with experiment. Reaction 6 is an example of adjacent C—C single and double bonds in a ring. However, this reaction also includes the effect of resonance stabilization in benzene. In addition to the Hartree–Fock methods, only INDO and, to a somewhat lesser extent, MINDO/2 account for this effect.

The third instance of single and double bond adjacency in hydrocarbons occurs in reactions 7 and 10, where the strained rings as well as the single–single C—C bond adjacencies tend to mask the effect. Comments on these reactions have been given above.

A triple bond and a single bond are more stable adjacent than they are separated. From reactions 3 and 9 all methods appear to account for this effect, with only INDO significantly exaggerating it.

On the other hand, the double bonds in allene (reaction 4) are experimentally less stable adjacent than they are separated. Only the extended *ab initio* method properly accounts for this, although the errors in MINDO/2, MINDO/3, and STO-3G are not large.

Reactions 11–15 illustrate the relative stabilities of adjacent and separated bonds in nonhydrocarbons. From reaction 13 it is seen that, like the C—C single bonds, the stabilization of adjacent O—C single bonds is underestimated by all methods, with the extended basis *ab initio* method giving the best agreement with experiment and INDO the worst. Of the semiempirical methods, MINDO/2 gives slightly better agreement than MINDO/3.

The stability of the adjacent single and double bonds (reactions 14 and 15) is fairly well reproduced by all methods, with the extended basis Hartree–Fock method, closely followed by MINDO/2, giving the best results for formamide and the minimal basis set Hartree–Fock method and INDO giving the best agreement with experiment for acetaldehyde. None of the calculated results are in serious error for these two reactions.

In contrast to allene, adjacent double bonds among the nonhydrocarbons are more stable experimentally than their separated counterparts. This effect appears to be fairly well accounted for by all methods, as seen in reactions 11 and 12. INDO, in particular, does surprisingly well here.

There are a number of conclusions that can be drawn from the results in Table 5. First, from reactions 5 and 14 in particular, it is clear that large differences can occur between the minimal and extended basis *ab initio* results, with the extended basis method almost always giving the best agreement with experiment. Second, in an apparent contrast to the results of Table 4, the strain energy of small rings, as defined in terms of bond separation energies, is *overestimated* by INDO and MINDO/3 and is fairly accurately accounted for by MINDO/2. This point will be discussed in Section 5.3.6. Finally, none of the methods gives consistently accurate results, although the overall accuracy of the extended basis and MINDO/2 calculations is impressive.

5.3.4. Bond Conversion Reactions

The reactions listed in Table 6 illustrate the relative stabilities of molecules containing single, double, and triple bonds. Reactions 1–3, for example, compare these bonds in hydrocarbons containing two carbon atoms. In comparing multiple with single bonds it is conceptually convenient to regard the molecular orbitals as localized. Thus, reaction 1 compares the stability of one of the two banana bonds in ethylene with the single bond in ethane. In this sense, reaction 1 is a bond separation reaction for a two-membered ring.

Inspection of reactions 1 and 2 shows that all methods overestimate the stability of a carbon–carbon single bond compared to half a double bond or one-third of a triple bond. INDO gives the worst result in each case, whereas the extended basis Hartree–Fock method, closely followed by MINDO/2, gives the best agreement. The same is true for reaction 3 (obtained from 1 and 2), which compares the double bond directly (i.e., without CH_4) with the average of the single and triple bonds.

Reactions 4–8 illustrate the effect of bond conversions and transfers in hydrocarbons containing three carbon atoms. Reactions 4 and 5, for example, show the small extra stabilities of double and triple bonds in C_3 as opposed to C_2 hydrocarbons. All methods except the extended basis *ab initio* method exaggerate this effect. Of the semiempirical methods, MINDO/3 best accounts for the double bond transfer, whereas MINDO/2 gives better results for the

Table 6. Bond Conversion Reactions[a]

Reaction	Exp.		Semiempirical			Ab initio	
	0°K[b]	298°K	INDO[66]	MINDO/2[44]	MINDO/3[17]	STO-3G[d]	Extended basis[g]
1. $CH_2=CH_2 + 2CH_4 \rightarrow 2CH_3CH_3$	-21.2	-17.2	-214.0	-30.8	-46.4	-53.2	-22.4
2. $CH\equiv CH + 4CH_4 \rightarrow 3CH_3CH_3$	-51.3	-43.2	-413.4	-67.5	-92.4	-96.6	-55.5
3. $CH\equiv CH + CH_3CH_3 \rightarrow 2CH_2=CH_2$	-9.1	-9.3	14.9	-7.9	0.4	9.1	-10.8
4. $CH_3CH_2CH_3 + CH_2=CH_2$ $\rightarrow CH_3CH=CH_2 + CH_3CH_3$	-3.5	-2.9	-8.4	-13.6	-6.0	-3.8	-2.8
5. $CH_3CH_2CH_3 + CH\equiv CH \rightarrow CH_3C\equiv CH + CH_3CH_3$	-5.7	-5.3	-20.7	-12.9	-16.1	-7.8	-7.6[h]
6. $CH_2CH_2CH_2 + CH_2=CH_2$ $\rightarrow CH=CHCH_2 + CH_3CH_3$	21.7	20.8	24.6	11.2	11.7	20.5	30.6[h]
7. $CH_2CH_2CH_2 \rightarrow \frac{3}{2}CH_2=CH_2$	8.2	6.1	204.1	30.2	20.3	34.7	7.3
8. $CH_3CH=CH_2 \rightarrow \frac{3}{2}CH_2=CH_2$	15.6	13.9	110.1	19.7	22.7	31.0	15.1
9. $CH_2CH_2CH_2CH_2 \rightarrow 2CH_2=CH_2$	—	18.0	349.7	35.0[c]	43.5[9]	79.0[e]	13.9[i]
10. $CH_2CH=CHCH_2 \rightarrow CH\equiv CH + CH_2=CH_2$	—	29.3	334.8	60.4[c]	43.9[9]	69.1[e]	22.2[i]
11. $CH_2CH_2CH_2CH_2 + CH_2=CH_2$ $\rightarrow CH_2CH=CHCH_2 + CH_3CH_3$	—	-2.0	0.0	-26.5[c]	-0.8[9]	0.9[e]	3.0[i]
12. $C_6H_6 \rightarrow 3CH\equiv CH$	151.1	143.1	668.3	153.3[c]	144.6[9]	209.7[f]	162.9[f]
13. $H_2C=C=O + CH_3CH_3 \rightarrow H_2C=O + CH_2=CHCH_3$	11.5	10.5	14.2	24.1	21.1	11.2[f]	8.8[k]
14. $O=C=O + CH_3CH_3 \rightarrow H_2C=O + CH_3CH=O$	49.4	48.6	47.6	47.3	46.2	45.4[f]	40.9[k]
15. $O=C=O + CH_3NH_2 \rightarrow H_2C=O + NH_2CH=O$	28.1	29.1	30.8	33.0	23.0	40.0[f]	19.8[k]
16. $O=C=O + CH_2=CH_2 \rightarrow H_2C=O + H_2C=C=O$	41.4	44.2	36.8	32.8	30.2	36.9[f]	39.4[k]

[a] Energies in kcal/mole. The best value among all methods is underlined. Unless otherwise noted, reference cited at column heading is reference for given column. All geometries are completely optimized unless otherwise indicated.

[b] Heats at 0°K corrected for zero-point vibrations. See Ref. 5.

[c] See Refs. 8 and 78. Results using partially optimized geometries.

[d] Results using optimized STO-3G geometries.[50]

[e] As footnote d for C_2 hydrocarbons. C_4 hydrocarbons partially optimized. See Ref. 55.

[f] Experimental geometries used. See Ref. 5.

[g] 6-31G* results using STO-3G optimized geometries. See Ref. 14.

[h] As footnote d, except 6-31G results using STO-3G optimized geometries.

[i] As footnote e, except results using 4-31G basis set.

[j] 6-31G* results using STO-3G optimized acetylene geometry and standard geometry for benzene.

[k] 4-31G results using experimental geometries. See Ref. 5.

triple bond. Reaction 6 can be interpreted as showing mainly the additional strain in cyclopropene compared to cyclopropane. Curiously, INDO and the *minimal* basis *ab initio* method best account for this effect. Reactions 7 and 8 are similar to isomerizations in that only two different molecules are involved, For reaction 7 only the extended basis *ab initio* result reasonably accounts for the stability of the single bond in the ring relative to the ethylene double bond. The same is true for 8, although the MINDO methods give better results here than for 7.

Reactions 9–11 involve four-membered rings. The pyrolyses of cyclobutane (9) and cyclobutene (10) are poorly accounted for by all but the extended basis method, with excess stability given to the rings. Reaction 11 is similar to 6 except that four-membered rings are involved. In contrast to 6, the double bond in the ring is the more stable situation. All methods but MINDO/2 accurately account for this.

In reaction 12 both MINDO methods account well for the stability of benzene relative to three acetylenes, with MINDO/3 giving extremely close agreement with experiment.

Reactions 13–16 compare C=C and C=O bonds in molecules involving two and three heavy atoms. In 13, for example, the stability of the adjacent double bonds in ketene compared to the separated C=C and C=O bonds is accounted for by all methods and particularly well by INDO and both *ab initio* methods. MINDO/2 and INDO best account for the stability of the double bonds in carbon dioxide in reactions 14 and 15. Reaction 16 measures the stability of the adjacent double bonds in carbon dioxide and the double bond in ethylene over the adjacent double bonds in ketene and the C=O bond in formaldehyde. Again, INDO is the best among the semiempirical methods for this reaction, although it does not do quite as well as the *ab initio* methods. It should be noted that for all of these nonhydrocarbon reactions, none of the methods fails seriously.

5.3.5. Hydrogenation Reactions

A small number of partial and complete hydrogenation reactions of hydrocarbons are listed in Table 7. These reactions are noteworthy because of the relative lack of agreement of any of the methods with each other or with experiment.

The errors in the extended basis Hartree–Fock calculations, while generally less than the other methods, are frequently between 15 and 35 kcal. The three cases in which MINDO/3 gives relatively good agreement with experiment are those reactions (5, 7, and 8) that contain no methanes. Moreover, the error in the MINDO/3 results tends to increase with the stoichiometric coefficient of methane, indicating that a large error can be associated with methane itself.[16] This also accounts for the relatively poor MINDO/3

Table 7. *Hydrogenation Reactions*[a]

	Exp.		Semiempirical			Ab initio	
Reaction	0°K[b]	298°K	INDO[c]	MINDO/2[d]	MINDO/3[e]	STO-3G[f]	Extended basis[h]
1. CH$_3$CH$_3$ + H$_2$ → 2CH$_4$	−18.1	−15.6	40.8	89.3	7.1	−18.8	−22.0
2. H$_2$C=CH$_2$ + 2H$_2$ → 2CH$_4$	−57.2	−48.2	−132.4	151.2	−32.0	−90.8	−66.3
3. HC≡CH + 3H$_2$ → 2CH$_4$	−105.4	−90.1	−290.7	214.2	−70.7	−153.6	−121.4
4. CH$_3$CH=CH$_2$ + 3H$_2$ → 3CH$_4$	−70.3	−58.5	−88.5	244.8	−25.4	−105.2	−84.4
5. 2CH$_3$CH=CH$_2$ + 3H$_2$ → 3CH$_3$CH$_3$	−86.3	−70.2	−299.4	221.7	−72.1	−154.0	−102.8
6. C$_6$H$_6$ + 9H$_2$ → 6CH$_4$	−164.8	−126.9	−203.9	765.1	−67.5	−258.9[g]	−201.3[i]
7. C$_6$H$_6$ + 6H$_2$ → 3CH$_3$CH$_3$	−110.5	−80.1	−326.3	497.2	−88.8	−202.5[g]	−135.3[i]
8. C$_6$H$_6$ + 3H$_2$ → 3CH$_2$=CH$_2$	6.8	17.7	193.3	311.5	28.5	13.5[g]	−2.4[i]

[a] Energies in kcal/mole. The best result among all methods is underlined. Unless otherwise indicated, reference cited at column heading is reference for the given column.
[b] Heats of reaction corrected for zero-point vibration.[66]
[c] Results using optimized geometries.
[d] Results using partially optimized geometries. See Refs. 8, 17, and 78.
[e] Results using optimized geometries. See Ref. 9.
[f] STO-3G results using optimized STO-3G geometries.[50]
[g] As footnote *f*, except standard equilibrium geometry used for benzene. See Ref. 5.
[h] 6-31G* results using optimized STO-3G geometries. See Ref. 14.
[i] As footnote *h*, except standard equilibrium geometries used for benzene. See Ref. 14.

energies for bond separation reactions. Moreover, a similar phenomenon exists for hydrogen in the MINDO/2 method, which gives relatively good agreement for reactions not containing hydrogen. As will be discussed in the next section, the unambiguous assignment of an absolute error to a single molecule requires that an appropriate zero level of energy be established.

5.3.6. Discussion of Errors

There are a number of difficulties encountered when attempting to analyze the source and nature of the errors that enter into the various energy calculations. These difficulties center about the choice of the origin of the energy scale. Both the magnitudes and the interpretations of the errors depend markedly on the particular choice of the zero of energy.

Seemingly acceptable choices would be to take as zero the energy of all the nuclei and electrons at rest and separated by infinity, or to compare calculated and experimental energies of atomization. However, these are unsatisfactory for the *ab initio* Hartree–Fock methods because of the large difference between the correlation energy of the closed-shell molecule and its constituent ions or open-shell atoms. The energy of a collection of isolated atoms is very poorly described by a closed-shell single-determinant wave function, be it *ab initio* or semiempirical. Computational difficulties with graphite similarly preclude the use of energies of formation.

These problems are avoided in the MINDO methods, which have a built-in energy scale. In MINDO/3, for example, the molecular energy is calibrated against experimental heats of atomization by calculating the left-hand side of the atomization reaction and using *experimental* atomic energies for the right-hand side. The adjustable parameters in the method are determined by a least squares fit of the calculated to the experimental heats of atomization using a variety of molecules. This parametrization procedure thus allows MINDO/3 to absorb the difference in correlation energy between the molecule and its constituent atoms, the effect of finite temperature, the molecular zero-point vibrational energy, and any effects due to the presence of the $1s$ electrons in the isolated heavy atoms and their absence in the molecule (valence shell approximation). For the purpose of presenting the results, the calculated heats of atomization are converted into heats of formation using the experimental heats of formation of the atoms. This approach is both straightforward and unbiased, and has the important advantage of allowing the assignment of an easily interpretable absolute error to the energy of a single molecule. As Dewar has illustrated,[16] one can easily point to those molecules that are well represented by the theory and those that are not and thereby determine which *reactions* will be properly accounted for and which will not.

As was seen earlier, the situation is not as simple for INDO and the *ab initio* methods. One cannot conveniently assign an absolute error to the energy of a given molecule, since only energies of calculated reactants relative to calculated products are obtained.

For the purpose of comparing the various methods, one could pick a list of molecules, construct all possible reactions involving those molecules, and then attempt to categorize the results and thereby determine the strengths and weaknesses of each method. This, however, would be a mammoth undertaking and would very likely yield far more information than could be easily digested.

The examples discussed in the previous sections represent a crude attempt along these lines. A much more convenient method of comparing results is to assign an energy and, hence, an error to each molecule. This can be done by choosing an energy scale which uses a small set of elementary molecules as a reference point for larger molecules, but, in contrast to using formation and atomization reactions, all the molecules in the reaction are calculated by closed-shell, single-determinant methods. One can easily define a number of such scales, the only question being which one should be used, since the errors in each method will depend upon the particular selection.

One obvious choice would be to use complete hydrogenation energies. This scale uniquely assigns an energy to each molecule relative to H_2, CH_4, NH_3, H_2O, and HF. From the results of Table 7, however, it appears that on this scale, all methods would show large errors. Moreover, an unfair bias against the MINDO methods would exist because of MINDO/2's poor value for the energy of hydrogen and MINDO/3's for methane.

For hydrocarbons, one can isolate methane by defining a scale of partial hydrogenation energies based on the reaction

$$C_nH_{2m} + \tfrac{1}{2}(3n - 2m)H_2 \rightarrow \tfrac{1}{2}n\,C_2H_6$$

which references molecular energies relative to hydrogen and ethane. On this scale, MINDO/3 gives superb agreement with experiment (except, of course, for methane) relative to all other methods.

One can also eliminate hydrogen by using bond separation reactions as an energy scale. For hydrocarbons the complete bond separation reaction (i.e., including reactions 1 and 2 of Table 6) is

$$C_n H_{2m} + (3n - 2m)CH_4 \rightarrow (2n - m)C_2H_6$$

The "complete bond separation" scale of energies fixes all molecular energies (including hydrogen) relative to ethane and methane. As seen in Tables 5 and 6, this scale strongly favors MINDO/2 over STO-3G and the other semiempirical methods.

One can, of course, devise other systematic procedures for comparing energies and they need not even reference a molecular energy to well-defined subunits. The point of the above discussion is that the choice of an energy scale can easily bias the comparisons.

Table 8. Error Analysis Using Heats of Formation[a]

Reaction	Exp.(298°K)	MINDO/2[44]	MINDO/3[9]
1. $\overset{\frown}{CH_2CH_2CH_2} \rightarrow 3H_2 + 3C$ (graphite)	−12.7	7.3 (−20.0)	−8.7 (−4.0)
2. $3H_2 + 3C$ (graphite) $\rightarrow CH_3CH{=}CH_2$	4.9	3.2 (1.7)	6.5 (−1.6)
3. $\overset{\frown}{CH_2CH_2CH_2} \rightarrow CH_3CH{=}CH_2$	−7.8	10.5 (−18.3)	−2.2 (−5.6)

[a] Energies in kcal/mole. Value in parentheses is difference from experiment (i.e., exp. − calc.). All geometries optimized.

The interpretation of the errors of a computed result can also depend dramatically on the choice of an energy scale. This is perhaps best illustrated by a specific example: the MINDO/2 and MINDO/3 calculations of the isomerization of cyclopropane to propene.

Table 8 shows how the heat of the isomerization is obtained from the heats of formation of cyclopropane and propene. The numbers in parentheses are the errors, taken always as the experimental minus the calculated energy quantity (in this case enthalpies at 298°K). The interpretation of Table 8 is as follows. MINDO/2 determines the heat of formation of cyclopropane to be too negative, while it accurately accounts for the heat of formation of propene. The two MINDO/2 errors for the formation reactions only partly cancel one another, leaving the rather large error for the isomerization. MINDO/3 also underestimates the heat of formation of cyclopropane, but only by 4 kcal. It also accurately accounts for the heat of formation of propane, but the error is in the opposite sense of MINDO/2's. Here, the two errors unfortunately add to give the overall error of 5.6 kcal. From these results it is clear that MINDO/2 and, to a much lesser extent, MINDO/3 underestimate the strain energy of the small ring since propene is well accounted for by both methods.[39]

Table 9 shows how the isomerization would be viewed using a bond separation rather than a heat of formation energy scale. The second reaction in Table 9 has been inserted to show the special separation of the double bond. The reverse of the "complete bond separation" reaction of propene can be obtained by simply adding reactions 2 and 3. The interpretation of Table 9 is as follows. MINDO/2 calculates the bond separation of cyclopropane to be too

Table 9. Error Analysis Using Bond Separation[a]

Reaction	Exp.(298°K)	MINDO/2[44]	MINDO/3[17]
1. $\overset{\frown}{CH_2CH_2CH_2} + 3CH_4 \rightarrow 3CH_3CH_3$	−19.6	−15.9 (−3.7)	−49.3 (29.7)
2. $2CH_3CH_3 \rightarrow CH_2{=}CH_2 + 2CH_4$	17.2	30.8 (−13.6)	46.4 (−29.2)
3. $CH_3CH_3 + CH_2{=}CH_2$ $\rightarrow CH_3CH{=}CH_2 + CH_4$	−5.3	−4.3 (−1.0)	0.5 (−5.8)
4. $\overset{\frown}{CH_2CH_2CH_2} \rightarrow CH_3CH{=}CH_2$	−7.8	10.5 (−18.3)	−2.2 (−5.6)

[a] Energies in kcal/mole. Value in parentheses is difference from experiment (i.e., exp. − calc.). All geometries optimized.

positive by a small amount. However, the bond separation energy of ethylene (reverse of 2) is too negative by 13.6 kcal. Although it accurately accounts for the (partial) bond separation energy of propene (reaction 3), the three errors unfortunately add to give the large error for the isomerization. MINDO/3, on the other hand, determines the bond separation energy of cyclopropane to be much too negative. A similar error appears in the bond separation energy of ethylene (i.e., MINDO/3 too negative) and to a lesser degree in the (partial) bond separation energy of propene. Thus the relatively good MINDO/3 result for the isomerization is due only to the near perfect cancellation of two large errors. Moreover, since the bond separation energy of a small ring is a measure of the ring strain energy,[5] we see that, in sharp contrast to the interpretation of Table 8, MINDO/2 only slightly underestimates the strain energy of the ring, whereas MINDO/3 *overestimates* this energy by 30 kcal.

The example given above is only meant to illustrate the pitfalls one can encounter when attempting to compare and interpret calculated reaction energies. Many other examples can easily be found involving all methods, *ab initio* as well as semiempirical, and other energy scales.

These difficulties are largely brought about by the nature and extent of the errors themselves. Thus, in Tables 4–7 there exists at least one reaction for each method in which the calculated result is in excellent agreement with experiment compared to the other methods. On the basis of the data given in these tables, one cannot decide whether or not a good calculated result is due to the fortuitous cancellation of errors or a poor result is due to an unfortunate combination of errors, such as the sum of individual errors, or the multiplication of an error in a given molecule by a large stoichiometric coefficient. In spite of these difficulties, a general estimation of the quality of the various methods can be ventured.

INDO generally gives poor and often spurious results. In particular, reactions involving ring compounds are most seriously in error, these errors often exceeding 100 kcal. Of all the reactions in Tables 4–7, only the bond separation energy of propene and bond transfer reactions among nonhydrocarbons are well accounted for by INDO.

At the other extreme, the extended basis set *ab initio* methods generally give good results. The worst error in the isomerizations of Table 5 is only 5 kcal, the largest bond separation errors are those for cyclopropene and cyclobutene (12.4 and 13.0 kcal, respectively), and the largest error of Table 6 is 11.8 kcal for the trimerization of acetylene. The rather larger errors for hydrogenation reactions are probably due to unfavorable stoichiometry. It is not difficult to show that if a method gives an error of $\gamma_{m,n}$ for the complete bond separation reaction of $C_n H_{2m}$ and an error of δ for the hydrogenation of ethane, then the error in the complete hydrogenation energy of $C_n H_{2m}$ must be $\gamma_{m,n} + (2n - m)\delta$. Thus, the 36 kcal error for the extended basis *ab initio* calculation of the complete hydrogenation of benzene (reaction 6 of Table 7) can be interpreted as being largely due to the relatively small 4 kcal error for the hydrogenation of ethane.

The other methods, MINDO/2, MINDO/3, and STO-3G, lie between the extremes of accuracy represented by INDO and the extended basis set methods. In Tables 4 and 6, STO-3G results are poorest for those reactions involving a ring compound as a reactant and not as product, the errors being as large as 60 kcal, with the reactant side being excessively favored. The errors for MINDO/2 and MINDO/3 are somewhat smaller (~20 kcal) for these reactions but are still noticeable. For the remaining reactions in these two tables the three methods are comparable. MINDO/2 and STO-3G are also comparable for the bond separation reactions of Table 5, both methods having a worst error of about 20 kcal with typical errors of about 10 kcal. MINDO/3 gives much larger errors for these reactions, the worst being about 60 kcal for benzene. This is mainly due to the relatively large MINDO/3 error for the heat of formation of methane.

In general, for heats of reaction, MINDO/2 and MINDO/3 appear to be comparable to or slightly better in quality than STO-3G. It is not clear from the scanty results listed here, however, that MINDO/3 represents a substantial improvement over MINDO/2.

5.4. Activation Parameters

With the development of efficient search techniques for the location of saddlepoint geometries, the calculation of activation parameters with the semiempirical methods becomes relatively straightforward. Since very few comparable calculations using *ab initio* methods have yet been performed, only selected results of MINDO/2 and MINDO/3 calculations are reported in Table 10.

As can be seen in Table 10, the quality of the calculated activation enthalpies is limited by the accuracy of the enthalpy of the overall reaction. Thus, even though the barriers to some reactions are in excellent agreement with experiment, those for the reverse reactions are often in error by as much as 30 kcal. There is some hope, however, that these types of calculations will give reasonable estimates of activation enthalpies in cases where the overall reaction enthalpy is correctly predicted. The MINDO/3 results for the Cope rearrangements are in excellent agreement with experiment, for example. Moreover, it has been noted in MINDO/2 calculations that when the geometry of the calculated transition state more closely resembles the reactant than the product, then the activation enthalpy calculated for the forward reaction is in better agreement with experiment than that of the reverse process. This was the case in the disrotatory closing of hexatriene to cyclohexadiene, for example.[32]

Table 10. *Activation Enthalpies and Entropies*[a]

Reaction	$\Delta H^{\dagger}_{exp}$ [b]	$\Delta H^{\dagger}_{M/2}$	$\Delta H^{\dagger}_{M/3}$ [b]	$\Delta S^{\dagger}_{exp}$	$\Delta S^{\dagger}_{M/2}$
Ethane rotation barrier	2.9[9]	1.9[44]	0.9[9]	—	—
Conrotatory opening	32.5	49.5[24]	48.9	1.8[24]	0.8[24]
Conrotatory closing	44.0	29.5[24]	51.0	—	—
Cope rearrangement: chair mechanism	33.5	22.3[13]	35.1	−4.0[13]	−3.8[13]
Cope rearrangement: boat mechanism	44.6	30.3[13]	41.4	−3.8[13]	−3.0[13]
Diels–Alder	26.4	39.4[c]	27.5	—	—
Retro Diels–Alder	66.9	96.0[c]	88.7	—	—
Pyrolysis of cyclobutane	62.5	—	62.0	—	—
Dimerization of ethylene	44.5	—	13.5	—	—
Disrotatory opening	44.4[32]	56.5[32]	—	—	—
Disrotatory closing	29.9[32]	24.6[32]	—	−7.0[32]	−6.0[32]
Cyclohexane inversion	10.8[13]	7.1[13]	—	6.8[13]	2.8[13]

[a] Enthalpies in kcal/mole, entropies in eu. M/2≡MINDO/2 and M/3≡MINDO/3.
[b] Unless otherwise noted, results from Ref. 38.
[c] Enthalpies for symmetric structure. Actual calculated values must be lower. See Refs. 30 and 31.

Also included in Table 10 are the MINDO/2 calculated entropies of activation for five of the reactions. These were calculated *a priori* by the methods described in Section 2. Note that in each case the sign and, often, the magnitude of the activation entropy are correctly given.

Although the energetic accuracy of these methods is limited, such calculations are useful in that, in many cases, they will give the correct order of the energies of alternative transition states and may hence be helpful in determining reaction mechanisms. Moreover, the overall quality of the entropies indicates that such calculations may also be useful in the understanding of kinetic isotope effects. Finally, transition state geometries calculated by semiempirical methods may serve as starting points for accurate *ab initio* determinations.

5.5. Vibrational Frequencies

The calculation of force constants and the associated vibrational frequencies provides a stringent test of the reliability of any potential function. The accurate prediction of the vibrational spectra of reactants and transition state is necessary if the barrier height and the entropy of activation are to be predicted. The vibrational frequencies are also needed to obtain the zero-point energies. Table 11 presents the results of a number of calculations on four hopefully representative molecules. The numerical methods described in Section 4.3 were used for these calculations. Before discussing some of the trends, it should be noted that neither the calculated nor experimental frequencies were corrected for anharmonicities. Thus, the errors apparent in the tables not only reflect computational deficiencies of each potential function, but also the neglect of anharmonicities. Finally, as noted earlier (Section 3.2), these calculations require that the MINDO energies be interpreted as potential energies.

There are several general trends in the calculated results. First, both the MINDO/2 and MINDO/3 results are generally in fair agreement with experiment. A closer inspection reveals that MINDO/2 gives better stretching than bending frequencies. The MINDO/3 C—H stretches would be in much better agreement were it not for a consistent error of about 450 cm^{-1} in all calculated C—H stretching modes. For both methods the bends are usually underestimated, with the out-of-plane bends and twists reproduced more accurately than the in-plane bends. There is also a tendency for both methods to reverse the order of some of the bands. For example, in ethylene, MINDO/2 predicts the high-frequency b_{3u} stretching frequency to be lower than the corresponding b_{1g} mode. Regarding overall agreement in all molecules studied, the MINDO/3 method provides a substantial improvement in the middle-frequency vibrations over those calculated with MINDO/2.

Table 11. *A Comparison of Calculated and Experimental Vibrational Frequencies*[a]

Symmetry	ν(obs)[3]	MINDO/2[79]	MINDO/3[17]
Ethylene			
a_g	1342	1114	1296
	1632	1684	1835
	3164	3019	3544
a_u	825	736	886
b_{1g}	1050	915	1024
	3272	3125	3538
b_{1u}	949	889	979
b_{2g}	943	787	811
b_{2u}	995	694	697
	3106	3145	3557
b_{3u}	1443	1113	1306
	2990	3134	3527
Acetylene			
π_g	612	508	488
π_u	729	776	885
Σ_u^+	1974	2085	2237
Σ_g^+	3374	3441	3827
Σ_u^+	3287	3381	3770
Propene			
a'	417	358	373
	919	764	840
	1043	836	951
	1224	1013	1158
	1278	1033	1209

Symmetry	ν(obs)[3]	MINDO/2[79]	MINDO/3[17]
	1399	1033	1306
	1416	1065	1317
	1448	1114	1400
	1647	1707	1872
	2852	3030	3429
	2916	3068	3452
	2979	3088	3459
	3012	3115	3530
	3081	3152	3545
a''	177	148	101
	578	496	510
	936	756	853
	996	804	900
	1166	859	973
	1472	1234	1309
	2960	3046	3450
Ketene			
a_1	1118	1005	1180
	1388	1329	1407
	2152	2366	2328
	3071	3215	3599
b_1	438	376	385
	977	749	850
b_2	3166	3231	3639
	528	423	383
	588	602	585

[a] Frequencies are given in cm^{-1}.

6. Conclusions and an Opinion

This chapter has been primarily concerned with the methodologies involved in extracting thermochemical and kinetic information from ground-state potential energy surfaces. The focus of attention has been the application of semiempirical molecular orbital theory to these problems.

In many ways, these methods are ideally suited to this type of application. Although current methods may fall short of large-basis-set Hartree–Fock methods in accuracy, the results are encouraging. Semiempirical methods frequently give fairly accurate geometries and their reaction energies are very often better than those given by small-basis-set *ab initio* methods. When one considers the computational ease of applying these methods, one must conclude that they can be extremely cost-effective.

It is, therefore, somewhat surprising that there is not more effort placed in the development and application of these methods. Dewar's group in Texas has almost single-handedly brought semiempirical molecular orbital theory to the present level. Although manpower and good computational resources are needed for developing these methods, this cannot be the principal reason for the neglect of this area. There are many research groups involved in developing and carrying out the very expensive *ab initio* calculations.

We believe that the reasons for the apathy and occasional hostility toward these methods may include some that fall outside of the domain of theoretical chemistry. When one reads the italicized Latin words, *ab initio*, one is struck with a sense of rigor. The very pronunciation of these words commands awe and respect. One does not deal lightly with *ab initio*. "Semiempirical," on the other hand, implies that all is not quite there. Does it mean semiaccurate, semireal, or does semi- mean pseudo? Webster defines "empirical" as "depending on experience or observation alone *without regard for science and theory*" (italics ours) and an "empiric" as a "quack, charlatan." The names of the individual methods themselves invite bias. STO-3G, 6-31G*, etc., mean business, whereas MINDO sounds like something from a childhood rhyme.

Whatever the reasons, we believe that these research areas should be more actively pursued, and that great care should be taken in naming the new methods that evolve.

References

1. M. J. S. Dewar, *Chem. in Britain* **11**, 97 (1975).
2. R. H. Schwendeman, *J. Chem. Phys.* **44**, 2115 (1966).
3. G. Herzberg, *Molecular Spectra and Molecular Structure. II. Infrared and Raman Spectra of Polyatomic Molecules*, D. Van Nostrand, New York (1945), pp. 501–530.

4. S. Glasstone, K. Laidler, and H. Eyring, *The Theory of Rate Processes*, McGraw-Hill, New York (1941).
5. W. J. Hehre, R. Ditchfield, L. Radom, and J. A. Pople, *J. Am. Chem. Soc.* **92**, 4796 (1970).
6. J. N. Murrell and G. L. Pratt, *Trans. Faraday Soc.* **66**, 1680 (1970).
7. E. W. Schlag, *J. Chem. Phys.* **38**, 2480 (1963).
8. M. J. S. Dewar and E. Haselbach, *J. Am. Chem. Soc.* **92**, 590 (1970).
9. R. C. Bingham, M. J. S. Dewar, and D. H. Lo, *J. Am. Chem. Soc.* **97**, 1285 (1975).
10. G. Klopman, *J. Am. Chem. Soc.* **86**, 4550 (1964).
11. J. A. Pople and G. A. Segal, *J. Chem. Phys.* **44**, 3289 (1966).
12. J. A. Pople, D. L. Beveridge, and P. A. Dobosh, *J. Chem. Phys.* **47**, 2026 (1967).
13. A. Komornicki and J. W. McIver, Jr., *J. Am. Chem. Soc.* **98**, 4553 (1976).
14. J. A. Pople, *J. Am. Chem. Soc.* **97**, 5306 (1975).
15. W. J. Hehre, *J. Am. Chem. Soc.* **97**, 5308 (1975).
16. M. J. S. Dewar, *J. Am. Chem. Soc.* **97**, 6591 (1975).
17. A. Komornicki, unpublished results.
18. J. W. McIver, Jr. and A. Komornicki, *Chem. Phys. Lett.* **10**, 303 (1971).
19. R. B. Woodward and R. Hoffmann, *Angew. Chem. Int. Ed. Engl.* **8**, 781 (1969).
20. J. A. Pople, *Int. J. Quantum Chem.* **S5**, 175 (1971).
21. A. R. Gregory and M. N. Paddon-Row, *Chem. Phys. Lett.* **12**, 552 (1972).
22. R. E. Stanton and J. W. McIver, Jr., *J. Am. Chem. Soc.* **97**, 3632 (1975).
23. M. J. S. Dewar and S. Kirschner, *J. Am. Chem. Soc.* **93**, 4291 (1971).
24. J. W. McIver, Jr. and A. Komornicki, *J. Am. Chem. Soc.* **94**, 2625 (1972).
25. R. Fletcher, *Optimization*, Academic Press, New York (1969).
26. J. Pancir, *Theor. Chim. Acta* **29**, 21 (1973).
27. M. J. S. Dewar, *Science* **187**, 1037 (1975).
28. B. A. Murtagh and R. W. H. Sargent, *Comput. J.* **13**, 185 (1970).
29. M. J. D. Powell, *Comput. J.* **7**, 303 (1965).
30. J. W. McIver, Jr., *J. Am. Chem. Soc.* **94**, 4782 (1972).
31. J. W. McIver, Jr., *Acc. Chem. Res.* **7**, 72 (1974).
32. A. Komornicki and J. W. McIver, Jr., *J. Am. Chem. Soc.* **96**, 5798 (1974).
33. T. A. Halgren, I. M. Pepperberg, and W. N. Lipscomb, *J. Am. Chem. Soc.* **97**, 1248 (1975).
34. J. N. Murrell and K. J. Laidler, *Trans. Faraday Soc.* **64**, 371 (1968).
35. J. W. McIver, Jr. and R. E. Stanton, *J. Am. Chem. Soc.* **94**, 8618 (1972).
36. P. Pulay, *Mol. Phys.* **17**, 197 (1969).
37. T. A. Halgren, D. A. Kleier, and W. N. Lipscomb, *Science* **190**, 591 (1975).
38. M. J. S. Dewar, *Science* **190**, 591 (1975).
39. R. C. Bingham, M. J. S. Dewar, and D. H. Lo, *J. Am. Chem. Soc.* **97**, 1294 (1975).
40. L. C. Snyder and H. Basch, *J. Am. Chem. Soc.* **91**, 2189 (1969).
41. P. C. Hariharan, W. A. Lathan, and J. A. Pople, *Chem. Phys. Lett.* **14**, 385 (1972).
42. J. D. Cox and G. Pilcher, *Thermochemistry of Organic and Organometallic Compounds*, Academic Press, New York (1970).
43. M. S. Gordon and J. A. Pople, *J. Chem. Phys.* **49**, 4643 (1968).
44. A. Komornicki, Doctoral Thesis, State Univ. of N.Y. at Buffalo (1974), unpublished.
45. M. D. Newton, W. A. Lathan, W. J. Hehre, and J. A. Pople, *J. Chem. Phys.* **52**, 4064 (1970).
46. D. E. Shaw, D. W. Lepard, and H. L. Welsh, *J. Chem. Phys.* **42**, 3736 (1965).
47. K. Kuchitsu, *J. Chem. Phys.* **44**, 906 (1966).
48. W. J. Lafferty and R. J. Thibault, *J. Mol. Spectrosc.* **14**, 79 (1964).
49. A. G. Maki and R. A. Toth, *J. Mol. Spectrosc.* **17**, 136 (1965).
50. L. Radom, W. A. Lathan, W. J. Hehre, and J. A. Pople, *J. Am. Chem. Soc.* **93**, 5339 (1971).
51. D. R. Lide, Jr. *J. Chem. Phys.* **33**, 1514 (1960).
52. D. R. Lide, Jr. and D. Christiensen, *J. Chem. Phys.* **35**, 1374 (1961).
53. A. Almenningen, I. M. Anfinisen, and A. Haaland, *Acta Chem. Scand.* **24**, 43 (1970).
54. W. Haugen and M. Trateberg, *Acta Chem. Scand.* **20**, 1726 (1966).
55. W. J. Hehre and J. A. Pople, *J. Am. Chem. Soc.* **97**, 6941 (1975).
56. O. Bastiansen, F. N. Fritsch, and K. Hedberg, *Acta Cryst.* **17**, 538 (1964).

57. P. H. Kasai, R. J. Myers, D. F. Eggers, Jr., and K. B. Wiberg, *J. Chem. Phys.* **30**, 512 (1959).
58. A. Langseth, B. P. Stoicheff, *Can. J. Phys.* **34**, 350 (1956).
59. J. F. Chiang, C. F. Wilcox, Jr., and S. H. Bauer, *J. Am. Chem. Soc.* **90**, 3149 (1968).
60. L. S. Bartell, K. Kuchitsu, and R. J. DeNui, *J. Chem. Phys.* **35**, 1211 (1961).
61. K. Kuchitsu, J. P. Guillory, and L. S. Bartell, *J. Chem. Phys.* **49**, 2488 (1965).
62. K. Kuchitsu and L. S. Bartell, *J. Chem. Phys.* **36**, 2460 (1962).
63. C. P. Courtney, *Ann. Soc. Sci. Bruxelles* **73**, 5 (1959).
64. K. Takagi and T. Oka, *J. Phys. Soc. Japan* **18**, 1174 (1963).
65. R. W. Kilb, C. C. Lin, and E. B. Wilson, Jr., *J. Chem. Phys.* **26**, 1695 (1957).
66. A. Komornicki and M. Flanigan, unpublished results.
67. P. Venkateswarlu and W. Gordy, *J. Chem. Phys.* **23**, 1200 (1955).
68. W. A. Lathan, L. A. Curtiss, W. J. Hehre, J. B. Lisle, and J. A. Pople, *Prog. Phys. Org. Chem.* **11**, 175(1974).
69. V. Blukis, P. H. Kasai, and R. J. Myers, *J. Chem. Phys.* **38**, 2753 (1963).
70. F. W. Dalby, *Can. J. Phys.* **36**, 1336 (1958).
71. P. A. Gigiure and I. D. Liu, *Can. J. Chem.* **30**, 948 (1952).
72. D. R. Lide, Jr., *J. Chem. Phys.* **27**, 343 (1957).
73. C. C. Costain, *J. Chem. Phys.* **29**, 864 (1958).
74. C. H. Chiang, R. F. Porter, and S. H. Bauer, *J. Am. Chem. Soc.* **92**, 5313 (1970).
75. R. H. Hunt, R. A. Leacock, C. W. Peters, and K. T. Hecht, *J. Chem. Phys.* **42**, 1931 (1965).
76. R. H. Hughes, *J. Chem. Phys.* **24**, 131 (1956).
77. A. Yamaguchi, I. Schishima, T. Shimanouchi, and S. Mizushima, *Spectro. Chim. Acta* **16**, 1471 (1960).
78. N. Bodor, M. J. Dewar, and D. H. Lo, *J. Am. Chem. Soc.* **94**, 5303 (1972).
79. G. Guzzardo, Jr., Masters Thesis, State Univ. of N.Y. at Buffalo (1972), unpublished.

2

Electronic Excited States of Organic Molecules

R. L. Ellis and H. H. Jaffé

1. Introduction

In this and earlier volumes of this series, many chapters deal in one way or another with the Schrödinger equation or solutions thereof. Since, however, in this chapter we will make use not only of the principal (electrostatic) term of the Hamiltonian operator but will be required to bring in various other terms as perturbations, we shall start by stating all terms which we shall need. We shall then focus on which terms of the operator we need in first order. The next step is to search for the proper wave functions of the problem. In this process we will ask the reader to forget what he believes he already knows about these functions, and to follow us in their derivation from first principles. In doing so we hope to provide a unified treatment, which is correct and which contains, through different sets of approximations, the starting points for various areas of research: spectroscopy, electronic as well as vibrational, molecular dynamics, and the detailed study of energy relaxation processes. We shall next introduce the procedures for setting up the trial functions and then the approximations we make in evaluating the integrals necessary for the solution.

A second section will deal with the interaction of light with matter and will lead to the development of equations for the transition moment. In the third section, we shall introduce previously neglected terms in the Hamiltonian operator as perturbations, provided they lead to spectroscopically interesting effects, and examine whether and to what extent we can refine the spectro-

R. L. Ellis • Department of Chemistry, University of Illinois, Urbana, Illinois, and *H. H. Jaffé* • Department of Chemistry, University of Cincinnati, Cincinnati, Ohio

scopic description of our states and transitions by perturbation theory. Finally, throughout the chapter, we shall cite examples of the applications of the methods developed from various parts of the literature and emphasize comparisons with experiment.

2. The Hamiltonian Operator

The Hamiltonian operator for systems of M nuclei and $2N$ electrons moving in the presence of an external electromagnetic field can be broken into a number of terms[1] as follows:

$$\mathcal{H} = H_0 + H_{so} + H_{ss} + H_F \tag{1}$$

In this equation the first term H_0 represents all the electrostatic and kinetic energies as indicated in the following equation*:

$$H_0 = \sum_{r=1}^{M} \left(-\frac{1}{2m_r} \nabla_r\right) - \sum_{i}^{2N} \frac{1}{2} \nabla_i - \sum_{r,i}^{M,2N} \frac{Z_r}{r_{ri}} + \sum_{i<j}^{2N,2N} \frac{1}{r_{ij}} + \sum_{r<s}^{M,M} \frac{Z_r Z_s}{r_{rs}} \tag{2}$$

where Z_r is the charge of nucleus r and m_r is its mass.

The next term, H_{so}, represents the coupling of the spin angular momentum of one electron with the orbital angular momentum of this same electron and with the spin orbital angular momenta of all other electrons in this system. This term has the form

$$H_{so} = -g\beta_e^2 \sum_{i} \sum_{j \neq i} r_{ij}^{-3} (2\mathbf{s}_i \cdot \mathbf{r}_{ij} \times \nabla_j + \mathbf{s}_j \cdot \mathbf{r}_{ij} \times \nabla_j) \tag{3}$$

The spin–spin interaction H_{ss} is given by

$$H_{ss} = -(g^2\beta_e^2/2) \sum_{i} \sum_{j} r_{ij}^{-5} [3(\mathbf{s}_i \cdot \mathbf{r}_{ij})(\mathbf{s}_j \cdot \mathbf{r}_{ij}) - r_{ij}^2 \mathbf{s}_i \cdot \mathbf{s}_j] \tag{4}$$

These are all the terms we will need to describe the electronic states of molecules in the absence of an external electromagnetic field. We could further include terms of coupling of electronic orbital and angular momenta with nuclear angular momenta (nuclear spins). These terms, however, are of no interest in electronic spectroscopy although they give rise to effects that are observed in NMR and ESR spectra.

---

*Throughout this chapter we shall use atomic units and adopt the following notation:

(a) For subscripts: $i, j, \ldots$ *electrons.* $\mu, \nu, \kappa, \lambda, \ldots$ *molecular orbitals.* $\rho, \sigma, \tau, \upsilon, \ldots$ atomic orbitals. $r, s, \ldots$ nuclei, atoms, centers. $n, m, o, \ldots$ states; configurations.

(b) For functions and operators: lower case, Greek or Roman symbols represent functions or operators dependent on a single electron (particle, coordinate) only; capital letters represent many-electron (many-particle, etc.) functions or operators. The letters ψ and Ψ will be reserved for *molecules*, and ϕ for *atomic* electron functions; χ and X will be used for vibrational and Φ for total nuclear coordinate functions; Ξ will be used for total electronic plus nuclear wave functions.

(c) The notation $\sum_{i<j}^{N,N}$ is used as shorthand for $\sum_{j=1}^{N} \sum_{i=1}^{j-1} \cdots$.

Finally, the last term, H_F, accounts for the interaction of an external electromagnetic field with the electrons and nuclei,

$$H_F = \sum_r^M \left(-\frac{Z_r}{m_r c i} \mathbf{A} \cdot \nabla_r + \frac{1}{2 m_r c^2} |A|^2 \right) + \sum_i^{2N} \left(\frac{1}{2ci} \mathbf{A} \cdot \nabla_i + \frac{1}{2c^2} |A|^2 \right) \tag{5}$$

3. The Zeroth-Order Approximation

The usual approach to the solution of the Schrödinger equation that arises from the Hamiltonian is to consider only the first term of Eq. (1). The Schrödinger equation in this form is given by

$$\left[\sum_r - \left(\frac{1}{2m_r} \nabla_r^2 \right) - \sum_i \frac{1}{2} \nabla_i^2 - \sum_{r,i} \frac{Z_r}{r_{ri}} + \sum_{i<j} \frac{1}{r_{ij}} + \sum_{r<s} \frac{Z_r Z_s}{r_{rs}} \right] \Xi(\mathbf{r}_r, \mathbf{r}_i) = E \Xi(\mathbf{r}_r, \mathbf{r}_i) \tag{6}$$

where $\mathbf{r}_r$ and $\mathbf{r}_i$ represent all the coordinates of the nuclei and of the electrons, respectively. To solve Eq. (6), we expand $\Xi(\mathbf{r}_r, \mathbf{r}_i)$ in a series as

$$\Xi(\mathbf{r}_r, \mathbf{r}_i; E) = \sum_n \Phi_n(\mathbf{r}_r) \Psi_n(\mathbf{r}_r, \mathbf{r}_i) \tag{7}$$

where now the functions $\Phi_n(\mathbf{r}_r)$ play the role of expansion coefficients. These coefficients themselves contain all the information about the translational, rotational, and vibrational motion of the system. Expansion such as Eq. (7) are *always* valid, provided the set Ψ_n is complete, but they are useful only if the series converges rapidly. Rapid convergence may be expected if we choose the Ψ_n to be the solutions to the following differential equation:

$$\left[\sum_i \left(-\frac{1}{2} \nabla_i^2 \right) + \sum_{i<j} \frac{1}{r_{ij}} - \sum_{ri} \frac{Z_r}{r_{ri}} + \sum_{r<s} \frac{Z_r Z_s}{r_{rs}} \right] \Psi_n(\mathbf{r}_r, \mathbf{r}_i) = \mathscr{E}_n(\mathbf{r}_r) \Psi_n(\mathbf{r}_r, \mathbf{r}_i) \tag{8}$$

If we now substitute Eqs. (7) and (8) into Eq. (6), we obtain a new differential equation:

$$\sum_n \left[E_n(\mathbf{r}_r) - \sum_r \frac{1}{2m_r} \nabla_r^2 \right] \Phi_n(\mathbf{r}_r) \Psi_n(\mathbf{r}_r, \mathbf{r}_i) = E \sum_n \Phi_n(\mathbf{r}) \Psi_n(\mathbf{r}_r, \mathbf{r}_i) \tag{9}$$

Multiplication from the left by $\Psi_m^*(\mathbf{r}_r, \mathbf{r}_i)$ and integration over all electronic coordinates $\mathbf{r}_i$ leads to the coupled set of differential equations (we assume that the functions $\Psi_i(\mathbf{r}_r, \mathbf{r}_i)$ are orthonormal)

$$\left[\mathscr{E}_m(\mathbf{r}_r) - \sum_r \frac{1}{2m_r} \nabla_r^2 - E \right] \Phi_m(\mathbf{r}_r)$$

$$= \sum_n \sum_r \left\{ \left[\int \Psi_m^*(\mathbf{r}_r, \mathbf{r}_i) \nabla_r \Psi_n(\mathbf{r}_r, \mathbf{r}_i) \, d\mathbf{r}_i \right] \cdot \frac{1}{m_r} \nabla_r \Phi_n(\mathbf{r}_r) \right.$$

$$\left. + \left[\int \Psi_m^*(\mathbf{r}_r, \mathbf{r}_i) \nabla_r^2 \Psi_n(\mathbf{r}_r, \mathbf{r}_i) \, d\mathbf{r}_i \right] \frac{1}{2m_r} \Phi_n(\mathbf{r}_r) \right\} \tag{10}$$

Equation (10) serves as the starting point for studies in molecular collisions and detailed discussions of its solution may be found in Volume 2 of this Series. The spectroscopist at this point generally introduces the Born–Oppenheimer approximation,[2] which simply sets the rhs of Eq. (10) equal to zero, with the result that the equation for $\Phi_m(\mathbf{r}_r)$ becomes

$$\left[-\sum_r \frac{1}{2m_r}\nabla_r^2 + \mathscr{E}_m(\mathbf{r}_r) - E\right]\Phi_m(\mathbf{r}_r) = 0 \tag{11}$$

The total wave function $\Xi(\mathbf{r}_r, \mathbf{r}_i)$ may now be written as

$$\Xi(\mathbf{r}_r, \mathbf{r}_i) = \Phi_m(\mathbf{r}_r)\Psi_m(\mathbf{r}_r, \mathbf{r}_i) \tag{12}$$

We may now interpret the $\Phi_m(\mathbf{r}_r)$ to represent functions which describe the motion of the nuclei in the effective potentials $\mathscr{E}_m(\mathbf{r}_r)$, which are established by the motion of the electrons. The functions $\Phi_m(\mathbf{r}_r)$ contain all the information about translational, rotational, and vibrational motion of the system. Most methods for the solution of Eq. (11) familiar to chemists assume a product form of translational, rotational, and vibrational wave functions. The final solution for each of these motions is still dependent on the electronic state [defined by $\psi_m(\mathbf{r}_r, \mathbf{r}_i)$ through the potential $\mathscr{E}_m(\mathbf{r}_r)$]. We have thus reduced the original differential equation (6), which was the equation of the total problem of nuclei and electrons, into two separate differential equations—Eq. (8), the equation for the electronic motion, and Eq. (11), the equation for the nuclear motion.

Equation (6) and the coupled sets of Eqs. (8) and (10) contain information about nonrelativistic behavior of the molecules. The present chapter will concentrate on the solution of Eq. (8) and its physical implications; the solution of Eq. (11) is the province of the infrared and microwave spectroscopist[3]; the coupled set, Eq. (10), contains information about "curve-crossing" and thus is important in the discussion of predissociation,[4,5] as well as many other problems in molecular scattering.[6,7]

4. The Electronic Wave Function

4.1. The All-Valence-Electron Approximation*

We shall now proceed to solve Eq. (8) for the electronic wave functions. Virtually all methods used today in quantum chemistry assume for the function Ψ of Eq. (8) the form of an antisymmetrized product of one-electron functions (orbitals). We shall at this point be interested only in molecules that have an even number of electrons and have a closed-shell ground state. In this case, the

---

*For convenience of indexing, the treatment that follows deals specifically only with elements of the first row and hydrogen. The essence of the proof, however, can readily be generalized for elements of other rows.

antisymmetrized product Ψ takes the form of a single Slater determinant:

$$\Psi_0 = [(2N)!]^{-1/2} |\psi_1(1)\bar{\psi}_1(2) \cdots \psi_N(2N-1)\bar{\psi}_N(2N)| \tag{13}$$

Since the present volume deals with semiempirical approximations, we must now also introduce the first of the approximations that are common to essentially all semiempirical methods today: The fact that we treat all *valence* electrons, and only these, explicitly in the wave function. In order to see what this approximation implies and how it changes the differential equation (8), we make the following choice for our ground-state Slater determinant:

$$\Psi_0 = [(2N)!]^{-1/2} |^1\phi_1(1) \,^1\bar{\phi}_1(2) \cdots \,^{M_c}\phi_{M_c}(2M_c-1) \,^{M_c}\bar{\phi}_{M_c}(2M_c)$$
$$\times \psi_1(2M_c+1)\bar{\psi}_1(2M_c+2) \cdots \psi_{N_v}(2N-1)\bar{\psi}_{N_v}(2N)| \tag{14}$$

Here we are assuming that there are $2N_c$ core electrons, $2N_v$ valence electrons, $2N$ total electrons $(2N = 2N_c + 2N_v)$, M nuclei, and $M_c = N_c$ nuclei other than hydrogen. In this determinant the ϕ represent the atomic functions of the core electrons centered on the nucleus that appears as a left superscript, and the ψ are generalized molecular orbitals accommodating the valence electrons. We now introduce this Slater determinant into Eq. (8), premultiply by its complex conjugate, and proceed to integrate with respect to the coordinates of all the electrons that are described by core functions.

In this integration we make use of the Schrödinger equation of the atomic problems describing the core electrons of atom r:

$$\left[-\frac{1}{2}(\nabla_1^2 + \nabla_2^2) - \frac{Z_r}{r_{r_1}} - \frac{Z_r}{r_{r_2}} + \frac{1}{r_{12}} - \mathscr{E}^r \right] |^r\phi_r(1) \,^r\phi_r(2)| = 0 \tag{15}$$

The resulting integral consists of $[(2N)!]^2$ terms, which occur in $(2N)!$ groups, each having $(2N)!$ members, but where each group, except for a permutation of indices, gives identical information. A typical such group, after integration over the coordinates of the "core electrons," gives

$$\Psi'^* \left\{ \sum_r^{M_c} \mathscr{E}^r - \sum_{r \neq s}^{M_c, N_c} \left\langle \,^s\phi_s(i) \left| \frac{2Z_r}{r_{ri}} \right| \,^s\phi_s(i) \right\rangle \right.$$

$$+ \sum_{r<s}^{M_c, M_c} \left[\left\langle \,^r\phi_r(i)^s\phi_s(j) \left| \frac{4}{r_{ij}} \right| \,^r\phi_r(i)^s\phi_s(j) \right\rangle - \left\langle \,^r\phi_r(i)^s\phi_s(j) \left| \frac{2}{r_{ij}} \right| \,^r\phi_r(j)^s\phi_s(i) \right\rangle \right]$$

$$+ \sum_r^{M_c} \sum_{j=2N_c+1}^{2N} \left\langle \,^r\phi_r(i) \left| \frac{2}{r_{ij}} \right| \,^r\phi_r(i) \right\rangle$$

$$\left. - \sum_{i=2N_c+1}^{2N} \left(\frac{1}{2}\nabla_i^2 + \sum_r^M \frac{Z_r}{r_{ir}} \right) + \sum_{i<j} \frac{1}{r_{ij}} + \sum_{r<s}^{M,M} \frac{Z_r Z_s}{r_{rs}} - \mathscr{E}_0(\mathbf{r}_r) \right\} \Psi'$$

$$= \sum_r^{M_c} \sum_\mu^{N_v} \Psi'^* \left\langle \,^r\phi_r(i) \left| \frac{1}{r_{ij}} \right| \psi_\mu(i) \right\rangle \frac{^r\phi_r(j)\Psi'}{\psi_\mu(j)} \tag{16}$$

where

$$\Psi' = \psi_1(2N_c+1)\bar{\psi}_1(2N_c+2) \cdots \psi_{N_v}(2N-1)\bar{\psi}_{N_v}(2N)$$

and Ψ'^* is equivalent to Ψ', except that all ψ_μ are replaced by ψ_μ^*.

The crux of the all-valence-electron approximation is that we neglect the rhs of Eq. (16), i.e., the exchange integrals between core and valence electrons. In Eq. (16), $\mathscr{E}^r$ is the energy of a pair of core electrons centered on nucleus r. The exchange integrals between electrons occupying core orbitals centered on different atoms, given by the second term in square brackets in Eq. (16) (κ_c), will not be treated further; at a given $\mathbf{r}_r$ it is a constant and will be included as such in the energy. We now define the total valence electron energy E_0^v as*

$$E_0^v(\mathbf{r}_r) = \mathscr{E}_0(\mathbf{r}_r) - \sum_r \mathscr{E}^r - \kappa_c \tag{17}$$

The remaining $(2N)!$ terms in this integration have exactly the same form. The three integrals appearing on the left-hand side of Eq. (16) represent the interaction energy of the charge distribution given by $2\,'\phi_r^*(i)\,'\phi_r(i)$ (a) with a nucleus, (b) with another such charge distribution, and (c) with one of the valence electrons, respectively. The charge distribution $2\,'\phi_r^*(i)\,'\phi_r(i)$ is always near nucleus r, spherically symmetric with respect to it, and at a distance from the charge with which it interacts which is large compared to its own effective diameter. Consequently, we can introduce the following approximation:

$$\int {}'\phi_r^*(i)\frac{2}{r_{in}}{}'\phi_r(i)\,d\tau_i = \frac{Z_r - Z_r^c}{r_{rn}} \tag{18}$$

This means, for the three integrals discussed, the following simplification:

$$\left\langle {}^s\phi_s(i)\left|\frac{2Z_r}{r_{ri}}\right|{}^s\phi_s(i)\right\rangle = \frac{Z_r^c Z_s}{r_{rs}} - \frac{Z_r Z_s^c}{r_{rs}} \tag{19a}$$

$$\left\langle {}'\phi_r(1)^s\phi_s(2)\left|\frac{4}{r_{12}}\right|{}'\phi_r(1)^s\phi_s(2)\right\rangle = \frac{Z_r Z_s}{r_{rs}} + \frac{Z_r^c Z_s^c}{r_{rs}} - \frac{Z_r Z_s^c}{r_{rs}} - \frac{Z_r^c Z_s}{r_{rs}} \tag{19b}$$

$$\left\langle {}'\phi_r(1)\left|\frac{2}{r_{12}}\right|{}'\phi_r(1)\right\rangle = \frac{Z_r - Z_r^c}{r_{r2}} \tag{19c}$$

In Eqs. (18) and (19), Z_r^c are the so-called core charges, i.e., the nuclear charge Z_r minus the number of core electrons, two for first row elements. Introduction of the approximations (19) into Eq. (16) leads to cancellation of all terms involving Z_r and Z_s. Finally, we must collect all our $(2N)!$ terms together and we find that they are all identical in form except for the indexing of the electrons in the equivalent of Ψ' and $\Psi^{*\prime}$. By collecting appropriate Ψ' and $\Psi^{*\prime}$ and eliminating redundant information, we can arrive at an expression with a single $N_v \times N_v$ determinant on the right and left. We can further eliminate the left

*In the remainder of this chapter, when we refer to the electronic energy, we will be implying E^v.

determinant and arrive at the Schrödinger equation for the valence electrons:

$$\left(\sum_{i}^{2N_v} -\frac{1}{2}\nabla_i + \sum_{i<j}^{2N_v}\frac{1}{r_{ij}} - \sum_{r,i}^{M,2N_v}\frac{Z_r^c}{r_{ri}} + \sum_{r<s}^{M,M}\frac{Z_r^c Z_s^c}{r_{rs}}\right)\Psi_0(\mathbf{r}_r,\mathbf{r}_i) = E_0^v(\mathbf{r}_r)\Psi_0(\mathbf{r}_r,\mathbf{r}_i) \quad (20)$$

with

$$\Psi_0(\mathbf{r}_r,\mathbf{r}_i) = [(2N_v)!]^{-1/2}|\psi_1(1)\bar{\psi}_1(2)\cdots\psi_{N_v}(2N_v-1)\bar{\psi}_{N_v}(2N_v)| \quad (21)$$

As indicated above, the derivation of the present section was carried out explicitly only for first row elements. It can readily be expanded to elements of other rows. However, it should be borne in mind that such expansion involves the approximations made, particularly in Eqs. (15) and (19); while these approximations seem readily justifiable for first row elements, they are obviously much more drastic when applied to heavier atoms.

Although Eq. (21) describes the ground state, and the above derivation has been carried through only for the ground state, we can derive equations corresponding to Eqs. (19) and (20), with different subscripts, for other types of states in a perfectly analogous manner. The only problem is that we must, in order to do the derivation, write down the trial function in an explicit form.

4.2. The SCF Procedure*

In Sections 3 and 4.1 we have developed an equation, Eq. (20), which has the form of an eigenvalue equation (secular equation). For the purpose of generality we have expressed our wave function Ψ_0 as a function of the nuclear coordinates $\mathbf{r}_r$ and as a result the energy E_0^v is also a function of $\mathbf{r}_r$. We cannot generally solve Eq. (20) in this form. Consequently, we now assign fixed values to $\mathbf{r}_r$, i.e., we fix the nuclear coordinates. Then Ψ_0 no longer depends explicitly on $\mathbf{r}_r$, but any function obtained implies the set that was assigned; this is called a parametric dependence and may be expressed by $\Psi_0(\mathbf{r}_i, \mathbf{r}_r)$.

Methods of solving Eq. (20) have been treated in other volumes of this series and in other chapters of the present volume; we shall not discuss these in any detail. However, we must reemphasize that exact solutions are practically never obtainable, and consequently, we always imply approximate solutions, generally considered obtained through use of the variational principle.

Most practical procedures involve specifying a trial function for Ψ_0, and we will discuss the nature of such trial functions in some detail in the next subsection. Here it will suffice to indicate that the *many-electron* function Ψ is almost always expressed as an (antisymmetric) product of *one-electron* functions (molecular orbitals) ψ—or as a linear combination of such products. The problem of solving Eq. (20) for Ψ then resolves into finding the set of ψ of which

*Detailed discussions of the SCF procedure can be found in many texts. A lucid treatment, specifically appropriate to semiempirical methods, is given by Pople and Beveridge.[40]

Ψ is composed. The process used, initiated by Hartree[8] and elaborated by Fock,[9] involves separating Eq. (20) into a set of coupled one-electron equations and solving theses separately. Since these equations are coupled, however, each equation depends on the solution to all others and accordingly requires an iterative process. When all the Ψ have been refined to the point where the potential calculated from them, and consequently the equations derived from them, gives back to the same ψ on solution, the set ψ is called self-consistent.

Today, Eq. (20) is always solved in a manner to obtain self-consistent solutions, at least at the level of the antisymmetric product function. We will imply such solutions whenever we discuss single-determinant functions.

4.3. The Trial Functions

We now wish to proceed to solve Eq. (20) by a variational procedure for various excited states. In order to make effective use of the variational theorem, we must specify trial functions. Virtually all trial functions commonly used have one property in common: They are either antisymmetrized products of a set of basis functions, or linear combinations of such antisymmetrized products.

4.3.1. The Ground State

Most work in semiempirical quantum chemistry in the recent past has used a single Slater determinant, as in Eq. (21), as the trial function on which energy minimization is based. Such functions, however, have occasionally been considered inadequate, particularly in conjunction with the description of excited states by configuration interaction including doubly excited states (*vide infra*). In that case, the single Slater determinant of Eq. (21) is replaced by an expansion over a series of Slater determinants. The first of these, the ground configuration, will be denoted as V_0 and is identical to the rhs of Eq. (21). Further members of the expansion are formed by "promoting" one (or more) electrons from orbitals occurring in V_0 to virtual orbitals, orthonormal to those in V_0, but unoccupied in the single-determinant ground state. Determinants so formed occur in groups because the electrons are indistinguishable. With one such "promotion" we obtain a *spin configuration* of the form

$$^{\mu\lambda'}V = [2(2N_v)!]^{-1/2}\{|\psi_1(1)\bar{\psi}_1(2) \cdots \psi_\mu(i)\bar{\psi}_{\lambda'}(i+1) \cdots \psi_{N_v}(2N_v-1)\bar{\psi}_{N_v}(2N_v)|$$

$$-|\psi_1(1)\bar{\psi}_1(1) \cdots \psi_{\lambda'}(i)\bar{\psi}_\mu(i+1) \cdots \psi_{N_v}(2N_v-1)\bar{\psi}_{N_v}(2N_v)|\} \qquad (22)$$

where an electron has been "promoted" from ψ_μ to $\psi_{\lambda'}$.* The configuration

---

*The left superscript on V will be used to denote these orbitals, with the symbol for the orbital, which in the ground configuration was virtual, primed.

$^{\mu\lambda'}V$ is called singly excited, since each of its determinants differs from V_0 by just one orbital.

Doubly (and higher) excited spin configurations can also be written readily: $^{\mu\nu\lambda'\kappa'}V$; the general expressions differ, depending on the existence of equalities between μ and ν and between λ' and κ':

$$^{\mu\mu\lambda'\lambda'}V = [(2N_v)!]^{-1/2}|\psi_1(1)\bar{\psi}_1(2) \cdots \psi_{\lambda'}(i)\bar{\psi}_{\lambda'}(i+1) \cdots \psi_{N_v}(2N_v - 1)$$
$$\times \psi_{N_v}(2N_v)| \tag{23}$$

$$^{\mu\mu\lambda'\kappa'}V = [2(2N_v)!]^{-1/2}\{|\psi_1(1)\bar{\psi}_1(2) \cdots \psi_{\lambda'}(i)\bar{\psi}_{\kappa'}(i+1) \cdots \psi_{N_v}(2N_v - 1)$$
$$\times \psi_{N_v}(2N_v)|$$
$$-|\psi_1(1)\bar{\psi}_1(2) \cdots \psi_{\kappa'}(i)\bar{\psi}_{\lambda'}(i+1) \cdots \psi_{N_v}(2N_v - 1)\psi_{N_v}(2N_v)|\} \tag{24}$$

$$^{\mu\nu\lambda'\lambda'}V = [2(2N_v)!]^{-1/2}\{|\psi_1(1)\bar{\psi}_1(2) \cdots \psi_\mu(i)\bar{\psi}_{\lambda'}(i+1) \cdots \psi_{\lambda'}(j)\bar{\psi}_\nu(j+1)$$
$$\times \cdots \psi_{N_v}(2N_v - 1)\psi_{N_v}(2N_v)| - |\psi_1(1)\bar{\psi}_1(2) \cdots \psi_{\lambda'}(i)\bar{\psi}_\mu(i+1) \cdots$$
$$\times \psi_\nu(j)\bar{\psi}_{\lambda'}(j+1) \cdots \psi_{N_v}(2N_v - 1)\bar{\psi}_{N_v}(2N_v)|\}, \qquad \mu \neq \nu \tag{25}$$

Finally, the spin configuration $^{\mu\nu\lambda'\kappa'}V$ with $\mu \neq \nu$ and $\lambda' \neq \kappa'$ requires six determinants; these can be combined in different combinations to give two linearly independent singlet spin configurations[10] (also three triplets and a quintet).

In principle, then, one should write the ground-state wave functions as

$$\Psi_0 = C_0V_0 + \sum_m C_{0m} {}^{\mu\lambda'}V_m + \sum_n C_{0n} {}^{\mu\nu\lambda'\kappa'}V_n + \cdots \tag{26}$$

where N is the size of the basis set, to be discussed in the next section. In general such expansions are difficult to handle as trial functions to Eq. (20); only in very recent years have wave functions of the type indicated, but with quite short expansions, been directly introduced into the variational problem, in a procedure known as multiconfiguration (MCSCF) SCF calculations.[11] More usually, the variational process is used to determine the ψ_μ occupied in V_0 and the associated virtual orbitals ψ_λ. Then, in configuration interaction (CI) the expansion coefficients of Eq. (26) are determined. Unfortunately, the wave functions resulting are no longer self-consistent.

Expansions such as Eq. (26), though finite, are extremely long, and in practice, for molecules, are always truncated. Different truncation schemes are in frequent use; they will be discussed below under the excited states. However, a special condition applies to ground-state expansions of the form of Eq. (26). Provided that the ψ_μ are self-consistent, i.e., that they have been obtained by solution of Eq. (20) with the trial function (21), the matrix elements $\langle V_0|H_0|^{\mu\lambda'}V\rangle$ *all* vanish according to Brillouin's theorem.[12–14] Consequently, if no terms in Eq. (2) beyond the first two are considered, Ψ_0 of Eq. (21) gives the lowest energy and all the C_{0m} of Eq. (26) are zero. However, when further terms are included in Eq. (26), this is no longer true, and *all* terms must be considered.

4.3.2. Excited States

Up to this point we have avoided any explicit consideration of electron spin. Throughout spectroscopy it is customary to define our wave functions in such a manner that they are, aside from being at least *approximate* eigenfunctions of $\mathcal{H}$ or H_0, *true* eigenfunctions of the total spin operator S^2 and of its z component S_z. These operators are defined by

$$S^2 = \sum_i \mathbf{s}_i \cdot \mathbf{s}_i; \qquad S_z = \sum_i s_{zi} \tag{27}$$

where $\mathbf{s}$ is the one-electron spin operator and s_z is its z component, both best defined in terms of their operation on the spin functions α and β: $s_x\alpha = \beta/2$; $s_x\beta = \alpha/2$; $s_y\alpha = i\beta/2$; $s_y\beta = -i\alpha/2$; $s_z\alpha = \alpha/2$; $s_z\beta = -\beta/2$. The spin configurations V in Eqs. (21)–(25) were so chosen as to be eigenfunctions of S^2 with eigenvalues of 0.* The eigenvalue of zero was chosen since we have restricted ourselves so far to the treatment of molecules with closed-shell (and therefore singlet) ground states. We shall also be dealing with doublet and triplet states, the wave functions of which will be eigenfunctions of S^2 with eigenvalues of $3/4$ and 2, respectively. The analogous expressions for spin configurations for triplets are

$$^{\mu\lambda'}T^1 = [(2N_v)!]^{-1/2}|\psi_1(1)\bar{\psi}_1(2) \cdots \psi_\mu(i)\psi_{\lambda'}(i+1) \cdots \psi_{N_v}(2N_v-1)\bar{\psi}_{N_v}(2N_v)|$$

$$^{\mu\lambda'}T^0 = [2(2N_v)!]^{-1/2}\{|\psi_1(1)\bar{\psi}_1(2) \cdots \psi_\mu(i)\bar{\psi}_{\lambda'}(i+1) \cdots \psi_{N_v}(2N_v-1)\bar{\psi}_{N_v}(2N_v)|$$

$$+ |\psi_1(1)\bar{\psi}_1(2) \cdots \psi_{\lambda'}(i)\bar{\psi}_\mu(i+1) \cdots \psi_{N_v}(2N_v-1)\bar{\psi}_{N_v}(2N_v)|\} \tag{28}$$

$$^{\mu\lambda'}T^{-1} = [(2N_v)!]^{-1/2}|\psi_1(1)\bar{\psi}_1(2) \cdots \bar{\psi}_\mu(i)\bar{\psi}_{\lambda'}(i+1) \cdots \psi_{N_v}(2N_v-1)\bar{\psi}_{N_v}(2N_v)|$$

where the three functions in Eq. (28) are the three components of the triplet, distinguished by eigenvalues of S_z, 1, 0, and -1, respectively, which are used as right superscripts.

We shall also be dealing with molecules with odd numbers of electrons, $2N_v+1$ valence electrons. In this case the ground states will generally be doublets, described by a spin configuration

$$D^{1/2} = [(2N_v+1)!]^{-1/2}|\psi_1(1)\bar{\psi}_1(2) \cdots \psi_{N_v}(2N_v-1)\bar{\psi}_{N_v}(2N_v)\psi_{N_v+1}(2N_v+1)|$$

$$\tag{29}$$

$$D^{-1/2} = [(2N_v+1)!]^{-1/2}|\psi_1(1)\bar{\psi}_1(2) \cdots \psi_{N_v}(2N_v-1)\bar{\psi}_{N_v}(2N_v)\bar{\psi}_{N_v+1}(2N_v+1)|$$

There are three types of singly excited spin configurations with respect to this

---

*We could always just define a single antisymmetrized product, and obtain the proper eigenfunction of S^2 by projection operator techniques.

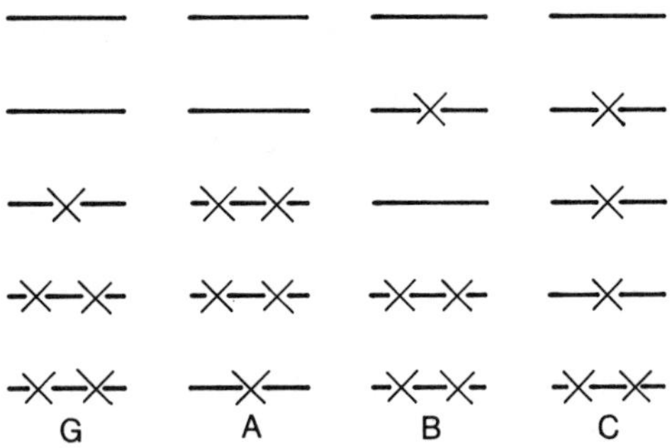

Fig. 1. The three types of excited configurations for doublet states.

G   A   B   C

doublet ground state, characterized in terms of the possible promotions in Fig. 1, and (we shall only give the Γ_+ component) for A- and B-type configurations,

$$^{\mu}D^{1/2} = [(2N_v+1)!]^{-1/2}|\psi_1(1)\bar{\psi}_1(2) \cdots \psi_\mu(i) \cdots \psi_{N_v+1}(2N_v)\bar{\psi}_{N_v+1}(2N_v+1)| \tag{30}$$

$$^{\lambda'}D^{1/2} = [(2N_v+1)!]^{-1/2}|\psi_1(1)\bar{\psi}_1(2) \cdots \psi_{N_v}(2N_v-1)\bar{\psi}_{N_v}(2N_v)\psi_{\lambda'}(2N_v+1)|$$

For C-type configurations we may write three determinants as eigenfunctions of S_z with eigenvalues of $+1/2$ (and three others with $-1/2$); all three are of the form

$$[(2N_v+1)!]^{-1/2}|\psi_1(1)\bar{\psi}_1(2) \cdots \psi_\mu(i)\psi_{\lambda'}(i+1) \cdots \psi_{N_v+1}(2N_v+1)| \tag{31}$$

with one of the three orbitals ψ_μ, ψ_{N_v+1} and $\psi_{\lambda'}$ barred (β spin). These three determinants may be combined into two linearly independent C-type spin configurations with $S^2 = 3/4$ (and one further one of quartet type, $S^2 = 6$).[15,16]

We can now proceed to define the various types of wave functions which have been used to define excited states of molecules.

a. VO Approximation. The simplest way of describing an excited state of a molecule is as a single spin configuration $\Psi_n = {}^{\mu\lambda'}V_n$ for a singlet state, and analogously for doublets, triplets, etc., with the specification that the ψ entering into $^{\mu\lambda'}V$ are the orbitals obtained in the solution for the ground state. This type of description is attractive because of its simplicity, and hence is extensively used in qualitative work; much notation, such as the concept of $n \to \pi^*$ and $\pi \to \pi^*$ transitions, derives from this approximation. Also, at this level of description, the concept of a singly (doubly) excited state is clearly defined, since the degree of excitation is that of the spin configuration.

Unfortunately, however, this type of description of excited states is frequently not sufficient. In many cases the first excited state of a molecule is reasonably well defined by a single spin configuration, but it is rare that the same is true for successive higher states.

b. VO-CI Approximations. Just as was indicated above for ground states, excited-state VO wave functions can be improved by *configuration interaction* (CI). For excited states, however, CI is of much greater importance since pairs of degenerate or nearly degenerate configurations frequently arise which interact strongly, leading to states which are far from degenerate.[17,18] Within the framework of the π-electron theories of spectra of aromatic hydrocarbons, this fact has long been recognized, related to the Coulson–Rushbrooke theorem,[19] as the pairing theorem.[18,22]

The CI expansion for singlet states leads to a wave function Ψ_n, fully analogous to Ψ_0 in Eq. (26). For triplets, the V_0 term is missing, and the configurations V are replaced by the T of Eq. (28). Only a single expansion is needed since the degeneracy of the three triplet components, at the present level of approximation (consideration of H_0 only), ensures expansion coefficients independent of the component used. Similarly, for doublets (ground and excited states) the V of Eq. (26) are replaced by the D of Eqs. (29)–(31). Unfortunately, however, Brillouin's theorem[12-14] does not hold for excited states (nor even for doublet ground states), and consequently more or less extensive CI is necessary for virtually all excited states.

The problem of truncation of the expansion now becomes important, and a number of solutions offer themselves. These can be classed into two major classes:

1. CI on singly excited configurations only. This seems to be the option most widely used. In it, the CI is extended only over spin configurations that are singly excited relative to the ground state. The preference for this way of proceeding probably lies more in its convenience than in any inherent theoretically validity. The total number of singly excited spin configurations corresponding to a ground state with $2N_v$ electrons and N basis members is $N_v(N-N_v) \le (N/2)^2$ (for a doublet ground state the corresponding number is $\le N^2$). Thus, it is relatively simple to calculate the energy of each configuration. Usually, however, the number of configurations is still too large to make it desirable to carry all in the full CI calculation, and further truncation is indicated. The most common procedure now is to truncate by configuration energy, carrying either a predetermined number of configurations, or all configurations of an energy below a given cutoff value.

2. CI is extended to doubly (and possibly even higher) excited spin configurations, the degree of excitation again being referred to the ground state. The number of such configurations is of the order of $(N/2)^4$ if N is the number of basis members! Consequently, a perturbation method[13,23] has been proposed; still, truncation schemes become even more important here.

(a) The simplest such scheme would appear to be to restrict the doubly excited spin configurations to those that differ from the desired excited state by only one orbital, i.e., that are singly excited with respect to it. Although computationally straightforward, this procedure has several drawbacks: It requires a different calculation for each excited state, thereby defeating one of the great advantages of the VO-CI process—that a single calculation provides a whole manifold of states. Second, it seems necessary for a reasonable spectro-scopic calculation to have both ground and excited states on the same basis, i.e., for each excited state a separate ground-state calculation is necessary, and the *various ground states are not necessarily equal.* Finally, the temptation to *restrict* the CI calculation to *only* those configurations singly excited to the leading term *must* be resisted vigorously: the most important configurational mixing usually occurs between configurations differing by two orbitals.[17,18]

(b) A second mode of truncation is analogous to the procedure described above for singly excited configurations. In view of the much larger number of configurations to be treated, a preset energy limit would appear most useful, since it allows immediate acceptance or rejection of configurations, without the lengthy sorting process.

(c) A third possible procedure involves the calculation of the matrix element between the doubly excited configuration and V_0, and elimination of any configuration for which the matrix element fails to exceed a cutoff value. Even more elaborate is the calculation of the matrix element, not only with V_0, but also with the leading spin configuration of the state or states of interest, and again elimination of configurations for which none of these matrix elements exceeds the cutoff. Finally, the most elaborate, and to our mind least satisfying, procedure is to print out all matrix elements between the ground and singly excited configurations on the one hand, and the doubly excited ones on the other, and to manually determine the configurations to be included.

Miraculously, it appears that most, if not all, of these procedures, and probably additional variations which have escaped us, have been used in the literature. In spite of this, we have been unable to find any work in the literature in which various procedures have been carefully compared. Furthermore, since we are dealing here with semiempirical methods, which are calibrated against experimental measurements or other theoretical calculations, use of a different procedure of truncation, unless shown to improve results in *all* cases tested, really requires recalibration of empirical parameters (cf. the following subsec-tion). The only procedure for which we have seen a systematic calibration is the CNDO/S method, involving VO-CI functions only to first excited configura-tions, and truncated to a preset number according to increasing energy.[24,22]

A final word of caution: In dealing with pure VO functions, we could readily distinguish singly and multiply excited *states.* As we proceed to VO-CI functions, of singly excited configurations only, this distinction retains its validity. As we begin to include higher excited configurations, the distinction begins to become fuzzy. However, as long as singly excited configurations predominate (as measured by the sum of the squares of the expansion coeffi-cients), the distinction retains some meaning. However, as we move up the manifold of excited states, configuration mixing tends to become so complex that these terms lose all meaning.

c. SCF (Direct) Method. In the methodology described to this point, we have assumed an SCF calculation of the ground state and are expressing or expanding the excited-state wave functions in terms of its occupied and virtual orbitals. Thus, the excited-state functions were *not* self-consistent; Eq. (20) had not been solved with excited-state functions as trial functions. If we wish to rectify these shortcomings, we may introduce the spin configurations, principally those in Eqs. (22), (28), and (30), as trial functions into Eq. (20) and solve this equation by one of the available methods.[25-31] As a consequence, although the *form* of the VO and SCF functions are the same, the underlying molecular orbitals ψ are *not*!

Direct SCF functions have many advantages over VO and VO-CI functions, and many disadvantages. They are self-consistent; they are the best single-configuration descriptions of the states attainable (within the limits of the basis set used). However, they do require a separate calculation for each state, and they fail to account for the most important term of the CI improvement, that required by the pairing theorem. In addition, normally they converge only on the lowest state of a given multiplicity manifold. Special techniques are available, e.g., to obtain convergence on the lowest state of a given symmetry within the manifold,[28] or of specific orbital makeup in terms of symmetry.[32] Further, since they are based on a different set of MOs (ψ), the evaluation of matrix elements connecting ground and excited state (or several excited states) is seriously complicated by the nonorthogonality of the two sets of MOs, to the point where we are not aware of any attempts to calculate such matrix elements.

Two basic procedures for resolving open-shell SCF procedures have been described: the *restricted*[28] and the *unrestricted*[27] Hartree–Fock schemes. The trial functions we have so far specified are of the restricted type. In these, the Hilbert space spanned by the electron coordinates is separated horizontally into a space of doubly occupied orbitals (closed shells) and a space of singly (or in general partially) occupied orbitals (open shells), and the two subspaces are separately minimized subject to orthonormality constraints. In the unrestricted Hartree–Fock scheme, the same space is subdivided vertically into an α-spin and a β-spin space. In this case we cannot use the configurations defined in Eqs. (22)–(25) and (28)–(31) but must, for singlet states, for instance, substitute configurations of the form of Eq. (32):

$$^{\mu\lambda'}V^u = [(2N_v)!]^{-1/2}|\psi_1^\alpha(1)\cdots\psi_\mu^\alpha(i)\cdots\psi_{N_v}^\alpha(N_v)\psi_1^\beta(N_v+1)\cdots\psi_{\lambda'}^\beta(N_v+i)$$
$$\times\cdots\psi_{N_v}^\beta(2N_v)| \tag{32}$$

where now the ψ_n^α and ψ_ν^β are distinct sets of orthonormal orbitals, each arising from minimization of its own subspace. Unfortunately, the wave functions obtained in this way are not eigenfunctions of S^2. They can be projected after the variational calculation is complete, the self-consistency is lost. Analytical projection prior to minimization has not been achieved. A partial resolution of this projection problem is contained in Smeyer's half-projected functions.[33] We shall, however, continue to base the present treatment on restricted Hartree–Fock functions, since most existing work is based on them.

Since all our wave functions are at best approximate eigenfunctions of H_0, it is really philosophically not justified to insist on true eigenfunctions of S^2. In other words, the unrestricted HF functions could be used as well as restricted ones. The spectroscopists' desire to classify states clearly in multiplicity manifolds has, however, led most calculations to be made by RHF methods.

d. Direct SCF-CI Functions. It would appear logical to extend the treatment by applying limited CI with the direct SCF functions as base. No serious new computational problems are encountered, provided the configurations are defined in terms of the self-consistent MOs of the excited state. Obviously self-consistency would be lost in the process. The potential advantage could only be a more rapidly converging CI expansion, i.e., that fewer terms might be required to obtain results of the same quality. In view of the considerable effort required to investigate such a procedure, it is not surprising that no one seems to have attempted it to date. In addition we would again like to emphasize, since we are dealing with semiempirical procedures, that a complete recalibration would be involved.

e. Multiconfiguration Functions. As previously indicated for ground states, one final technique is available to describe trial functions for minimization by Eq. (20). In this technique, virtually unexplored in semiempirical methods (but cf. Hinze[34]), the trial function is written as a linear combination of antisymmetrized products. The method appears particularly desirable for introduction, from the start, of the interactions due to the pairing theorem. Considerable computational problems arise in the use of such trial functions but there is little doubt that these can be overcome.[29,30,34–39] To date no calculations for excited states of molecules by an MC SCF method appear to have been reported.

4.4. The ZDO Approximation

In virtually all semiempirical work on spectra in the last few years, the zero-differential-overlap (ZDO) approximation has been invoked either rigorously or with minor exceptions. Since this approximation has been dealt with in other chapters of this Treatise, we only reiterate that it assumes that the differential charge distribution $\phi_\rho^*(1)\phi_\sigma(1)\,d\tau_1$ for $\rho \neq \sigma$ is neglected everywhere in space, so that all integrals involving this charge distribution are neglected.

The principal exception made to this rule lies in the evaluation of some one-center integrals; in the INDO method, the differential charge distribution $\phi_\rho^{r*}(1)\phi_\sigma^r(1)\,d\tau_1$, i.e., for atomic orbitals ρ and σ both centered on atom r, is systematically retained. Even in the strictest ZDO method, CNDO, one-center integrals of the form

$$\langle \phi_\rho^r(1)|\mathbf{r}_1|\phi_\sigma^r(1)\rangle$$

occurring in the calculation of dipole and transition moments, are systematically retained, although they should be neglected by strict application of ZDO.

Finally, an *apparent* contradiction exists when, in ZDO-based methods, certain integrals $\beta_{\rho\sigma}$ are approximated as proportional to overlap integrals $S_{\rho\sigma}$. Formally, $S_{\rho\sigma}$, or even the underlying differential overlap, does not enter into the derivation of $\beta_{\rho\sigma}$; only empirical experience has shown that properly computed values of $\beta_{\rho\sigma}$ correlate well with $S_{\rho\sigma}$, and hence the approximations to be discussed in the next section.

4.5. The Semiempirical Approximations

At this point we must further specify the form of our orbital ψ. We shall, throughout, use the customary form of the LCAO expansion, i.e., expand ψ in terms of a basis set of atomic orbitals ϕ:

$$\psi_\mu = \sum_\rho c_{\mu\rho}\phi_\rho \tag{33}$$

Throughout semiempirical quantum chemistry, with rare exceptions, it is customary to use the minimum basis sets, which include only the s and p orbitals of first row elements. Substitution of these LCAO expansions into the eigenvalue problem produces what is known as the Fock matrix, which then must be diagonalized. The elements of the Fock matrix have been discussed in many places for the ground state.[40,41] They are, in semiempirical theory, expressed in terms of a series of integrals; some of these are treated as adjustable parameters, others are treated as systematic parameters, still others are evaluated theoretically, and yet others are neglected according to the ZDO scheme or some variant thereof. The different methods, CNDO, INDO, MINDO, etc., differ principally in the approximations made at this point, although a complete specification of a "method" should also include the particular trial functions used.

The various basic semiempirical techniques, such as CNDO/2, INDO, MINDO, etc., have been described in other chapters of this treatise. One method has been developed specifically for spectroscopic work, the CNDO/S method.[24] It is a derivative of the CNDO/2 method of Pople and co-workers[42] and its parametrization is best defined by comparison to that method.[43] Aside from the changes in parametrization, the CNDO/S method implies application of limited CI over a truncated set of singly excited configurations. In the range used (30–80 configurations), the total number does not seem to seriously affect the results.

4.5.1. Gammas

The original CNDO/2 method approximates all Coulomb repulsion integrals (gammas) between different orbitals as a $\Gamma_{\mu\nu}$ between a pair of $2s$ orbitals, and calculates theoretically the value of this quantity from Slater functions. In the CNDO/S method, the one-center Γ's have been approximated by the well-known method introduced by Pariser[44] in the Pariser–Parr–Pople method for π electrons in the following form:

$$\Gamma_{\mu\mu} = \tfrac{1}{2}(I_\mu + A_\mu) \tag{34}$$

where I_μ and A_μ are valence state ionization potential and electron affinity, respectively. The two-center Γ's are again approximated similar to the PPP method, in which two approximations have been current. One is the Pariser–Parr approximation[45] in which, at long distances (>0.35 nm), Γ is expressed as the repulsion of two charged spheres and, at short distances, an interpolation formula between the monocentric term and the long-range terms is performed. The other approximation, due to Nishimoto and Mataga,[46] is given by the formula

$$\Gamma_{\mu\nu} = \frac{e^2}{[r_{rs} + 2e^2/(\Gamma_{\mu\mu} + \Gamma_{\nu\nu})]} \tag{35}$$

4.5.2. Betas

The beta terms of CNDO/2 have also been modified significantly. CNDO/2 defines this term as

$$\beta_{\mu\nu} = \tfrac{1}{2} S_{\mu\nu}(\beta_r^0 + \beta_s^0) \tag{36}$$

where orbital μ is centered on atom r, and ν on s, and the β^0 are adjustable parameters which have been calibrated to best reproduce *ab initio* calculations on diatomic molecules. In the CNDO/S method the calculation of the overlap integral S of Eq. (36) has been modified by decomposing it into two parts, a σ and a π component. The σ component is calculated in the normal way from the Slater orbitals just as in CNDO/2. The π component is also calculated in this manner, but is multiplied by a factor κ, for which the optimum value of 0.585 has been determined. Thus, $\beta_{\mu\nu}$ becomes

$$\beta_{\mu\nu} = \tfrac{1}{2}(\beta_r^0 + \beta_s^0)(S_{\mu\nu}^\sigma + \kappa S_{\mu\nu}^\pi) \tag{37}$$

Introduction of this approximation in no way affects the rotational invariance built into the original CNDO/2 method, nor does it distinguish between "σ and π electrons." As a consequence, the method is perfectly applicable to all kinds of molecules, nonplanar as well as planar, without any restriction. Finally, in the CNDO/S method, the parameters β^0 of Pople have been reoptimized.

The parametrization of any semiempirical method involves the assignment of numerical values to certain arbitrary parameters. These values are chosen in general to fit certain properties and are always chosen in connection with the use of a certain type of trial function. Thus, CNDO/2 was parametrized for agreement with *ab initio* ground-state wave functions of diatomic molecules, using single Slater determinants as trial functions. CNDO/S, on the other hand, was parametrized on spectroscopic transitions using VO-CI trial functions with an expansion of 30–80 singly excited spin configurations. It has since been demonstrated that the CNDO/S method also gives good results with SCF trial functions in those cases where it has been tested. It should be noted that a parametrization of a semiempirical method is valid only with the set of trial functions with which it has been calibrated. Any use with a different set of trial functions requires complete reparametrization, or at least verification that the calibration is valid within the particular approximation used.

4.6. Comparison of Various Methods

4.6.1. Closed-Shell Methods

Although a few papers foreshadowed the development,[47] the announcement of CNDO/1[48,49] may be taken as the beginning of the era of the ZDO all-valence-electron methods. The beginning of this new era has, of course, not terminated the use of the π-electron methods, in particular the Pariser–Parr–Pople method[20,21]; however, new theoretical and methodological developments in recent years in this area are few and the method is well documented[50]; consequently, we shall not discuss it in this chapter.

As might have been anticipated, the ZDO methods, once they became widely known, were rapidly applied to spectroscopic problems. With the large variety of these methods available it is surprising that the literature contains few systematic studies of comparison among the various choices; the reason for this will soon become apparent.

One of the earliest systematic examinations of the usefulness of ZDO methods to the calculation of electronic spectra was the application by Clark and Ragle[51] of CNDO/1 and CNDO/2 to ethylene and benzene. They found it necessary to reparametrize extensively in order even to interpret the benzene spectrum, and then required reinterpretation of the spectrum. Similar results were obtained by Kroto and Santry.[52] Another early critical study was the systematic evaluation of the importance of configuration interaction in the calculation of electronic spectra by Giessner-Prettre and Pullman.[53] They found that, short of an extensive reparametrization which they were unwilling to undertake, no amount of CI (involving the inclusion of singly, doubly, and triply excited configurations) is sufficient to properly reflect electronic spectra.

The situation changed drastically in 1968 when the CNDO/S method was introduced,[24] replacing theoretical Coulomb repulsion integrals by semiempirical values, and modifying the approximations for $F_{\mu\nu}$ (cf. Section 4.5). As an example of the effect of this reparametrization, Table 1 shows spectra of benzene calculated by various methods. Simultaneously with this work and independent of it, Clark[54] proposed similar modifications of $\Gamma_{\mu\nu}$ and a different change of the $F_{\mu\nu}$ approximation, which, although unexplored, should serve the same purpose. Unfortunately, while Clark's work was successfully applied to cyclopropane, ethylene oxide, and ethylenimine, it does not seem to have been further tested.

Table 1, which is typical for the appearance of the results in other compounds, shows the three shortcomings of the CNDO/2 parametrization for spectroscopic calculations: (1) All excitation energies are vastly exaggerated. (2) States arising from $\sigma \to \pi^*$ and $\pi \to \sigma^*$ transitions are completely mixed in with $\pi \to \pi^*$ states. (3) The relative spacings of $\pi-\pi^*$ states are not well reproduced. The same table further shows the effect of the various possible interpolation formulas for the two-center Coulomb integrals: For the ^1L_b and ^1B states in benzene the Pariser–Parr[45] and Mataga–Nishimoto[46] approximations give essentially identical results, but for ^1L_a and ^3L_a the results differ widely. The same effect has been observed in many molecules and has led to acceptance of the Pariser–Parr approximation for triplets and of the Mataga–Nishimoto approximation for singlets.[55]

The reasons that CNDO-type methods require reduction (scaling) of the Coulomb repulsion integrals are undoubtedly the same as in the PPP method: corrections for the integrals neglected because of the ZDO approximation,

Table 1. Comparison of the Benzene Spectrum Calculated by the
CNDO/2 and CNDO/S Methods[a]

State[b]	CNDO/2	CNDO/S Mataga	CNDO/S Pariser	Exp.
$^3L_a(^3B_{1u})$	9.19	2.55	3.43	3.6
$^3B(^3E_{1u})$	9.12	3.96	4.36	—
$^3L_b(^3B_{2u})$	9.98	4.84	4.85	—
$^1L_b(^1B_{2u})$	10.04	4.84	4.85	4.7
$^1L_a(^1B_{1u})$	10.40	6.10	5.20	6.1
$^1B(^1E_{1u})$	13.00	6.79	6.88	6.9
$^1A_{1g}(\sigma-\pi^*)$	10.35	6.61	7.24	—
$^1A_{2u}(\sigma-\pi^*)$	10.16	6.68	7.20	—
$^1A_{2u}(\sigma-\pi^*)$	9.98	6.90	7.36	—
$^1E_{2g}(\sigma-\pi^*)$	12.35	7.98	7.84	—
$^1E_{2u}(\sigma-\pi^*)$	12.86	7.62	8.30	—

[a] In all cases, configuration interaction over 60 states. All numbers represent energies (in eV). All states are listed for which the calculated energy lies below the ^1B state (as calculated by the same method).
[b] Platt and symmetry notation.

and for correlation.[50] The need for correcting the $F_{\mu\nu}$ is surely related to the definition of β^0 in the CNDO method as an *atomic* parameter, and thus the lack of distinction of s and p orbitals, an approximation which was originally introduced to attain hybridizational invariance.[48] A comparison of results from CNDO/S calculations with those from extended Hückel and PPP calculations has been made on some large organic compounds.[56]

Some more recent attempts[57–59] to use INDO methods[60] in the original parametrization have been no more successful than the use of CNDO in its original parametrization, and again apparently for the same reasons.

A number of further attempts either to improve the CNDO/S method or its parametrization have naturally been made. From the authors' indications, most of these attempts appear to be basically successful; however, none of them seem to have been tested in a sufficient range of compounds or classes of compounds to seriously establish their general utility.

Thus, for instance, an all-valence-electron method without CI seems to have worked reasonably in representing the spectra of formaldehyde, HCN, and CO_2.[61,62] The possibility of avoiding the inclusion of CI in these cases probably is due to the small size of the molecules, so that the number of interacting configurations is small, and the excited states happen to be configurationally pure or nearly so. Other reparametrizations of CNDO/CI techniques have also been proposed,[63–65] but without more extensive tests such attempts are not convincing. Not surprisingly, when the changes that transform CNDO/2 into CNDO/S are applied to INDO,[66–68] and the one-center integrals G^1 and F^2, specific to INDO, are properly scaled, a method is obtained which gives results paralleling those of CNDO/S; however, it has the advantage of properly reflecting singlet–triplet splitting in $n–\pi^*$ transitions, which is given by an integral systematically neglected in CNDO/S.

4.6.2. Open-Shell Methods

With the advent of the ZDO SCF all-valence-electron techniques, investigators soon attempted to develop open-shell methods to directly calculate excited states of molecules. Among the early attempts were those of Dixon,[69] Kroto and Santry,[70] and Song.[71] These methods suffered from the same difficulties as the closed-shell CNDO/2 calculations.

Once the CNDO/S method with its configuration interaction was introduced, the pressure for open-shell methods was greatly reduced, since a CI calculation not only provides the energy of open-shell states in the singlet manifold, but can equally serve to provide triplet states.

The new recent interest in open-shell methods seems to have come largely from concern for doublet states, i.e., the spectroscopy of free radicals. Work in this direction has gone on simultaneously and independently in a number of different laboratories, almost all representing a more or less direct application of the CNDO/S method.[72–76] Since photoelectron spectra also give information about the relative energies of doublet states in radical cations, application of open-shell methods to these data followed immediately.[77,78]

The simplest route to free radical and radical ion spectra is an open-shell calculation of the ion ground state, again followed by limited CI with singly excited configurations.[16] Some new problems arise here, since convergence is not always easily obtained,[79] and degenerate ground states occur when the partially occupied orbitals are degenerate.[80]

More elaborate treatments are available in terms of direct SCF calculations of various excited states.[75] Such methods, to date, are somewhat limited with regard to the states accessible to them. They have, however, also been worked out for triplet states.[81] They are effective only for molecules with some symmetry.

5. The Interaction of Matter and Electromagnetic Fields

In the presence of an electromagnetic field, a system of particles may undergo a transition from a state of energy E_o to a state of energy E_f. The states are characterized by the function $\Xi(\mathbf{r}_r, \mathbf{r}_i; E)$ of Eq. (6). Time-dependent perturbation theory is generally used to treat this process.[82] In Eq. (1) the term H_F represents the energy due to the interaction of the electromagnetic field with our assembly of particles. Generally, we are interested only in "weak" electromagnetic fields, and hence we choose as perturbation operator the terms in H_F that are linear in A. The details of determining an expression for the transition probability have been discussed by many authors.[83,84] We simply state here the final result for a linearly polarized field averaged over all orientations:

$$\text{trans. prob.} = \frac{64\pi^3}{3\omega_{fi}^2}\rho(\omega_{fo})|\mathbf{M}|$$

where

$$\mathbf{M} = \int \Xi(\mathbf{r}_r, \mathbf{r}_i; E_f)\left|\sum_r -\frac{Z_r}{2m_r}\exp\left(i\frac{\omega_{fo}}{c}\hat{\mathbf{n}}\cdot\mathbf{r}_r\right)\nabla_r\right.$$

$$\left.+\sum_i \frac{1}{2}\exp\left(i\frac{\omega_{fo}}{c}\hat{\mathbf{n}}\cdot\mathbf{r}_i\right)\nabla_i\right|\Xi(\mathbf{r}_r, \mathbf{r}_i; E_o)\,d\mathbf{r}_r\,d\mathbf{r}_i \tag{38}$$

$$\omega_{fo} = E_f - E_o$$

and $\hat{\mathbf{n}}$ is a unit vector in the direction of propagation of the radiation. The integration in Eq. (38) is over both electronic and nuclear coordinates.

5.1. Transition Moments

The quantity $\mathbf{M}$ in Eq. (38) is generally referred to as the *transition moment*; the nature of the molecule and of the upper and lower states affect the transition probability through $\mathbf{M}$, and hence we must proceed to evaluate this quantity.

In the present discussion, we are interested in optical transitions; so the transition frequency ω_{fo} will be in the optical region of the spectrum and, consequently, the function $\exp(i\omega_{fo}\hat{\mathbf{n}}\cdot\mathbf{r}_r/c)$ will oscillate much more rapidly than $\int \Xi(\mathbf{r}_r, \mathbf{r}_o; E_f)\nabla_r\Xi(\mathbf{r}_r, \mathbf{r}_i; E_o)\, d\mathbf{r}_i$ and the first term of $\mathbf{M}$ may be ignored. Then the transition moment for an electronic transition has the form

$$\mathbf{M} \approx \int \Xi_f(\mathbf{r}_r, \mathbf{r}_i; E_f)\left|\sum_j \frac{1}{2}\exp\left(i\frac{\omega_{fo}}{c}\hat{\mathbf{n}}\cdot\mathbf{r}_j\right)\nabla_j\right|\Xi_o(\mathbf{r}_r, \mathbf{r}_i; E_o)\, d\mathbf{r}_r\, d\mathbf{r}_i \quad (39)$$

To obtain the value of $\mathbf{M}$ for a particular electronic transition, two very basic assumptions are made.

First we assume that the exponential in Eq. (39) may be expanded and only the first term retained. The approximation is generally referred to as the dipole approximation*:

$$\mathbf{M} \approx \iint \Xi(\mathbf{r}_r, \mathbf{r}_i; E_f)\sum_j \nabla_j \Xi(\mathbf{r}_r, \mathbf{r}_i; E_o)\, d\mathbf{r}_r\, d\mathbf{r}_i \quad (40)$$

Second, we assume that Eq. (6) has been solved in the Born–Oppenheimer approximation. This leads to expressions for the initial and final states of the form

$$\Xi(\mathbf{r}_r, \mathbf{r}_i; E_o) \approx \Phi_o(\mathbf{r}_r)\Psi_o(\mathbf{r}_r, \mathbf{r}_i)$$
$$\Xi(\mathbf{r}_r, \mathbf{r}_i; E_f) \approx \Phi_f(\mathbf{r}_r)\Psi_f(\mathbf{r}_r, \mathbf{r}_i) \quad (41)$$

Substitution of Eq. (41) into Eq. (40) gives

$$\mathbf{M}(E_f \leftarrow E_o) \approx \int \Phi_f(\mathbf{r}_r)\mathbf{M}_\nabla(\mathbf{r}_r)\Phi_o(\mathbf{r}_r)\, d\mathbf{r}_r \quad (42)$$

where

$$\mathbf{M}_\nabla(\mathbf{r}_r) = \int \Psi_f(\mathbf{r}_r, \mathbf{r}_i)\sum_j \nabla_j\Psi_o(\mathbf{r}_r, \mathbf{r}_i)\, d\mathbf{r}_i \quad (43)$$

Provided that the $\Psi(\mathbf{r}_r, \mathbf{r}_i)$ are exact solutions of Eq. (7), Eq. (43) has an analog which is perfectly equivalent to it:

$$(i\omega_{fo})^{-1}\mathbf{M}_\nabla(\mathbf{r}_r) = \mathbf{M}_r(\mathbf{r}_r) \quad (44)$$

where

$$\mathbf{M}_r(\mathbf{r}_r) = \int \Psi_f(\mathbf{r}_r, \mathbf{r}_i)\sum_i \mathbf{r}_i\Psi_o(\mathbf{r}_r, \mathbf{r}_i)\, d\mathbf{r}_i \quad (45)$$

---

*The second term in this expansion produces the terms generally referred to as the magnetic dipole and electric quadrupole terms.[85] These will not be further discussed here.

Substitution of Eq. (44) into Eq. (42) gives

$$(i\omega_{fo})^{-1}\mathbf{M}(E_f \leftarrow E_o) \approx \int \Phi_f(r_r)\mathbf{M}_r(\mathbf{r}_r)\Phi_o(\mathbf{r}_r)\,d\mathbf{r}_r \qquad (46)$$

In discussing electronic transition in molecules, it is common to consider the motion of the electrons and nuclei in a coordinate system fixed at the center of mass of the molecule and rotating with it. This, in effect, ignores the rotational and translational motion of the system. In such a coordinate system Eq. (46) may be written in terms of a set of coordinates Q_α which are appropriate for the description of the vibrational motion of the system. The wave functions which describe this motion depend on sets of quantum numbers (v for the lower, w for the upper state) and are represented by $X_f^w(Q_\alpha)$ and $X_o^v(Q_\alpha)$.[86] The energies E_f and E_o now may be identified as the total electronic and vibrational energies and Eq. (46) becomes

$$(i\omega_{fo})^{-1}\mathbf{M}(E_f \leftarrow E_o) \approx \int X_f^w(Q_\alpha)\mathbf{M}_r^{fo}(Q_\alpha)X_o^v(Q_\alpha)\,dQ_\alpha \qquad (47)$$

Although Eq. (47) is extremely useful for the discussion of selection rules of electronic transitions, the difficulties which arise in the calculation of $\mathbf{M}_r(Q_\alpha)$ make quantitative evaluation impossible in large molecules. In order to evaluate $\mathbf{M}(E_f \leftarrow E_o)$ we must introduce an approximation, called the Condon approximation.[87]

In this approximation we expand $\mathbf{M}_r(Q_\alpha)$ about some fixed point Q_α°,*

$$\mathbf{M}_r^{fo}(Q_\alpha) = \mathbf{M}_r^{fo}(Q_\alpha^\circ) + \sum_\alpha \left(\frac{\partial \mathbf{M}_r^{fo}(Q_\alpha)}{\partial Q_\alpha}\right)_{Q_\alpha} Q_\alpha + \cdots \qquad (48)$$

where

$$\mathbf{M}_r^{fo}(Q_\alpha^\circ) = \int \Psi_f(Q_\alpha^\circ, \mathbf{r}_i)\sum_j \mathbf{r}_j\Psi_o(Q_\alpha^\circ, \mathbf{r}_i)\,d\mathbf{r}_i \qquad (49)$$

and

$$\frac{\partial \mathbf{M}_r^{fo}(Q_\alpha)}{\partial Q_\alpha} = \int \frac{\partial \Psi_f(Q_\alpha, \mathbf{r}_i)}{\partial Q_\alpha}\sum_j \mathbf{r}_j\Psi_o(Q_\alpha, \mathbf{r}_i)\,d\mathbf{r}_i$$

$$+ \int \Psi_f(Q_\alpha, \mathbf{r}_i)\sum_j \mathbf{r}_j\frac{\partial \Psi_o(Q_\alpha, \mathbf{r}_i)}{\partial Q_\alpha}\,d\mathbf{r}_i \qquad (50)$$

The Condon approximation consists in terminating the expansion (48) after the first term and substituting into Eq. (47), with the result

$$(i\omega_{fo})^{-1}\mathbf{M}(E_f \leftarrow E_o) \approx \mathbf{M}_r^{fo}(Q_\alpha^\circ)F_{wv} \qquad (51)$$

where

$$F_{wv} = \int X_f^w(Q_\alpha)X_o^v(Q_\alpha)\,dQ_\alpha \qquad (52)$$

*The point generally chosen is the equilibrium position of the ground electronic state.

The term F_{wv}^2, multiplied by a Boltzmann factor determining the distribution among the v in the initial state, determines the relative intensities of the vibronic components $v \to w$.[88] If one is interested only in the total probability of an electronic transition, the proper procedure is to sum over all vibrational quantum numbers of both initial and final states (the sets v and w). It can be shown that the sum over the probabilities for each vibronic component gives a factor of unity. Consequently, the total probability depends only on

$$|\mathbf{M}_r^{fo}(Q_\alpha)|^2$$

The CNDO/S wave functions will now be used to develop an expression for $\mathbf{M}_r^{fo}$. Within the framework of the CNDO/S method the function $\Psi_o(Q_\alpha', \mathbf{r}_i)$ is taken to be a single Slater determinant and $\Psi_f(Q_\alpha', \mathbf{r}_i)$ is taken as a linear combination of spin configurations as in Eq. (28). The dipole integral is expressed in the CNDO/S method, invoking the LCAO Expansion, in terms of atomic orbitals, as the sum of the three terms:

$$\mathbf{M}_r^{fo}(Q_\alpha') = \sqrt{2} \sum_n \sum_s \sum_\rho^{(s)} C_{fn}^* c_{\nu\rho} \left\{ c_{\lambda'\rho}^* \langle \phi_\rho | \mathbf{r} | \phi_\rho \rangle \right.$$

$$\left. + \sum_{\sigma \neq \rho}^{(s)} c_{\lambda'\sigma}^* \langle \phi_\sigma | \mathbf{r} | \phi_\rho \rangle + \sum_{r \neq s}^{(r)} \sum_\sigma c_{\lambda'\sigma}^* \langle \phi_\sigma | \mathbf{r} | \phi_\rho \rangle \right\} \qquad (53)$$

Here the superscript on the summations over ρ and σ indicates summation over all AOs centered on the atom defined by the index.

The third term of Eq. (53) vanishes as a consequence of the ZDO approximation. Strict application of this approximation also requires the second term to vanish; however, since this frequently is the leading term, rather than neglecting it, we shall use its theoretical value; this is readily done since it is an atomic integral

$$\langle \phi_\rho | \mathbf{r} | \phi_\sigma \rangle = 5a_o / 2(3\zeta_s)^{1/2} \qquad (54)$$

if ϕ_ρ or ϕ_σ is an s electron, and the other is a p electron. If both are s or both are p electrons, the integral vanishes. The first term of Eq. (53) is readily evaluated. We decompose the radius vector $\mathbf{r}$ to the electron i into the radius vector from the (arbitrary) origin to the nucleus s on which i is centered, $\mathbf{r}_s$, and the radius vector of the electron relative to nucleus s, $\mathbf{r}_{si}$,

$$\mathbf{r} = \mathbf{r}_s + \mathbf{r}_{si}$$

The vector $\mathbf{r}_s$ does not contain electronic coordinates, and the AOs ϕ_ρ, as spherical harmonics, integrate over $\mathbf{r}_{si}$ to zero. Thus

$$\langle \phi_\rho^s | \mathbf{r} | \phi_\rho^s \rangle = \mathbf{r}_s \qquad (55)$$

There is a discussion among practitioners of semiempirical methods as to which of the two formulations, Eq. (42) or Eq. (46), provides the better approximation within the semiempirical framework. Again, let us stress that if the wave functions were exact eigenfunctions, the problem would not arise; the two formulations would be completely equivalent. Only when the wave functions are approximate, as is the case in a semiempirical method is there a question. We can contribute some insight into this controversy by inspection of the two equations, (43) and (45). Since the operation of ∇_i reduces the order of the r dependence in Eq. (43), while the operation of r_i increases this dependence [in Eq. (45)], it is clear that the latter equations are the better approximation if we have confidence in the "tails" of the wave function, while the former equations are more useful for approximate solutions that have been parametrized to give good energies without any particular attention to the tails. Since the latter is the case for the majority of semiempirical functions, it would appear that, if one wishes to completely evaluate the integrals which arise in the LCAO expansion, it would be better to use the former.

However, it appears to us that the use of either of these equations is unnecessarily complicated and cumbersome. The semiempirical methods we have discussed introduce many serious approximations. Equations (53)–(55) give us computationally trivial expressions without violating formally, or at least in spirit, the approximations made. The contribution of the first two terms of Eq. (53) is independent of which of the forms, Eq. (43) or (45), is used since we *are* dealing with eigenfunctions of the appropriate atomic problem. Thus the only question remaining is whether we are justified in neglecting the last term of Eq. (53) altogether, or how we should approximate it. We maintain that it is best neglected, and cite the results obtained with the CNDO/S method as proof.

Finally, we cannot finish this discussion without introducing the measure of transition probability most commonly used by spectroscopists, the oscillator strength f. This quantity is readily defined in terms of the transition moment:

$$f = (2\omega_{fo}/3)|\mathbf{M}_r(Q'_\alpha)|^2 \tag{56}$$

5.2. Photoelectron Cross Sections

A quantity which is closely related to the transition moment discussed in the preceding subsection is the cross section for photoionization. This quantity has particular importance since it determines the intensity of the bands observed in photoelectron spectra.

The photoionization process can be treated in a manner quite analogous to the absorption process, except that the final state described by Ψ_f has some special properties. It is made up of the antisymmetric product of the wave function of a free electron and the proper wave function of the remaining molecular ion. The latter is readily obtained; if the ionization occurs from a closed-shell neutral molecule M, the desired function is a doublet state of the appropriate ion M^+. The wave function of the free electron may be approximated by a plane wave. The resulting Ψ_f may then be substituted in Eq. (46).

The evaluation of these transition moments is reasonably straightforward, although sufficiently lengthy that we shall not reproduce it here. It was first outlined by Lohr and Robin,[90] rederived and first applied by Thiel and Schweig,[89] and further extended by Ellison.[91]

6. Spin–Orbit and Spin–Spin Coupling

Throughout this chapter we have insisted, with the spectroscopist, that all our wave functions be eigenfunctions of the spin operator S^2 (and S_z). This has permitted us to classify the electronic states calculated strictly into multiplicity manifolds, and all transitions between states in different manifolds are strictly forbidden. However, we know experimentally that such transitions occur, both in radiative and in radiationless manner. Thus, phosphorescence and singlet–triplet absorption are well-known and well-documented examples of transitions between different manifolds, while the population of triplet manifolds by intersystem crossing and the nonradiative relaxation of triplet states to singlet ground states are examples of radiationless processes involving changes between multiplicity manifolds.[92]

The problem lies in the fact that the spin angular momentum (i.e., the expectation value of the spin operator S^2) is not a good constant of motion (although approximately so, hence the usefulness of our treatment in other sections); the true constant of motion is the *total* angular momentum. This, within the Born–Oppenheimer approximation, i.e., the neglect of the rhs of Eq. (10) and hence the uncoupling of nuclear [Eq. (11)] and electronic motion [Eq. (8)], is the *total* electronic angular momentum, composed of orbital and spin components. To take account of these facts requires the inclusion in the Hamiltonian operator of relativistic terms, in particular H_{so} and H_{ss} [Eqs. (3) and (4)].

6.1. Spin–Orbit Coupling

We shall use first-order perturbation theory to account for the inclusion of H_{so}, Eq. (3), in the Hamiltonian. In connection with the phenomena listed, we are not concerned with the (generally negligible) perturbation *energies*, but with *transition probabilities*. Consequently, we will focus attention here on the perturbed wave functions, which are needed to calculate such probabilities. Under the effect of the perturbation, the spin-pure wave functions of multiplicity α, $^\alpha\Psi_n$, mix to form new, spin-impure wave functions U. However, since the mixing between manifolds generally is small, we shall designate a function derived from a perturbation of $^{\alpha}\Psi_n$ as $^{'\alpha'}U_n$ to indicate the spin-impure function of multiplicity near α. The admixture to $^{'\alpha'}U_n$ of other spin-pure function $^{\beta}\Psi_m$ is governed by the matrix element $\langle^{\beta}\Psi_m|H_{so}|^{\alpha}\Psi_n\rangle$, which we shall now discuss.

6.1.1. Spin–Orbit Matrix Elements

The general expression for a perturbed wave function $^{'\alpha'}U_n$ is

$$^{'\alpha'}U_n^a = {}^{\alpha}\Psi_n^a + \sum_{\beta}\sum_m\sum_b \frac{\langle^{\beta}\Psi_m^b|H_{so}|^{\alpha}\Psi_n^a\rangle}{E_{n\alpha}-E_{m\beta}}{}^{\beta}\Psi_m^b(1-\delta_{mn}\delta_{\alpha\beta})$$

$$= {}^{\alpha}\Psi_n^a + \sum_m\sum_b^{\alpha} \frac{\langle^{\alpha}\Psi_m^b|H_{so}|^{\alpha}\Psi_n^a\rangle}{E_{n\alpha}-E_{m\beta}}{}^{\alpha}\Psi_m^b(1-\delta_{mn})$$

$$+ \sum_{\beta}\sum_m\sum_b^{\beta} \frac{\langle^{\beta}\Psi_m^b|H_{so}|^{\alpha}\Psi_n^a\rangle}{E_{n\alpha}-E_{m\beta}}{}^{\beta}\Psi_m^b(1-\delta_{\alpha\beta}) \qquad (57)$$

where α and β specify the multiplicity, and a and b the components of a multiplet (in terms of eigenvalues of S_z or any other appropriate index desired). The first term on the rhs of (57) represents the zeroth-order function. The middle term is of no interest for the phenomena treated in this section since it reflects mixing of states of *equal* multiplicity; hence this term will be neglected.

The general expression for H_{so} given in Eq. (3) may be rewritten as

$$H_{so} = \sum_i \zeta'(\mathbf{r}_i)\mathbf{l}_i \cdot \mathbf{s}_i - \frac{1}{c}\sum_i\sum_{j\neq i}\left[\frac{(\mathbf{r}_i-\mathbf{r}_j)\times(\mathbf{v}_i-\mathbf{v}_j)}{r_{ij}^3} - \frac{1}{2}\frac{(\mathbf{r}_i-\mathbf{r}_j)\times\mathbf{v}_i}{r_{ij}^3}\cdot\mathbf{s}_i\right] \qquad (58)$$

where $\mathbf{l}_i$ and $\mathbf{s}_i$ are the electronic orbital and spin angular momenta of electron i, $\mathbf{r}_i$ and $\mathbf{v}_i$ are the position and velocity vectors of this electron, and r_{ij} is the distance between electrons i and j. The quantity $\zeta'(\mathbf{r}_i)$ in Eq. (56) accounts for the interaction of an electron with the field produced by the nuclear cores which make up the molecule. The other terms in Eq. (58) account for the interaction of each electron with the field produced by the other electrons in the system. By rearranging the second term in Eq. (58) we can express H_{so} in terms of the products $\mathbf{l}_i \cdot \mathbf{s}_i$ and $\mathbf{l}_i \cdot \mathbf{s}_j$. In the treatment which follows we shall neglect terms with $\mathbf{l}_i \cdot \mathbf{s}_j$ (spin–other orbit coupling), and write H_{so} as

$$H_{so} = \sum_i\sum_r \zeta_r(\mathbf{r}_{ri})\mathbf{l}_i \cdot \mathbf{s}_i \qquad (59)$$

Unlike the operator $\zeta'(\mathbf{r}_i)$ in Eq. (58), the operator $\zeta_r(\mathbf{r}_{ri})$ accounts for the interaction of the electron with a field by the nuclei as well as the other electrons,[93] and has been expressed in terms of the electron–nuclear distance r_{ri}.

We then must proceed to evaluate the matrix elements $\langle^{\beta}\Psi_m^b|H_{so}|^{\alpha}\Psi_n^a\rangle$, with $\beta \neq \alpha$. Although the operators $\mathbf{l}_i$ and $\mathbf{s}_i$ may be written as differential operators,[94] we need only characterize them with respect to their operation on the appropriate functions, **s** with respect to the one-electron spin functions α and β, **l** with respect to the electronic wave functions; since the latter will eventually be expanded in atomic orbitals, **l** is specified adequately by its operation on s, p, d, etc., atomic orbitals.[93]

The operator (59), as a sum of one-electron operators, can change only the spin of one electron at a time. Consequently, it can readily be shown that (59) vanishes unless the multiplicity $\beta = \alpha \pm 2$ (except for the neglected $\beta = \alpha$). This greatly reduces the expansion of the wave functions in (57).

The matrix elements of Eq. (59) can, of course, be evaluated for any function Ψ; however, the algebra and arithmetic become messy unless the *same set* of molecular orbitals is used to construct both $^\alpha\Psi_m$ and $^\beta\Psi_n$ as antisymmetrized products (or linear combinations thereof). Consequently, most recent work on spin–orbit coupling has been based on VO-CI functions as base for the perturbation.

As we introduce such functions [cf. Eqs. (22)–(31)] into the spin–orbit matrix element, we obtain for each an expansion in spin configurations; e.g., for the interaction of a singlet, $^1\Psi_n$, and the m_s component of a triplet, $^3\Psi_m^{m_s}$, and carrying the expansion only to singly excited configurations, we have

$$\langle {}^1\Psi_n | H_{so} | {}^3\Psi_m^{m_s} \rangle = \sum_{k,l} C_{nk}^* C_{ml} \langle {}^{\mu\lambda'} V_k | H_{so} | {}^{\nu\kappa'} T_l^{m_s} \rangle$$

$$+ \sum_l C_{n0} C_{ml} \langle V_0 | H_{so} | {}^{\nu\kappa'} T_l^{m_s} \rangle \qquad (60)$$

The integration in the Dirac brackets of (60) over the closed-shell portion (i.e., over the coordinates of all electrons described by doubly occupied MOs) can be carried out immediately[95]; following such integration, spin and space functions can be separated, giving

$$\langle {}^{\mu\lambda'} V | H_{so} | {}^{\nu\kappa'} T \rangle = \langle (\mu\lambda' + \lambda'\mu) \, {}^1\omega | h_1 + h_2 | (\nu\kappa' - \kappa'\nu) \, {}^3\omega_{m_s} \rangle \qquad (61)$$

where

$$h_i = \sum_r \zeta_r(i) \mathbf{l}_i \cdot \mathbf{s}_i \qquad (62a)$$

and

$$(\mu\lambda' + \lambda'\mu) = (\psi_\mu(1)\psi_{\lambda'}(2) + \psi_{\lambda'}(1)\psi_\mu(2)) \qquad (62b)$$

and the $^3\omega_{m_s}$ are the standard eigenfunctions of S_z. It is convenient, at this point, to form linear combinations of $^3\omega_1$ and $^3\omega_{-1}$, because the new functions $^3\omega_+$ and $^3\omega_-$ transform as rotations about the magnetic axes. For molecules in which all three principal axes lie in symmetry elements, the magnetic and principal axes coincide, and thus the new spin functions transform like molecular rotations. After this transformation, Eq. (61) must be modified by replacing the indices m_s by the arbitrary index a, which may take the values $+$, 0, or $-$.

The transformed spin function are given by

$$^1\omega = (1/\sqrt{2})(\alpha\beta - \beta\alpha), \qquad ^3\omega_- = (1/\sqrt{2})(\beta\beta - \alpha\alpha)$$

$$^3\omega_0 = (1/\sqrt{2})(\alpha\beta + \beta\alpha), \qquad ^3\omega_+ = (i/\sqrt{2})(\beta\beta + \alpha\alpha)$$

$$(63)$$

Since the operator h_i [cf. Eq. (62)] is a dot product of two operators, one containing only space coordinates, the other only spin coordinates, Eq. (61) can be rewritten as

$$\langle^{\mu\lambda'}V|H_{so}|^{\nu\kappa'}T^a\rangle = \sum_{i=1}^{2}\left[\langle(\mu\lambda' + \lambda'\mu)|\sum_r\zeta_r(i)\mathbf{l}_i|(\nu\kappa' - \kappa'\nu)\rangle \cdot \langle^1\omega|\mathbf{s}_i|^3\omega_a\rangle\right]$$

$$(64)$$

The operation of

$$\mathbf{s}_i = \hat{\mathbf{i}}s_{xi} + \hat{\mathbf{j}}s_{yi} + \hat{\mathbf{k}}s_{zi}$$

on the various spin functions is readily performed,* leading to

$$\langle^{\mu\lambda'}V|H_{so}|^{\nu\kappa'}T^a\rangle = \langle(\mu\lambda' + \lambda'\mu)|\sum_r[\zeta_r(1)\mathbf{l}_1 - \zeta_r(2)\mathbf{l}_2] \cdot \hat{\mathbf{a}}_a|(\nu\kappa' - \kappa'\nu)\rangle$$

$$(65)$$

where $\hat{\mathbf{a}}_a$ is $\hat{\mathbf{i}}, \hat{\mathbf{j}}$, or $\hat{\mathbf{k}}$ for a representing the $-$, $+$, and 0 spin function, respectively.

Next, we proceed to the integration over the space coordinates. First, we recognize that, since $\mathbf{l}$ is a one-electron operator, the integral (64) vanishes unless either $\mu = \nu$ or $\lambda' = \kappa'$. We then introduce the LCAO expansion, together with the ZDO approximation, and collect terms:

$$\langle^{\mu\lambda'}V|H_{so}|^{\nu\kappa'}T^a\rangle = 2\sum_\rho\sum_\sigma[c_{\mu\rho}^*c_{\nu\sigma}\delta_{\lambda'\kappa'} - c_{\lambda'\rho}^*c_{\kappa'\sigma}\delta_{\mu\nu}]\langle\phi_\rho|\sum_r\zeta_r\mathbf{l} \cdot \hat{\mathbf{a}}_a|\phi_\sigma\rangle$$

$$(66)$$

As a result of the ZDO approximation, the operation of $\mathbf{l}$ on an atomic orbital ϕ_σ^s centered on atom s can only lead to another orbital centered on the same atom s (or zero). Also, terms with ϕ_ρ and ϕ_σ not centered on the same atom vanish according to the ZDO approximation. Thus, we are led to

$$\langle^{\mu\lambda'}V|H_{so}|^{\nu\kappa'}T^a\rangle = 2\sum_r\sum_\rho\sum_\sigma[c_{\mu\rho}^*c_{\nu\sigma}\delta_{\lambda'\kappa'} - c_{\lambda'\rho}^*c_{\kappa'\sigma}\delta_{\mu\nu}]\langle\phi_\rho^r|\zeta_r\mathbf{l} \cdot \hat{\mathbf{a}}_a|\phi_\sigma^r\rangle \quad (67)$$

Next, we evaluate the dot product $\mathbf{l} \cdot \hat{\mathbf{a}}_a$:

$$\mathbf{l} \cdot \hat{\mathbf{a}}_- = l_x, \qquad \mathbf{l} \cdot \hat{\mathbf{a}}_+ = l_y, \qquad \mathbf{l} \cdot \hat{\mathbf{a}}_0 = l_z$$

Operation of any of the components of $\mathbf{l}$ on an s orbital yields zero; similarly, operation of the component l_q on the orbital p_q, where q is x, y, or z, $l_q p_q$ yields zero.[96] Finally, carrying out the operation of $\mathbf{l}$, we obtain

$$\langle\zeta_r\rangle = \langle\phi_\rho^r|\zeta_r\mathbf{l} \cdot \hat{\mathbf{a}}_a|\phi_\sigma^r\rangle = \langle\phi_\rho^r|Z_r/r_{ri}^3|\phi_\tau^r\rangle\delta_{\rho\tau}$$

$$(68)$$

*Cf. Table I of Ref. 96.

We neglect these if $\rho \neq \tau$ (according to the ZDO approximation), and use empirical atomic spin–orbit constants[97] ζ_r for the nonvanishing integrals. We can thus express any matrix element over H_{so}, and consequently the perturbed wave function describing any spin-impure state, $^{'\alpha'}U_n$.

6.1.2. Phosphorescence Lifetimes and Singlet–Triplet Absorption Intensities

Many problems can be solved with the help of the spin-impure functions of the preceding section. Probably most interesting among these are the lifetime of phosphorescence (i.e., the natural radiative lifetime of the emission from the lowest triplet state of molecules with singlet ground states), and the intensity of absorption from a singlet ground state to a triplet state. Both of these quantities can readily be expressed in terms of the transition moment $\mathbf{M}(^{'1'}U_0, {}^{'3'}U_m)$ [cf. Eq. (51)], where, for phosphorescence, $m = 1$. Substituting the wave functions from Eq. (53) into the expression for $\mathbf{M}$ and truncating after the first-order terms leads to the expression

$$\mathbf{M}(^{'3'}U_1^\alpha, {}^{'1'}U_0) = \left\langle {}^3\Psi_1^\alpha \left| \sum_i e\mathbf{r}_i \right| {}^1\Psi_0 \right\rangle + \sum_m^{\text{singlets}} \frac{\langle {}^1\Psi_m | H_{so} | {}^3\Psi_1^a \rangle}{E_1 - E_m} \left\langle {}^1\Psi_0 \left| \sum_i e\mathbf{r}_i \right| {}^1\Psi_m \right\rangle$$

$$+ \sum_k^{\text{triplets}} \sum_b^{3} \frac{\langle {}^3\Psi_k^b | H_{so} | {}^1\Psi_0 \rangle}{E_0 - E_k} \left\langle {}^3\Psi_k^b \left| \sum_i e\mathbf{r}_i \right| {}^3\Psi_1^a \right\rangle \tag{69}$$

The first term in Eq. (69) vanishes. All other terms consist of a product of a transition moment between two spin-pure (unperturbed) states, readily evaluated as indicated in Section 5, and a spin–orbit matrix element, the evaluation of which was indicated in the preceding section.

6.2. Spin–Spin Coupling

In the preceding section we have dealt with the mixing of electronic states in different multiplicity manifolds, and some of its spectroscopic consequences, through the operation of the spin–orbit coupling term in the Hamiltonian, Eq. (3). In this section we shall examine the spin–spin coupling operator, Eq. (4), and in particular the lifting of the degeneracy of the three components of triplet states which it induces.

We shall again describe the triplet wave function as a VO-CI function of the form

$$^3\Psi_n^{m_s} = \sum_m C_{nm}{}^{\mu\lambda'} T_m^{m_s} \tag{70}$$

The triplet spin configurations $^{\mu\lambda'}T_m^{m_s}$ are given in Eq. (28), and are eigenfunctions of S_z with eigenvalues $m_s = +1, 0, -1$. As evaluated in zeroth order, under the Hamiltonian H_o, the three components are degenerate with energy

^3E_n. However, the spin–spin coupling operator H_{ss}[Eq. (4)] and the spin–orbit coupling operator H_{so} [Eq. (3)] split this degeneracy and destroy S_z as a good quantum number. The resulting energies of the triplet component $^3\Psi_n^a$ are given by[98]

$$^3E_n^a = {^3E_n} + \lambda_a \tag{71}$$

where the λ_a are the roots of the 3×3 secular equation

$$|H_{m_s m_{s'}} - \lambda \delta_{m_s m_{s'}}| = 0 \tag{72}$$

with the matrix elements

$$H_{m_s m_{s'}} = \langle {^3\Psi_n^{m_s}} | H_{ss} | {^3\Psi_n^{m_{s'}}} \rangle - \sum_k \frac{\langle {^3\Psi_n^{m_s}} | H_{so} | {^1\Psi_k} \rangle \langle {^1\Psi_k} | H_{so} | {^3\Psi_n^{m_{s'}}} \rangle}{^1E_k - {^3E_n}} \tag{73}$$

The second term in Eq. (73), of second order in the spin–orbit coupling operator, is frequently neglected and the matrix element $H_{m_s m_{s'}}$ is then evaluated only in terms of the spin–spin coupling operator.

In the situation mentioned in the preceding section when the symmetry is sufficient to guarantee coincidence between magnetic and principal axes,[99] the same transformation of spin coordinates [Eq. (63)] made there brings Eq. (72) into diagonal form, and all that remains is the evaluation of the diagonal elements H_{aa}, which, for this simple case, are given in Eq. (83), below. However, the development of full matrix elements is unnecessarily clumsy, and we shall not make this transformation of spin coordinates at this time. Instead, we shall develop the full matrix in terms of the usual eigenfunctions of S_z with eigenvalues m_s.[100]

The evaluation of $H_{m_s m_{s'}}$ proceeds in the same manner as the evaluation of the matrix elements over H_{so} in the preceding section. The integration over the closed shell reduces the matrix element to an expansion over two-electron integrals:

$$H_{m_s m_{s'}} = \sum_l \sum_k C_{nl}^* C_{nk} \langle (\mu\lambda' - \lambda'\mu)\, {^3\omega_{m_s}} | \hat{H}_{ij} | (\nu\kappa' - \kappa'\nu)\, {^3\omega_{m_{s'}}} \rangle \tag{74}$$

where $\hat{H}_{ij}$ is given by (the q and p representing the Cartesian axes x, y, and z)

$$\hat{H}_{ij} = g^2\beta_e^2 \left[\sum_q s_{qi}s_{qj}(r_{ij}^2 - 3q_{ij}^2) - \sum_{p>q} 3(s_{qi}s_{pj} + s_{pi}s_{qj})q_{ij}p_{ij} \right] r_{ij}^{-5} \tag{75}$$

It is convenient to express the two-electron operator $\hat{H}_{ij}$ in terms of the components of the two-spin operators S_q defined by

$$S_q = s_{qi} + s_{qj} \tag{76}$$

and their squares

$$\hat{H}_{ij} = \frac{1}{2}g^2\beta_e^2 \sum_q \left[S_q^2(r_{ij}^2 - 3q_{ij}^2) - \sum_p 3(S_qS_p + S_pS_q)q_{ij}p_{ij} \right] r_{ij}^{-5} \tag{77}$$

Substitution of (76) into Eq. (74) gives

$$H_{m_s m_{s'}} = \sum_l \sum_k C^*_{nl} C_{nk} \left\langle {}^3\omega_{m_s} \left| \sum_{p,q} D_{pq} S_p S_q \right| {}^3\omega_{m_{s'}} \right\rangle \tag{78}$$

with

$$D_{pp} = \tfrac{1}{2} g^2 \beta_e^2 \langle (\mu\lambda' - \lambda'\mu) | (r_{ij}^2 - 3p_{ij}^2) r_{ij}^{-5} | (\nu\kappa' - \kappa'\nu) \rangle \tag{79}$$

$$D_{pq} = \tfrac{1}{2} g^2 \beta_e^2 \langle (\mu\lambda' - \lambda'\mu) | -3 p_{ij} q_{ij} r_{ij}^{-5} | (\nu\kappa' - \kappa'\nu) \rangle, \qquad p \neq q$$

Introducing the LCAO expansion then leads to

$$D_{pq} = \tfrac{1}{2} g^2 \beta_e^2 \sum_\rho \sum_\sigma \sum_\tau \sum_\upsilon (c^*_{\mu\rho} c^*_{\lambda'\sigma} - c^*_{\lambda'\rho} c^*_{\mu\sigma}) c_{\nu\tau} c_{\kappa'\upsilon} I(\rho\sigma\tau\upsilon) \tag{80}$$

with

$$I_{pp}(\rho\sigma\tau\upsilon) = \langle \phi_\rho(i)\phi_\sigma(j) | (r_{ij}^2 - 3p_{ij}^2) r_{ij}^{-5} | \phi_\tau(i)\phi_\upsilon(j) \rangle$$

$$I_{pq}(\rho\sigma\tau\upsilon) = \langle \phi_\rho(i)\phi_\sigma(j) | -3 p_{ij} q_{ij} r_{ij}^{-5} | \phi_\tau(i)\phi_\upsilon(j) \rangle, \qquad p \neq q \tag{81}$$

In addition, the ZDO approximation makes I_{pq}, and hence D_{pq}, vanish unless $\rho = \tau$, $\sigma = \upsilon$. Thus,

$$D_{pq} = \tfrac{1}{2} g^2 \beta_e^2 \sum_\rho \sum_\sigma (c^*_{\mu\rho} c^*_{\lambda'\sigma} - c^*_{\lambda'\rho} c^*_{\mu\sigma}) c_{\nu\rho} c_{\kappa'\sigma} I(\rho\sigma\rho\sigma) \tag{82}$$

The values of the integrals which appear in Eq. (81) have been evaluated and appear in the literature.[101,102] The results of the operation of the spin operators $S_p(S_q)$ in Eq. (78) and the subsequent integration over the spin coordinate are straightforward and have recently been tabulated by McGlynn et al.[100] The matrix elements $H_{m_s m_{s'}}$ are now completely specified within the framework of the CNDO/S method. All that remains is to bring the matrix in Eq. (73) into diagonal form, thus obtaining the λ_a's.

Since the matrix of Eq. (73) is symmetric, it may be diagonized by a proper unitary transformation. Such a transformation is equivalent to finding a set of rectangular coordinate axis x, y, z and spin functions such that the expectation value of H_{ss} is diagonal. As pointed out above, in molecules with sufficient symmetry, the proper coordinate axes coincide with the principal axes and the spin functions are given in Eq. (63). Therefore, it is immediately possible to write down an expression for the λ's in such molecules. The result is

$$\lambda_x = -D_{xx} = \tfrac{1}{3} D - E, \qquad \lambda_y = -D_{yy} = \tfrac{1}{3} D + E, \qquad \lambda_z = -D_{zz} = -\tfrac{2}{3} D \tag{83}$$

where, from (73) and (82),*

$$D = \tfrac{3}{4} g^2 \beta_e^2 \sum_m \sum_l C_{nm}^*(\mu\lambda') C_{nl}(\nu\kappa')$$

$$\times \sum_\rho \sum_\rho (c_{\mu\rho}^* c_{\lambda'\sigma} - c_{\lambda'\rho}^* c_{\mu\sigma}) c_{\nu\rho} c_{\kappa'\sigma} I_{zz}(\rho\sigma\rho\sigma) \qquad (84)$$

and

$$E = \tfrac{3}{4} g^2 \beta_e^2 \sum_m \sum_\lambda C_{nm}^*(\mu\lambda') C_{nl}(\nu\kappa')$$

$$\times \sum_\rho \sum_\sigma (c_{\mu\rho}^* c_{\lambda'\sigma} - c_{\lambda'\rho}^* c_{\mu\sigma}) c_{\nu\rho} c_{\kappa'\sigma} \tfrac{1}{3}(I_{yy} - I_{xx}) \qquad (85)$$

Therefore, the effect of H_{ss} is to break the degeneracy of the triplet states, forming three new states with energy $^3E_0 + \lambda_x$, $^3E_0 + \lambda_y$, and $^3E_0 + \lambda_z$ and characterized by the spin functions $^3\omega_-$, $^3\omega_+$, and $^3\omega_0$, respectively, of Eq. (63).

7. Vibrationally Induced Transitions

Electronic transitions between states (o and f) of a molecule for which the transition moment, Eq. (51), vanishes are called *forbidden*, and, according to the theory of Section 5, should not be observed. However, the experimental observation of forbidden transitions is not at all unusual and is well documented.[103] Two theoretical methods have established themselves in the literature for the treatment of this phenomenon; one, first proposed by Herzberg and Teller,[104] is generally referred to by their names (HT). The other, originally discussed by Born and Oppenheimer[2] as arising from a breakdown of their approximation [cf. Section 3, particularly Eq. (10)] will be referred to as the Born–Oppenheimer-breakdown (BOB) theory.

7.1. Herzberg–Teller Theory

The theory most frequently invoked in the spectroscopic literature,[105–108] and recently formulated using CNDO/S wave function, is the HT theory.[109,110] However, the use of HT theory has recently been criticized.[111] The criticism, however, is based more on the manner of application of HT theory than on the theory itself.

HT theory seeks to consider the second term on the rhs of Eq. (48) in the series expansion of $\mathbf{M}_r^{fo}(Q_\alpha)$. First, the total electronic Hamiltonian $H(Q_\alpha, \mathbf{r}_i)$ is

---

*In Eqs. (84) and (85) the coefficients $C_{on}(\mu\lambda')$ and $C_{om}(\nu\kappa')$ are the expansion coefficients for the configuration, where $\mu\lambda'$ and $\nu\kappa'$ are the single occupied MO's.

expanded about the position Q_α°

$$H(Q_\alpha, \mathbf{r}_i) = H(Q_\alpha^\circ, \mathbf{r}_i) + \sum_\alpha \left(\frac{\partial H(Q_\alpha, \mathbf{r}_i)}{\partial Q_\alpha}\right)_{Q_\alpha^\circ} Q_\alpha + \cdots \tag{86}$$

We shall treat the second term on the rhs of Eq. (86) as a perturbation. The perturbed wave function $\Psi_0(Q_\alpha, \mathbf{r}_i)$ is expressed in terms of the unperturbed functions $\Psi_m(Q_\alpha^\circ, \mathbf{r}_i)$:

$$\Psi_o(Q_\alpha, \mathbf{r}_i) = \Psi_o(Q_\alpha^\circ, \mathbf{r}_i) + \sum_{m \neq o} \sum_\alpha H_{mo} Q_\alpha \Psi_m(Q_\alpha^\circ, \mathbf{r}_i) \tag{87}$$

where

$$H_{mo} = \frac{\int \Psi_m(Q_\alpha^\circ, \mathbf{r}_i)[\partial H(Q_\alpha, \mathbf{r}_i)/\partial Q_\alpha]_{Q_\alpha^\circ} \Psi_o(Q_\alpha^\circ, \mathbf{r}_i)\, d\mathbf{r}_i}{E_o(Q_\alpha^\circ) - E_m(Q_\alpha^\circ)} \tag{88}$$

In Eq. (88), $E_o(Q_\alpha^\circ)$ and $E_m(Q_\alpha^\circ)$ represent the *electronic* energies associated with $\Psi_o(Q_\alpha^\circ, \mathbf{r}_i)$ and $\Psi_m(Q_\alpha^\circ, \mathbf{r}_i)$, respectively. Expressions analogous to (87) and (88) may, of course, be written for the final perturbed wave function $\Psi_f(Q_\alpha, \mathbf{r}_i)$. Substitution of these equations for $\Psi_o(Q_\alpha, \mathbf{r}_i)$ and $\Psi_f(Q_\alpha, \mathbf{r}_i)$ in Eq. (50) and evaluation of the derivative at Q_α° leads to

$$\sum_\alpha \left(\frac{\partial \mathbf{M}_r^{fo}(Q_\alpha)}{\partial Q_\alpha}\right)_{Q_\alpha^\circ} = \sum_{n \neq f} \sum_\alpha H_{nf} \mathbf{M}_r^{no}(Q_\alpha^\circ) + \sum_{m \neq o} \sum_\alpha H_{mo} \mathbf{M}_r^{mf}(Q_\alpha^\circ) \tag{89}$$

Finally, the substitution of Eqs. (89) and (48) into Eq. (47) provides the HT result for the transition moment:

$$(i\omega_{fo})^{-1} \mathbf{M}(E_f \leftarrow E_o) \approx \sum_\alpha \int X_f^w(Q_\alpha) Q_\alpha X_o^v(Q_\alpha)\, dQ_\alpha$$

$$\times \left[\sum_{n \neq f} H_{nf} \mathbf{M}_r^{no}(Q_\alpha^\circ) + \sum_{m \neq o} H_{mo} \mathbf{M}_r^{mf}(Q_\alpha^\circ)\right] \tag{90}$$

Equation (90) is the general expression for the transition moment in the HT theory. The matrix elements H_{nf} and H_{mo} are often found expressed in terms of the operator $\partial/\partial Q$ rather than $(\partial H/\partial Q_\alpha)_{Q_\alpha^\circ}$. The relationship between these operators arises through the Hellmann–Feynman theorem,[112,113] which states

$$\int \Psi_k(Q_\alpha^\circ, \mathbf{r}_i)\left(\frac{\partial H(Q_\alpha, \mathbf{r}_i)}{\partial Q_\alpha}\right)_{Q_\alpha^\circ} \Psi_l(Q_\alpha^\circ, \mathbf{r}_i)\, d\mathbf{r}_i$$

$$= [E_l(Q_\alpha^\circ) - E_k(Q_\alpha^\circ)]\left[\int \Psi_k(Q_\alpha, \mathbf{r}_i)\frac{\partial}{\partial Q_\alpha}\Psi_l(Q_\alpha, \mathbf{r}_k)\, d\mathbf{r}_i\right]_{Q_\alpha^\circ} + \left(\frac{\partial E_l(Q_\alpha)}{\partial Q_\alpha}\right)_{Q_\alpha^\circ} \delta_{kl}$$

$$\tag{91}$$

and the matrix elements become

$$H_{mo} = \int \Psi_m(Q_\alpha, \mathbf{r}_i) \left[\frac{\partial}{\partial Q_\alpha} \Psi_o(Q_\alpha, \mathbf{r}_i) \right]_{Q_\alpha^\circ} d\mathbf{r}_i$$

$$H_{nf} = \int \Psi_n(Q_\alpha, \mathbf{r}_i) \left[\frac{\partial}{\partial Q_\alpha} \Psi_f(Q_\alpha, \mathbf{r}_i) \right]_{Q_\alpha^\circ} d\mathbf{r}_i$$

(92)

In the application of Eq. (90) the second sum in the square brackets is often ignored on the basis of the argument that H_{mo} is normally very small. This particular approximation, which is not necessary for HT theory, is the prime target of the recent attacks on this method.[111] Whereas this last approximation is not essential to the theory, others that have to be made to make the calculation feasible, or at least reasonable, are at least as severe:

1. The assumption that the vibrations are harmonic, i.e., that X_f^w and X_o^v are eigenfunctions of a harmonic oscillator.
2. The assumption that the system is at low temperature, and at equilibrium, so that the entire set of quantum numbers v consists of all zeros.
3. The assumption that X_f^w and X_o^v are functions of identical normal coordinates, undistorted and undisplaced relative to one another.

On the basis of these three assumptions it is concluded that the only transitions that have nonvanishing intensity are those in which the sets w and v differ only in a single quantum number, and in this one only by one. Since all v are zero, all w but one are zero. With these simplifications, Eq. (90) becomes

$$(i\omega_{fo})^{-1} \mathbf{M}(E_f \leftarrow E_o) \approx \sum_\alpha (2\mu_\alpha \omega_\alpha)^{-1/2} \left[\sum_{n \neq f} H_{nf} \mathbf{M}_r^{no}(Q_\alpha^\circ) + \sum_{m \neq o} H_{mo} \mathbf{M}_r^{mf}(Q_\alpha^\circ) \right]$$

(93)

where μ_α is the reduced mass corresponding to the normal mode α and ω_α is the associated frequency.[114]

Using only the first term in Eq. (93), Roche and Jaffé have recently calculated transition probabilities for the $^1B_{2u} \leftarrow {}^1A_{1g}$ transition in benzene.[109] H_{nf} given in Eq. (88) may be written as

$$H_{nf} = \left[\frac{\partial}{\partial Q_\alpha} \int \Psi_n(Q_\alpha^\circ, \mathbf{r}_i) H(Q_\alpha, \mathbf{r}_i) \Psi_f(Q_\alpha^\circ, \mathbf{r}_i) \, d\mathbf{r}_i \right]_{Q_\alpha^\circ}$$

(94)

The integral which appears in Eq. (94) is simply the matrix element of the electronic Hamiltonian between two different electronic states n and f, and in the CNDO/S approximation can be written in a simple form.[109] The taking of the derivative $\partial/\partial Q_\alpha$ is straightforward. When Eq. (94) was evaluated using wave functions for the states n and f determined at Q_α°, the results were not in agreement with experiment. This lack of agreement was traced to the fact that when both the LCAO coefficients and the LCAO bases were held fixed at Q_α°

Table 2. The Calculated and Experimental Oscillator Strengths
for Transition to the $^1B_{2u}$ and $^1B_{1u}$ States in Benzene

State	$f \times 10^2$		
	Ref. 109	Ref. 110	Exp.[a]
$^1B_{2u}$	0.09	0.08	0.14
$^1B_{1u}$	23.0	9.1	9.4

[a] See Ref. 110.

the field exerted on the nuclei was too strong. To overcome this difficulty the LCAO coefficients are taken as they are calculated in the geometry Q_α°, but the LCAO basis set is allowed to move with the nuclei.[107,115] This is known as the orbital following approximation. Strictly speaking, the wave functions constructed in this manner are not the $\Psi_n'(Q_\alpha^\circ, \mathbf{r}_i)$ which appear in Eq. (45) but are closely related, and give excellent results for transitions in benzene.

Ziegler and Albrecht[110] have recently studied this same transition. Their formulation of the HT theory is very different from the development given above; however, it is consistent with the central theme of the theory, namely the inclusion of the second term of Eq. (48) into Eq. (47). Their calculations include the equivalent of both terms in Eq. (90), and they conclude that the second term is significantly smaller than the first. Some results for the benzene system are given in Table 2.

7.2. Born–Oppenheimer Breakdown Theory

The BOB theory begins from the viewpoint that the separation of the molecular wave function $\Xi(Q_\alpha, \mathbf{r}_i)$ into a product of a nuclear function $\Phi_m(Q_\alpha)$ and an electronic function $\Psi(Q_\alpha, \mathbf{r}_i)$ [cf. Eq. (7)] is in error because we have neglected the rhs of Eq. (10). BOB theory as recently formulated thus asserts that the nuclear kinetic energy should be treated as a perturbation.[111] In what follows we shall treat the problem completely and show that both the HT and the kinetic energy terms arise separately and that they are additive.

For a molecule with total energy E_o (electronic plus vibrational), the correct wave function Ξ' may be expanded in terms of the approximate set Ξ_n of Eq. (7):

$$\Xi'(Q_\alpha, \mathbf{r}_i; E_o) = \Xi(Q_\alpha, \mathbf{r}_i; E_o) + \sum_{m \neq o} \sum_\alpha H_{mo} \Xi(Q_\alpha, \mathbf{r}_i; E_m)$$

where

$$H_{mo} = \frac{\iint \Xi(Q_\alpha, \mathbf{r}_i; E_m)[-(\partial^2/\partial Q_\alpha^2)\Xi(Q_\alpha, \mathbf{r}_i; E_m)]\, dQ_\alpha\, d\mathbf{r}_i}{E_o - E_m} \qquad (95)$$

and analogously for the final state with energy E_f.

The transition moment for the transition $o \to f$ is

$$(i\omega_{fo})^{-1}\mathbf{M}(E_f \leftarrow E_o) \approx \iint \Xi(Q_\alpha, \mathbf{r}_i; E_f) \sum_j \mathbf{r}_j \Xi(Q_\alpha, \mathbf{r}_i; E_o) \, dQ_\alpha \, d\mathbf{r}_i$$

$$+ \sum_{n \neq f} \sum_\alpha H_{nf} \iint \Xi(Q_\alpha, \mathbf{r}_i; E_n) \sum_j \mathbf{r}_j \Xi(Q_\alpha, \mathbf{r}_i; E_o) \, dQ_\alpha \, d\mathbf{r}_i$$

$$+ \sum_{m \neq o} \sum_\alpha H_{mo} \iint \Xi(Q_\alpha, \mathbf{r}_i; E_f) \sum_j \mathbf{r}_j \Xi(Q_\alpha, \mathbf{r}_i; E_m) \, dQ_\alpha \, d\mathbf{r}_i \tag{96}$$

Substitution of Eqs. (41) and (45) into Eq. (96) gives

$$(i\omega_{fo})^{-1}\mathbf{M}(E_f \leftarrow E_o) \approx \int \Phi_f(Q_\alpha)\mathbf{M}_r^{fo}(Q_\alpha)\Phi_o(Q_\alpha) \, dQ_\alpha$$

$$+ \sum_{n \neq f} \sum_\alpha H_{nf} \int \Phi_n(Q_\alpha)\mathbf{M}_r^{no}(Q_\alpha)\Phi_o(Q_\alpha) \, dQ_\alpha$$

$$+ \sum_{m \neq o} \sum_\alpha H_{mo} \int \Phi_f(Q_\alpha)\mathbf{M}_r^{fm}(Q_\alpha)\Phi_m(Q_\alpha) \, dQ_\alpha \tag{97}$$

Equation (97) is the *complete* statement, to first order, of the transition moment as expanded in Eq. (46). The problem is then to evaluate the coefficients H_{nf} and H_{mo} as well as the integrals appearing in Eq. (97). The evaluation of these integrals reduces to the problem we have already solved for Eq. (47). Using the Condon expansion Eq. (48), Eq. (97) becomes

$$(i\omega_{fo})^{-1}\mathbf{M}(E_f \leftarrow E_o) \approx \mathbf{M}_r^{fo}(Q_\alpha^\circ) + \sum_\alpha \left(\frac{\partial \mathbf{M}_r^{fo}(Q_\alpha)}{\partial Q_\alpha}\right)_{Q_\alpha^\circ} \int X_f^w(Q_\alpha)Q_\alpha X_o^v(Q_\alpha) \, dQ_\alpha$$

$$+ \sum_{n \neq f} \sum_\alpha H_{nf}\left[\mathbf{M}_r^{no}(Q_\alpha^\circ) \int X_n^u(Q_\alpha)X_o^v(Q_\alpha) \, dQ_\alpha\right.$$

$$\left. + \left(\frac{\partial \mathbf{M}_r^{no}(Q_\alpha)}{\partial Q_\alpha}\right)_{Q_\alpha^\circ} \int X_n^u(Q_\alpha)Q_\alpha X_o^v(Q_\alpha) \, dQ_\alpha\right]$$

$$+ \sum_{m \neq o} \sum_\alpha H_{mo}\left[\mathbf{M}_r^{fn}(Q_\alpha^\circ) \int X_f^w(Q_\alpha)X_m^u(Q_\alpha) \, dQ_\alpha\right.$$

$$\left. + \left(\frac{\partial \mathbf{M}_r^{fm}(Q_\alpha)}{\partial Q_\alpha}\right)_{Q_\alpha^\circ} \int X_f^w(Q_\alpha)Q_\alpha X_m^u(Q_\alpha) \, dQ_\alpha\right] \tag{98}$$

In the case of symmetry-forbidden transitions, $\mathbf{M}_r^{fo}(Q_\alpha^\circ)$ vanishes; however, the same is not necessarily true for $\mathbf{M}_r^{no}(Q_\alpha^\circ)$ and $\mathbf{M}_r^{fm}(Q_\alpha^\circ)$ appearing in Eq. (98), which, when we retain only the first nonvanishing terms, may be rewritten as

$$(i\omega_{fo})^{-1}\mathbf{M}(E_f \leftarrow E_o) \approx \sum_\alpha \left(\frac{\partial \mathbf{M}_r^{fo}(Q_\alpha)}{\partial Q_\alpha}\right)_{Q_\alpha^\circ} \int X_f^w(Q_\alpha)Q_\alpha X_o^v(Q_\alpha)\, dQ_\alpha$$

$$+ \sum_{n\neq f}\sum_\alpha H_{nf}\mathbf{M}_r^{no}(Q_\alpha^\circ)\int X_n^u(Q_\alpha)X_o^v(Q_\alpha)\, dQ_\alpha$$

$$+ \sum_{m\neq o}\sum_\alpha H_{mo}\mathbf{M}_r^{fm}(Q_\alpha^\circ)\int X_f^w(Q_\alpha)X_m^u(Q_\alpha)\, dQ_\alpha \qquad (99)$$

Finally, we have to evaluate the perturbation coefficients H_{nf} and H_{mo} of Eq. (95). Using Eq. (7), the H_{mo} becomes

$$H_{mo} = (E_o - E_m)^{-1}\int X_m^u(Q_\alpha)G_{mo}(Q_\alpha)\left[\frac{\partial}{\partial Q_\alpha}X_o^v(Q_\alpha)\right]dQ_\alpha$$

with

$$G_{mo}(Q_\alpha) = \int \Psi_m(Q_\alpha, \mathbf{r}_i)\left[\frac{\partial}{\partial Q_\alpha}\Psi_o(Q_\alpha, \mathbf{r}_i)\right]d\mathbf{r}_i$$

and analogously for H_{nf}. We now expand $G_{mo}(Q_\alpha)$ about the position Q_α° and retain only the leading term; thus H_{mo} becomes

$$H_{mo} = (E_o - E_m)^{-1}\left[\int \Psi_m(Q_\alpha, \mathbf{r}_i)\frac{\partial}{\partial Q_\alpha}\Psi_o(Q_\alpha, \mathbf{r}_i)\, d\mathbf{r}_i\right]_{Q_\alpha^\circ}$$

$$\times \int X_m^u(Q_\alpha)\left(-\frac{\partial}{\partial Q_\alpha}X_o^v(Q_\alpha)\right)dQ_\alpha$$

and similarly for H_{nf}. Substitution of these results in Eq. (99) yields

$$(i\omega_{fo})^{-1}\mathbf{M}(E_f - E_o)$$

$$\approx \sum_{n\neq f}\sum_\alpha \left[\int \Psi_n(Q_\alpha, \mathbf{r}_i)\frac{\partial}{\partial Q_\alpha}\Psi_f(Q_\alpha, \mathbf{r}_i)\, d\mathbf{r}_i\right]_{Q_\alpha^\circ}\mathbf{M}_r^{no}(Q_\alpha^\circ)\int X_f^w(Q_\alpha)Q_\alpha X_o^v\, dQ_\alpha$$

$$+ \sum_{m\neq o}\sum_\alpha \left[\int \Psi_m(Q_\alpha, \mathbf{r}_i)\frac{\partial}{\partial Q_\alpha}\Psi_o(Q_\alpha, \mathbf{r}_i)\, d\mathbf{r}_i\right]_{Q_\alpha^\circ}\mathbf{M}_f^{fm}(Q_\alpha^\circ)$$

$$\times \int X_f^w(Q_\alpha)Q_\alpha X_o^v(Q_\alpha)\, dQ_\alpha$$

$$+ \sum_{n\neq f}\sum_\alpha (E_f - E_n)^{-1}\left[\int \Psi_n(Q_\alpha, \mathbf{r}_i)\frac{\partial}{\partial Q_\alpha}\Psi_f(Q_\alpha, \mathbf{r}_i)\, d\mathbf{r}_i\right]_{Q_\alpha^\circ}$$

$$\times \int X_n^u(Q_\alpha)\left(-\frac{\partial}{\partial Q_\alpha}X_f^w(Q_\alpha)\right)dQ_\alpha\mathbf{M}_r^{no}(Q_\alpha^\circ)\int X_n^u(Q_\alpha)X_o^v(Q_\alpha)\, dQ_\alpha$$

$$+ \sum_{m\neq o}\sum_\alpha (E_o - E_m)^{-1}\left[\int \Psi_m(Q_\alpha, \mathbf{r}_i)\frac{\partial}{\partial Q_\alpha}\Psi_o(Q_\alpha, \mathbf{r}_i)\, dQ_\alpha\right]_{Q_\alpha^\circ}$$

$$\times \int X_m^u(Q_\alpha)\left(-\frac{\partial}{\partial Q_\alpha}X_o^v(Q_\alpha)\right)dQ_\alpha\mathbf{M}_r^{fm}(Q_\alpha^\circ)\int X_f^w(Q_\alpha)X_n^u(Q_\alpha)\, dQ_\alpha \qquad (100)$$

The first two terms in Eq. (100) are just the result obtained from HT theory in the preceding subsection. The last two terms arise from the nuclear kinetic energy perturbation, and thus represent the BOB correction. In that sense, then, the HT and BOB corrections prove to be additive. However, it should be clear that Eq. (97) is a *complete* statement of the effect of the breakdown of the Born–Oppenheimer approximation; Eq. (100) arose only after application of the Condon approximation.

A general evaluation of the relative importance of the various terms in Eq. (100) is difficult. One generalization, however, can readily be made: The BOB (kinetic energy) correction terms are similar in form to the HT terms, except for a weighting factor involving the reciprocal of an energy difference. This suggests that, in general, the HT terms should predominate. This conclusion seems to be borne out by the results for benzene (Table 2).

8. Application of ZDO Methods

8.1. Simple Organic Compounds

Many of the applications of the CNDO/S and the various other ZDO methods parametrized for spectroscopy have been to molecules for which spectral calculations had been, or could be, made by π-electron methods such as Pariser–Parr–Pople.[20,21] Examples are studies of benzene, the azines, and their substitution products,[24,116–120] of five-membered heterocycles,[121,122] of quinones,[123] of polycycles,[22,124] and many others.[125] One may ask what advantages these methods, which, because of their greatly expanded basis sets must be substantially more costly, have over the earlier methods. A number of these present themselves:

1. Because of the inclusion of all electrons explicitly in the calculation, ZDO methods directly and automatically give information about $n-\pi^*$ transitions; such information is essentially unobtainable in π-electron methods without the introduction of additional highly empirical approximations which usually have only very limited ranges of validity. Some of the early work in the ZDO methods was specifically directed to this aim.

2. In the π-electron methods workers have long been plagued with problems of multiple parameters for atoms of a single element, depending on the distribution of their electrons between the σ and π populations. In particular, the parametrization of N and O has been subject to this uncertainty and an entire literature exists on the systematic choice of such parameters. In contrast, the all-valence-electron methods, since they make no distinction between σ and π orbitals, are completely free of this problem.

3. The treatment of ionic compounds in π-electron methods again is a major problem; the semiempirical parameters are based on a neutral core, and the distribution of the charge has to be carefully adjusted. In the CNDO/S method numerous ionic compounds have been dealt with without any special precautions.[126,127]

4. Strictly, π-electron methods are applicable only to planar molecules. Much effort has gone into attempts to treat adequately even such small deviations from planarity as the effect of a methyl substituent, and many different approximation treatments have been proposed: treatment only by perturbation of the parameters of the unsaturated atoms, and treatment in a true hyperconjugative fashion,[128] or as a single pseudoatom with an electron pair. In the ZDO methods, in principle, departure from planarity produces no problem. In practice, the resulting reduction in symmetry makes the assignment of orbitals to preconceived chemical notions difficult, but does not affect the spectroscopic results of the calculation. The proper treatment does require the assumption of specific conformations; however, in our experience, transition energies and intensities have been insensitive to the particular conformer assumed.

The fact that the majority of applications of the methods under discussion has been to planar and pseudoplanar π-electron systems is mostly due to the fact that most of the carefully investigated compounds fall in this class. An exhaustive bibliography of such applications is well beyond the scope of the present chapter; a few more examples shall suffice to indicate the range: cyclopropane, ethylene oxide, ethylene imine,[54] ethylene oximes,[129] toluic acids,[130] β-γ unsaturated ketones,[131] α-dicarbonyls,[132] aliphatic cyclic ketones,[133] nucleic acid bases,[65] and porphyrin derivatives.[134]

The methods permit, however, the treatment of completely nonplanar systems and application to spectra that have nothing to do with π electrons; thus, spectra of alkanes have been dealt with,[135] but most such applications will fall into later, more specialized sections. A particularly elegant use of results of CNDO/S calculations is presented in a review and reinterpretation of the spectra of polar aromatics.[136]

8.2. Inorganic Compounds

Most authors have been unwilling to extend the parametrization beyond the first row of the periodic system; this has naturally greatly restricted the use of ZDO methods to inorganic compounds. Still, quite a number of such applications have been reported; we again give only a few random examples: NO_3^- and boron trihalides[137]; the series from nitrite to nitrate through nitromethane, nitramide, nitric acid, and methylnitrate[138]; $N(C(CN)_2)_2$[139]; cyanogen and cyanoacetylen[140]; $CO(CN)_2$.[141]

8.3. Interacting Nonplanar π-Electron Systems

There are of course many molecules in which two or more separate π electron systems interact to a lesser or greater degree. A borderline case of such a system is biphenyl, which has been treated with some success by π-electron methods,[142] but which has been reexamined by CNDO/S.[64] More characteristic of this type of system are the spiropyranes[143] and tris-[2,2,2]-paraxylylene.[144]

8.4. Triplet States

Although the bulk of work on spectroscopy of organic molecules for many years has been concerned with singlet states (absorption and fluorescene spectroscopy), interest in triplet states has grown rapidly in recent years. The ZDO methods consequently are coming into play in providing information about such states. Triplet energies are readily available through the virtual orbitals of the ground state and CI,[22,24,55,145] as well as by direct SCF calculation.[81,145–148] Intensities for singlet–triplet absorption and lifetimes for phosphorescence also are becoming available through spin–orbit coupling calculations (cf. Section 6).

The availability of triplet energies also permits prediction of transitions within the triplet manifold.[149–151] CNDO and INDO wave functions have recently been applied to the calculation of Davydov splitting of the lowest $\pi-\pi^*$ triplets of hydrocarbons and the $n-\pi^*$ triplet of pyrazine.[152–154]

8.5. Free Radicals and Doublet States; Photoelectron Spectra

The recently developed methods for doublet states are rapidly finding application in the calculation of optical spectra of free radicals and radical ions. Thus, for example, the old problem of the spectra of hydrocarbon anion and cation radicals has been reexamined,[155] and a number of small inorganic radicals have been treated,[72,75] The difficult problem of the treatment of radicals with partially occupied degenerate orbitals was resolved[80] and calculations on cyclopropenyl, cyclopentenyl, tropyl, and phenyl anion and cation radicals have been reported.[156]

Closely related to the problem of the optical spectra of free radicals and radical ions is the treatment of photoelectron spectra. Most work in this area has been based on Koopmans' theorem,[157] coupled with any of the ZDO methods. A problem that arises immediately is that virtually all of these methods give ionization potentials which are too high generally by about 1 eV; the ordering and spacing within the PE spectrum are generally much better represented. A special parametrization for PE spectra (SPINDO) has also been proposed.[158]

However, Koopmans' theorem is, at best, a rather crude approximation; a better representation of the ordering and spacing within the PE spectrum is given by a calculation of the doublet states of the radical cation in question, be it by a VO-CI calculation of all states or by SCF calculation of individual ones.[77,78]

Intensities of photoelectron spectra should become available through the cross-section calculations outlined in Section 5.2; however, only preliminary results are so far available.[89-91]

8.6. Rydberg Transitions

All electronic transitions so far discussed in this chapter are transitions within the valence shell of the molecules concerned. Further transitions may, of course, be expected, in which electrons are excited into orbitals of the higher shells of the atoms, the Rydberg transitions,[159] which occur generally in the vacuum-ultraviolet region. Since the calculations so far discussed in this chapter employ only minimum basis sets, such transitions are not included. We know of only a single attempt to deal with Rydberg states and transitions to them.[160] By expanding the basis set to the $n = 2$ shell for hydrogen and the $n = 3$ shell for carbon, Rydberg spectra of a number of aliphatic hydrocarbons were calculated with good results. Unfortunately, the size of the computation is greatly expanded in the process; in particular, to deal with successive members of a Rydberg series the expansion of the basis set must go to higher and higher shells, and the empirical estimation of integrals becomes more and more questionable. Consequently, it does not seem likely that semiempirical methods will be pushed much further in this direction.

8.7. Treatment of *d* Orbitals

One of the oldest controversies in chemistry relates to the type of binding in compounds of elements of the second row of the periodic system (Si, P, S, and Cl); in particular to the question of whether $3d$ orbitals need to be included in the basis set, and if so, how. This question has again been raised in conjunction with the ZDO methods. Some early calculations with the Wolfsberg–Helmholz method[161] gave encouraging results.[162] Santry and Segal[163] provided an analysis of possible extensions of CNDO/2 to ground states of molecules involving *d* orbitals of elements in the second row, which, however, failed to yield a simple answer or a generally applicable method. However, their "method" has been applied in the literature to excited states with varying success.[164] The main problem appears to lie in the description of the $3d$ orbitals, which, depending on the charge of the second row atom, may be very diffuse or more or less contracted. In some cases it may, of course, be possible to ignore d orbitals completely. Thus, a study of sulfur heterocycles without intervention of d orbitals has given eminently successful results.[165]

Some attempts have also been made to apply ZDO methods to transition metal compounds, but not enough systematic work has appeared in the literature to give a meaningful evaluation. See also Chapter 4.

8.8. Geometry of Excited States

The CNDO method was originally conceived in part to permit the calculation of the geometries of molecules, and some rather sophisticated algorithms have been developed to achieve the energy minimization.[166] A number of attempts have been made to similarly obtain geometries of excited states. The simplest of these appears to be to infer the excited-state geometry from static parameters (charge densities and bond orders) calculated by the CNDO/S method for the state in question.[167] Crude as it is, this method appears to yield reasonable results. Attempts to do direct CNDO/2 calculations on the states[168,169] suffer from the usual difficulties that the method does not provide reasonable wave functions for excited states. On the other hand, use of the CNDO/S method involves problems with electron repulsion integrals, which are not the proper functions of interatomic distance. Initial attempts to minimize the excited-state energy as a sum of ground-state energy calculated by CNDO/2 plus excitation energy calculated by CNDO/2 have shown considerable promise.[151,170]

8.9. Spin–Orbit, Spin–Spin, and Vibronic Coupling

Relatively little work has yet been reported using these perturbation methods, although it appears that they will receive increasing use.

The most straightforward application of the spin–orbit coupling theory developed in Section 6.1 is to a calculation of phosphorescence lifetimes and singlet–triplet absorption intensities.[95,96,148,171] More complicated applications involve the use of SOC to calculate the splitting of first-order degeneracies in molecular ions.[172,173] Another potential application is in the calculation of intersystem crossing probabilities,[174,175] but no such applications have yet been reported.

Spin–spin coupling leads, as shown in Section 6.2, directly to the zero–field splitting observed in ESR spectra, and it appears that a number of groups are working in this area.[176]

Several groups have dealt with vibronic coupling in formaldehyde[109] and in benzene[109,110] using various formulations of Herzberg–Teller theory (cf. Section 7).

8.10. Ionization Potentials

Once an SCF calculation has been made of the ground state of a molecule, Koopman's theorem[157] may be used to obtain ionization potentials.[177] Although the CNDO/S method was not specifically parametrized for this purpose, it appears that it does actually somewhat better in calculation of ionization potentials than CNDO/2,[178] probably because the parametrization is specific to energy *differences*. A particularly interesting use of such ionization potentials to photoelectron spectra is discussed in Section 8.5. An alternate approach to ionization potentials is, of course, equally feasible, though more expensive—the separate calculation of the two states. This approach is further discussed in Section 8.5.

Ionization potentials for excited states are obtainable in analogous ways. although Koopmans' theorem was originally postulated for closed shells, and Roothaan[28] questions its applicability in his open-shell method, it has been shown that a trivial transformation of the Roothaan equations leads to an extended Koopmans' theorem for open shells.[179] Some recent CNDO/S calculations confirm that the ionization potentials so obtained are only slightly larger than values obtained as differences between two SCF calculations.[180]

8.11. Dipole Moments and Polarizabilities

Once a molecular wave function has been calculated, the dipole moment is readily available.[181] Thus excited-state dipole moments may be calculated readily.[182] The results obtained compare favorably with experimental values.

Recently, interest in molecular polarizabilities has been renewed. Diagonal elements α_{qq} of the polarizability tensor α have been calculated by variational perturbation methods; SCF calculations are carried out including terms for external applied electromagnetic fields; the induced dipole moment is calculated and extrapolated to zero field.[183–186] A much simpler procedure giving essentially similar results has been proposed.[187] It involves the direct calculation of α by perturbation theory.[188] The set of VO/CI functions are used as basis functions for the perturbation. Unfortunately, the expansion does not appear to converge rapidly. These methods have also been applied to excited states.[183,187,189]

8.12. Miscellaneous Studies

ZDO methods have been used to help shed light on a number of other problems. Thus, Smith and co-workers, in their studies of circular dichroism, have used CNDO/S to help locate excited singlet states and to obtain magnetic transition moments.[190] Similarly, the method has been used to identify the upper states in two-photon absorption processes.[191]

A largely unexplored area, of prime importance in solution spectroscopy, is the treatment of solvent effects. Some initial attempts[187,192] are encouraging but no generally useful method has yet evolved.

9. Conclusions

We have attempted in this chapter to review the semiempirical methods currently available to obtain energies and wave functions for excited states of molecules. In the process we have discovered that reasonable values for energies (relative to experimental measurements) can be obtained by ZDO CNDO-like methods, provided the method differs from CNDO in three crucial points: (1) it should include a reasonably long configuration-interaction expansion, (2) it should involve scaled electronic repulsion integrals, and (3) it should distinguish properly between s amd p orbitals in the β's.

The associated wave functions have been shown to provide a basis for reasonable expectation values leading to such properties as transition moments, and to serve as basis sets for perturbation calculations of such diverse properties as intensities (or lifetimes) for forbidden transitions, zero-field splitting, and excited-state geometries.

Whereas the CNDO/S method is by far the most widely adopted and most extensively documented method we have encountered, it appears that wave functions obtained by many variants are equally acceptable.

References

1. R. McWeeny and B. T. Sutcliffe, *Methods of Molecular Quantum Mechanics*, Academic Press, New York (1969), pp. 287–298.
2. M. Born and J. R. Oppenheimer, *Ann. Phys. (Paris)* **84**, 457 (1927).
3. E. B. Wilson, J. C. Decius, and P. C. Cross, *Molecular Vibrations*, McGraw-Hill, New York, (1955), Chapters 1–3 and 11.
4. G. Herzberg, *Molecular Spectra and Molecular Structure III. Electronic Spectra and Electronic Structure of Polyatomic Molecules*, Van Nostrand Reinhold, New York (1966), pp. 442–482.
5. O. K. Rice, *Phys. Rev.* **35**, 1538, 1551 (1930); *J. Chem. Phys.* **9**, 258 (1941).
6. M. S. Child, *Molecular Collision Theory*, Academic Press, New York (1974).
7. R. D. Levine, *Quantum Mechanics of Molecular Rate Processes*, Oxford University Press, Oxford (1969).
8. D. R. Hartree, *Proc. Camb. Phil. Soc.* **24**, 89 (1928).

9. V. Fock, *Z. Physik.* **61**, 126 (1930).
10. R. Daudel, R. Lefebvre, and C. Moser, *Quantum Chemistry Methods and Applications*, Interscience, New York (1959).
11. J. Hinze and C. C. J. Roothaan, *Prog. Theor. Phys.* **40**, S37(1967).
12. L. Brillouin, *Les Champs 'self consistents' de Hartree et de Fock*, Hermann, Paris (1934), p. 19.
13. C. Møller and M. S. Plesset, *Phys. Rev.* **40**, 618 (1934).
14. M. Cohen and A. Delgarno, *Proc. Phys. Soc. (Lond.)* **77**, 748 (1961).
15. J. Fajer, B. H. J. Bielski, and R. H. Felton, *J. Phys. Chem.* **72**, 1281 (1968).
16. R. Zahradník and P. Čársky, *J. Phys. Chem.* **74**, 1235 (1970).
17. M. J. S. Dewar and H. C. Longuet-Higgins, *Proc. Phys. Soc. (Lond.)* A **68**, 81 (1954).
18. R. Pariser, *J. Chem. Phys.* **24**, 250 (1956).
19. C. A. Coulson and G. S. Rushbrooke, *Proc. Camb. Phil. Soc.* **36**, 193 (1940).
20. R. Pariser and R. G. Parr, *J. Chem. Phys.* **21**, 466 (1953).
21. J. A. Pople, *Trans. Faraday Soc.* **49**, 1375 (1953).
22. R. L. Ellis and H. H. Jaffé, *J. Mol. Spec.* **50**, 474 (1974).
23. P. Claverie, S. Diner, and J. P. Malrieu, *Theor. Chim. Acta* **8**, 390, 404 (1967); **10**, 467 (1968).
24. J. DelBene and H. H. Jaffé, *J. Chem. Phys.* **48**, 1807 (1968).
25. R. Lefèbvre, *J. Chim. Phys.* **54**, 168 (1957).
26. S. Huzinaga, *Phys. Rev.* **120**, 866 (1960); **122**, 131 (1962).
27. J. A. Pople and R. N. Nesbet, *J. Chem. Phys.* **22**, 571 (1954).
28. C. C. J. Roothaan, *Rev. Mod. Phys.* **32**, 179 (1960).
29. W. J. Hunt, T. H. Dunning, Jr., and W. A. Goddard, III, *Chem. Phys. Lett.* **3**, 606 (1969).
30. W. A. Goddard, III, T. H. Dunning, Jr., and W. J. Hunt, *Chem. Phys. Lett.* **4**, 231 (1969).
31. G. A. Segal, *J. Chem. Phys.* **33**, 360 (1970).
32. H. H. Jaffé and R. L. Ellis, *J. Comput. Phys.* **16**, 20 (1974).
33. Y. G. Smeyers and L. Doreste-Suarez, *Int. J. Quantum Chem.* **7**, 687 (1973).
34. C. W. Eaker and J. Hinze, *J. Am. Chem. Soc.* **96**, 4084 (1974).
35. G. Das and A. C. Wahl, *J. Chem. Phys.* **44**, 87 (1966).
36. G. Das, *J. Chem. Phys.* **46**, 1568 (1967).
37. A. Veillard and E. Clementi, *Theor. Chim. Acta* **7**, 13 (1967).
38. S. Huzinaga, *Prog. Theor. Phys.* **41**, 307 (1969).
39. M. H. Wood and A. Veillard, *Mol. Phys.* **26**, 595 (1973).
40. J. A. Pople and D. L. Beveridge, *Approximate Molecular Orbital Theory*, McGraw-Hill, New York (1970), Ch. II.
41. H. H. Jaffé, *Acc. Chem. Res.* **2**, 136 (1969).
42. J. A. Pople and G. A. Segal, *J. Chem. Phys.* **44**, 3289 (1966).
43. J. A. Pople and D. L. Beveridge, *Approximate Molecular Orbital Theory*, McGraw-Hill, New York, (1970), Ch. III.
44. R. Pariser, *J. Chem. Phys.* **21**, 568 (1953).
45. R. Pariser and R. G. Parr, *J. Chem. Phys.* **21**, 767 (1953).
46. K. Nishimoto and N. Mataga, *Z. Physik. Chem. (Frankfurt)* **12**, 335; **13**, 140 (1957).
47. G. Klopman, *J. Am. Chem. Soc.* **86**, 4550 (1964).
48. J. A. Pople, D. F. Santry, and G. A. Segal, *J. Chem. Phys.* **43**, S129 (1965).
49. J. A. Pople and G. A. Segal, *J. Chem. Phys.* **43**, S136 (1965).
50. R. G. Parr, *Quantum Theory of Molecular Electronic Structure*, Benjamin, New York (1964), p. 3.
51. P. A. Clark and J. L. Ragle, *J. Chem. Phys.* **46**, 4235 (1967).
52. H. W. Kroto and D. P. Santry, *J. Chem. Phys.* **47**, 792 (1967).
53. G. Giessner-Prettre and A. Pullman, *Theor. Chim. Acta* **13**, 265 (1969); **17**, 120 (1970); **18**, 14 (1970); **20**, 378 (1971).
54. D. T. Clark, *Theor. Chim. Acta* **10**, 111 (1968).
55. R. L. Ellis, G. Kuehnlenz, and H. H. Jaffé, *Theor. Chim. Acta* **26**, 131 (1972).
56. J. Arriau, J. P. Campillo, J. Deschamps, G. Tarrago, and R. Jacquier, *Bull. Soc. Chim. Fr.* **1973**, 1403.

57. F. A. VanCatledge, *J. Am. Chem. Soc.* **95**, 1173 (1973).
58. Z. Yoshida and T. Kobayashi, *J. Chem. Phys.* **58**, 334; **59**, 3444 (1973).
59. W. R. Wadt and W. A. Goddard, III, *J. Am. Soc.* **96**, 5997 (1974).
60. J. A. Pople, D. L. Beveridge, and P. A. Dobosh, *J. Chem. Phys.* **47**, 2026 (1967).
61. I. Absar, C. S. Lin, and K. L. McEwen, *Can. J. Chem.* **50**, 646 (1972).
62. I. Absar and K. L. McEwen, *Can. J. Chem.* **50**, 653 (1972).
63. M. Rajzman, G. Pouzard, and L. Bouscasse, *Compte. Rend. C* **273**, 595 (1971).
64. A. Tajiri, S. Takagi, and M. Hatano, *Bull. Chem. Soc. Japan* **46**, 1067 (1973).
65. W. Hug and I. Tinoco, *J. Am. Chem. Soc.* **95**, 2803 (1973).
66. S. Y. Chen and R. M. Hedges, *Theor. Chim. Acta* **31**, 275 (1973).
67. J. Ridley and M. Zerner, *Theor. Chim. Acta* **32**, 111 (1973).
68. R. Lake and H. H. Jaffé, unpublished results.
69. R. N. Dixon, *Mol. Phys.* **12**, 83 (1967).
70. H. W. Kroto and D. P. Santry, *J. Chem. Phys.* **47**, 2736 (1967).
71. P. S. Song, *J. Phys. Chem.* **72**, 536 (1968).
72. R. Zahradník and P. Čársky, *Theor. Chim. Acta* **27**, 121 (1972).
73. H. M. Chang and H. H. Jaffé, *Chem. Phys. Lett.* **23**, 146 (1973).
74. J. M. Howell and P. Jørgensen, *J. Amer. Chem. Soc.* **95**, 2813 (1973).
75. H. M. Chang, H. H. Jaffé, and C. A. Masmanidis, *J. Phys. Chem.* **79**, 1118 (1975).
76. P. Čársky, W. Machcek, and R. Zahradník, *Coll. Czech. Chem. Comm.* **38**, 3067 (1973).
77. R. Zahradník, P. Čársky, and Z. Slanina, *Coll. Czech. Chem. Comm.* **38**, 1886 (1973).
78. R. L. Ellis, H. H. Jaffé, and C. A. Masmanidis, *J. Am. Chem. Soc.* **96**, 2623 (1974).
79. H. H. Jaffé, H. M. Chang, and C. A. Masmanidis, *J. Comput. Phys.* **14**, 180 (1974).
80. J. Kuhn, P. Čársky, and R. Zahradník, *Theor. Chim. Acta* **33**, 263 (1974).
81. H. M. Chang, H. H. Jaffé, and C. A. Masmanidis, *J. Phys. Chem.* **79**, 1109 (1975).
82. L. Pauling and E. B. Wilson, *Introduction to Quantum Mechanics*, McGraw-Hill, New York (1935).
83. W. Heitler, *The Quantum Theory of Radiation*, 3rd ed., Clarendon Press, Oxford (1954).
84. E. Merzbacher, *Quantum Mechanics*, Wiley, New York (1961).
85. G. Herzberg, *Molecular Spectra and Molecular Structure. III. Electronic Spectra and Electronic Structure of Polyatomic Molecules*, Van Nostrand Reinhold, New York (1966), p. 134.
86. G. Herzberg, *Molecular Spectra and Molecular Structure. II. Infrared and Raman Spectra of Polyatomic Molecules*, Van Nostrand Reinhold, New York (1945), Ch. II.
87. E. U. Condon, *Phys. Rev.* **32**, 858 (1928).
88. G. Herzberg, *Molecular Spectra and Molecular Structure. III. Electronic Spectra and Electronic Structure of Polyatomic Molecules*, Van Nostrand Reinhold, New York (1966), Ch. II.
89. W. Thiel and A. Schweig, *Chem. Phys. Lett.* **12**, 49 (1971); **16**, 409 (1972); **21**, 541 (1973).
90. L. L. Lohr and M. B. Robin, *J. Am. Chem. Soc.* **92**, 7241 (1970).
91. F. Ellison, *J. Chem. Phys.* **61**, 507 (1974).
92. S. P. McGlynn, T. Asumi, and M. Kinoshita, *Molecular Spectroscopy of the Triplet State*, Prentice-Hall, New York (1969).
93. E. U. Condon and G. H. Shortley, *The Theory of Atomic Spectra*, 7th ed., Cambridge Univ. Press (1967).
94. F. L. Pilar, *Elementary Quantum Chemistry*, McGraw-Hill, New York (1968), p. 133.
95. C. A. Masmanidis, H. H. Jaffé, and R. L. Ellis, *J. Phys. Chem.* **79**, 2052 (1975).
96. R. L. Ellis, R. Squire, and H. H. Jaffé, *J. Chem. Phys.* **55**, 3499 (1971).
97. D. S. McClure, *J. Chem. Phys.* **17**, 905 (1949).
98. L. Pauling and E. B. Wilson, *Introduction to Quantum Mechanics*, McGraw-Hill, New York (1935), pp. 165–372.
99. K. W. H. Stevens, *Prog. Roy. Soc. (Lond.)* **214A**, 237 (1952).
100. S. P. McGlynn, L. G. Vanquickenborne, M. Kinoshita, and D. G. Carroll, *Introduction to Applied Quantum Chemistry*, Holt, Rinehart, and Winston, New York (1972), Ch. 12.
101. J. B. Lounsbury, *J. Chem. Phys.* **42**, 1549 (1965); **46**, 2193; **47**, 1566 (1967).
102. L. B. Lounsbury and G. W. Barry, *J. Chem. Phys.* **44**, 4367 (1966).

103. G. Herzberg, *Molecular Spectra and Molecular Structure. III. Electronic Spectra and Electronic Structure of Polyatomic Molecules*, Van Nostrand Reinhold, New York (1966), pp. 137–141.
104. G. Herzberg and E. Teller, *Rev. Mod. Phys.* **13**, 75 (1947).
105. J. N. Murrell and J. A. Pople, *Proc. Phys. Soc. (Lond.) A* **69**, 245 (1956).
106. J. A. Pople and J. W. Sidman, *J. Chem. Phys.* **36**, 1588 (1958).
107. A. D. Liehr, *Can. J. Phys.* **35**, 1123 (1957); **36**, 1588 (1958).
108. A. C. Albrecht, *J. Chem. Phys.* **33**, 156 (1960).
109. M. Roche and H. H. Jaffé, *J. Chem. Phys.* **60**, 1193 (1974).
110. L. Ziegler and A. C. Albrecht, *J. Chem. Phys.* **60**, 3558 (1974).
111. G. Orlandi and W. Siebrand, *Chem. Phys. Lett.* **15**, 465 (1972).
112. H. Hellmann, *Einführung in die Quantenchemie*, Franz Deuticke, Leipzig (1937).
113. R. P. Feynman, *Phys. Rev.* **56**, 340 (1939).
114. E. B. Wilson, J. C. Decius, and P. C. Cross, *Molecular Vibrations*, McGraw-Hill, New York (1955), pp. 289–91.
115. A. D. Liehr, Ph.D. Dissertation Harvard University (1958).
116. J. DelBene and H. H. Jaffé, *J. Chem. Phys.* **49**, 1221 (1968).
117. G. W. King and A. A. VanPutte, *J. Mol. Spec.* **42**, 514; **44**, 286 (1972).
118. J. S. Yadav, P. G. Mishra, and D. K. Rai, *Spec. Lett.* **5**, 471 (1972).
119. J. S. Yadav, P. C. Mishra, and D. K. Rai, *J. Mol. Struct.* **13**, 253 (1972).
120. J. R. Huber and J. E. Adams, *Ber.* **78**, 217 (1974).
121. J. DelBene and H. H. Jaffé, *J. Chem. Phys.* **48**, 4050 (1968).
122. M. Cignitti and L. Paoloni, *Theor. Chim. Acta* **25**, 277 (1972).
123. P. E. Stevenson, *J. Phys. Chem.* **76**, 2424 (1972).
124. T. Bluhm, H. H. Perkampus, and J. V. Knop, *Ber. Bunsenges. Phys. Chem.* **76**, 1256 (1972).
125. R. Zahradník and P. Čársky, *Coll. Czech. Chem. Comm.* **38**, 1876 (1973).
126. O. Chalvet, H. H. Jaffé, and E. De La Serna, *J. Phys. Chem.* **79**, 2543 (1975).
127. O. Chalvet, H. H. Jaffé, and J. C. Rayez, *Photochem. Photobiol.*, in press.
128. R. S. Mulliken, C. A. Rieke, and W. G. Brown, *J. Am. Chem. Soc.* **63**, 41 (1941).
129. B. Gonbeau and H. Sauvaitre, *J. Mol. Struct.* **14**, 235 (1972).
130. C. Sieiro and J. I. Fernandez-Alonso, *An. Quím.* **70**, 484 (1974).
131. K. N. Houk, D. J. Northington, and R. E. Duke, Jr., *J. Am. Chem. Soc.* **94**, 6233 (1972).
132. J. F. Arnett, G. Newkome, W. L. Mattice, and S. P. McGlynn, *J. Am. Chem. Soc.* **76**, 4385 (1974).
133. E. N. Svendsen and M. T. Vala, *Acta Chim. Scand.* **26**, 3475 (1972).
134. G. M. Maggiora and L. J. Weimann, *Chem. Phys. Lett.* **22**, 297 (1973).
135. H. Ohmichi, A. Tajiri, and T. Nakajuma, *Bull. Chem. Soc. Japan* **45**, 3026 (1976).
136. C. J. Seliskar, O. S. Khalil, and S. P. McGlynn, in *Excited States* (E. C. Lim, ed.), Academic Press, New York (1974), Vol. I, pp. 231ff.
137. H. J. Maria, J. R. McDonald, and S. P. McGlynn, *J. Am. Chem. Soc.* **95**, 1050 (1973).
138. L. E. Harris, *J. Chem. Phys.* **58**, 5615 (1973).
139. C. Leibovici, *J. Mol. Struct.* **20**, 429 (1974).
140. R. E. Connors, J. L. Roebber, and K. Weiss, *J. Chem. Phys.* **60**, 5011 (1974).
141. C. H. Warren, *Theor. Chim. Acta* **30**, 1 (1973).
142. H. Suzuki, *Electronic Absorption Spectra and Geometry of Organic Molecules*, Academic Press, New York (1967).
143. B. Tinland, R. Guglimetti, and O. Chalvet, *Tetrahedron* **29**, 665 (1973).
144. F. Imashiro, Z. Yoshida, and I. Tabushi, *Tetrahedron* **29**, 3521 (1973).
145. P. Gacoin and J. M. Leclercq, *J. Chem. Phys.* **59**, 4351 (1973).
146. H. H. Jaffé, C. A. Masmanidis, H. M. Chang, and R. L. Ellis, *J. Chem. Phys.* **60**, 1696 (1974).
147. H. H. Jaffé and C. A. Masmanidis, *Chem. Phys. Lett.* **24**, 416 (1974).
148. L. V. Orlovska, *Opt. Spektr.* **36**, 878 (1974).
149. J. J. Mikula, R. W. Anderson, L. E. Harris, and E. W. Stuebing, *J. Mol. Spec.* **42**, 350 (1972).
150. P. Crozet, *Chem. Phys. Lett.* **25**, 114 (1974).
151. R. L. Ellis and H. H. Jaffé, unpublished.

152. A. Tiberghien and G. Delacote, *J. Phys. (Paris)* **31**, 637 (1970); *Chem. Phys. Lett.* **8**, 88 (1971); **14**, 184 (1972).
153. A. Tiberghien, G. Delacote, and P. Devaux, *Chem. Phys. Lett.* **9**, 642 (1971).
154. J. L. Fave, G. Delacote, and A. Tiberghien, *Mol. Phys.* **26**, 17 (1973).
155. P. Jørgensen and J. C. Poulsen, *J. Phys. Chem.* **78**, 1420 (1974).
156. J. Kuhn, P. Čársky, and R. Zahradník, *Coll. Czech. Chem. Comm.* **39**, 2175 (1974).
157. T. Koopmans, *Physica* **1**, 104 (1933).
158. C. Fridh, L. Asbrink, and E. Lindholm, *Chem. Phys. Lett.* **15**, 282, 408 (1972).
159. G. Herzberg, *Molecular Spectra and Molecular Structure. III. Electronic Spectra and Electronic Structure of Polyatomic Molecules*, D. Van Nostrand Co., Inc., Princeton, N.J. (1966), p. 340.
160. D. R. Salahub and C. Sandorfy, *Theor. Chem. Acta* **20**, 227 (1971).
161. M. Wolfsberg and L. Helmholz, *J. Chem. Phys.* **20**, 837 (1952).
162. S. D. Thompson, D. G. Carroll, F. Watson, M. O'Donnell, and S. P. McGlynn, *J. Chem. Phys.* **45**, 1367 (1966).
163. D. P. Santry and G. A. Segal, *J. Chem. Phys.* **47**, 158 (1967).
164. C. Guimon, D. Gonbeau, and G. Pfister-Guillouzo, *Tetrahedron* **29**, 3399, 3599 (1973).
165. G. Pfister-Guillouzo, D. Gonbeau, and J. Deschamps, *J. Mol. Struct.* **14**, 81, 95 (1972).
166. J. W. McIver and A. Komornicki, *Chem. Phys. Lett.* **10**, 303 (1971).
167. J. S. Yadav, P. C. Mishra, and D. V. Rai, *Mol. Phys.* **26**, 193 (1973).
168. G. H. Kirby and K. Miller, *Chem. Phys. Lett.* **3**, 643 (1969).
169. A. Albinati, F. Maraschi, and M. Zocchi, *J. Chem. Soc. (Fz)*, **69**, 798 (1973).
170. H. H. Jaffé and G. Koser, *J. Org. Chem.* **40**, 3082 (1975).
171. G. Lancelot, *Mol. Phys.* **29**, 1099 (1975).
172. B. F. Minaev, *Opt. Spektr.* **32**, 22 (1972).
173. J. L. Berkovsky, F. O. Ellison, T. H. Lee, and J. W. Rabalais, *J. Chem. Phys.* **59**, 5342 (1973).
174. G. W. Robinson, *J. Mol. Spec.* **6**, 58 (1961).
175. M. A. El-Sayed, *J. Chem. Phys.* **36**, 573 (1962); **38**, 2834 (1963).
176. C. A. Masmanidis and H. H. Jaffé, to be published.
177. J. A. Pople and D. L. Beveridge, *Approximate Molecular Orbital Theory*, McGraw-Hill, New York (1970), p. 36.
178. J. DelBene and H. H. Jaffé, *J. Chem. Phys.* **50**, 563 (1969).
179. W. G. Laidlaw and F. W. Bliss, *Theor. Chim. Acta* **2**, 181 (1964).
180. J. A. Singerman and H. H. Jaffé, *J. Phys. Chem.* **80**, 1928 (1976).
181. J. A. Pople and D. L. Beveridge, *Approximate Molecular Orbital Theory*, McGraw-Hill, New York (1970), p. 87.
182. G. W. Kuehnlenz, C. A. Masmanidis, and H. H. Jaffé, *J. Mol. Struct.* **15**, 445 (1973).
183. R. Mathies and A. C. Albrecht, *J. Chem. Phys.* **60**, 2500 (1974).
184. H. Meyer and A. Schweig, *Theor. Chim. Acta* **29**, 375 (1973).
185. N. S. Hush and M. L. Williams, *Chem. Phys. Lett.* **5**, 507 (1970).
186. D. Rinaldi and J. L. Rivali, *Theor. Chim. Acta* **32**, 243 (1974).
187. F. T. Marchese and H. H. Jaffé, *Theor. Chim. Acta*, in press.
188. E. Merzbacher, *Quantum Mechanics*, Wiley, New York, (1961), Section 17.4.
189. H. Meyer, K. W. Schulte, and A. Schweig, *Chem. Phys. Lett.* **31**, 187 (1975).
190. H. E. Smith, R. K. Orr, and F. M. Cheng, *J. Am. Chem. Soc.* **97**, 3126 (1975).
191. N. Y. C. Chu and K. Weis, *Chem. Phys. Lett.* **27**, 567 (1974).
192. P. Cremaschi, A. Gamba, and M. Simonetta, *Theor. Chim. Acta* **31**, 155 (1973).

3

Photochemistry

Josef Michl

1. Introduction

Within the framework of the Born–Oppenheimer approximation, the theoretician's task in describing photochemical processes naturally falls into two parts, a static and a dynamic part. Computation of molecular wave functions and construction of potential energy hypersurfaces is the static part of the problem, while the study of molecular dynamics on these surfaces represents the dynamic part. In the present treatise, the two topics are separated. Molecular dynamics is discussed in Volumes 1 and 2, and the use of quantum chemical methods for the construction of potential energy hypersurfaces for excited states is the subject of the present chapter. Such methods vary in sophistication from simple qualitative arguments concerning the molecular orbitals and states involved, which shall not be discussed here (for such discussions see Refs. 1–7; for a review see Ref. 8), to calculations by semiempirical methods such as PPP,[9] CNDO,[10] etc., on which we shall concentrate here, to *ab initio* studies by methods discussed in Volumes 3 and 4 of this series,* which will not be mentioned. In addition to being restricted to a discussion of semiempirical treatments of static aspects of photochemistry, the chapter also concentrates exclusively on studies of organic molecules. Theoretical as well as experimental experience with photochemistry of inorganic and organometallic compounds is presently more limited.

---

*For a recent review of *ab initio* calculations of triplet hypersurfaces see Ref. 11a. For an illuminating qualitative discussion of the results of a series of *ab initio* calculations on both singlet and triplet surfaces see Ref. 11b.

Josef Michl • Department of Chemistry, University of Utah, Salt Lake City, Utah

After a brief description of the basic qualitative features of the most important photochemical and photophysical processes, which will permit us to summarize the desirable attributes of the theoretical treatments to be used, we shall proceed to a discussion of the nature of the available parametrized computational schemes. This will be done in two steps: first, listing the model Hamiltonians used in the existing methods, and second, listing the various approximate ways of finding solutions to the Hamiltonians. In each case, we shall not concentrate on a detailed description of the semiempirical methods themselves, which can be found in Volume 7 and in the literature cited, but on an evaluation of their applicability to the problems of photochemistry.

Subsequently, we shall discuss specific results for two selected types of photochemical processes. First, the somewhat trivial case of proton-shift equilibria in excited states will be taken up briefly. This will be followed by a discussion of semiempirical studies in which an attempt was made to understand the course of electrocyclic reactions, first those in which an attempt was made to map the relevant part of the potential energy hypersurface, then those in which calculations were performed only at the starting geometry and conclusions drawn from state ordering and slopes in the direction of an assumed reaction coordinate.

The chapter will conclude with a brief summary of the present state of affairs and some guesses concerning future developments.

The literature search ran up to Spring 1975.

2. Photochemical Processes

Reactions discussed in this chapter are initiated by excitation of an organic molecule into a relatively low-lying electronic excited state, either by photon absorption or by energy transfer. Processes involving photoionization, electron transfer in excited states, or highly energetic excitation will not be considered. Our interest will end with the formation of the first thermally equilibrated product in ground electronic state. In practice, this is often not the product actually isolated, but the steps following the photochemical process proper can be understood in terms of ground-state reactivity discussed in Chapter 1 in this volume.

The absorption event is a rather complicated phenomenon, and its description requires an explicit consideration of the radiation field as well as the molecule. The outcome of the interaction may depend on the detailed nature of the field, such as pulse duration, and the details are still a matter of controversy (for recent discussions and leading references, see Ref. 12). Regardless of these details of the excitation process, it is possible to state in general that one of two things happens subsequently extremely rapidly in a large organic molecule:

The motions of the nuclei soon become governed either by the lowest excited state (typically one of the same multiplicity as the initial state) or the ground state. The excess of kinetic energy of such motion over that corresponding to thermal equilibrium with the reaction medium is rapidly lost, so that in about 10^{-10}–10^{-12} sec the molecule ends up thermally equilibrated in one or another minima in its lowest excited singlet (S_1) or triplet (T_1) state or in its ground state (S_0). The thermal equilibration may take longer in media of very low thermal conductivity (low-pressure gas, low-temperature inert gas matrix). In such "isolated molecule" experiments, a large molecule acts as its own temperature bath and distributes its excess energy over all degrees of freedom.* Then, no vibrational mode will contain very many quanta of excitation, and the situation still need not be very different from the one described above.

The relaxation from higher excited states to the lowest one (internal conversion) is almost always very rapid, presumably because the states lie close together and may also share points of touching or near touching. Similarly rapid internal conversion all the way to S_0 is less common and can be expected if the molecule passes through a "funnel" in a region of geometries at which the $S_0 - S_1$ difference is very small, particularly if the S_0 and S_1 surfaces touch (or if a touching is intended but weakly avoided). Some cases of slower relaxation from higher excited states, revealed usually by emission or photochemical product formation from such states, are also known.[8]

In an approximate but often useful way, one can imagine the loss of excess energy from the initially prepared excited molecule on its way to thermal equilibrium in a minimum in S_1, T_1, or S_0 as a complicated motion of a point on a sloping surface, generally in the direction of the steepest descent, with interruptions due to collisions, and "jumps" in which kinetic energy is suddenly increased as the point starts to obey the dictates of a new, lower hypersurface. The horizontal distance covered by the point by the time equilibrium has been reached may be very small ("vertical internal conversion"), but it may also be sizeable, and the point may pass above barriers in S_1, T_1, or S_0, which would have been prohibitively high, at the same temperature of the medium, if the initial excitation had been less energetic. If a barrier in S_1 (T_1) has been passed before thermal equilibrium in S_1 (T_1) was achieved, wavelength-dependent ("hot excited state") photochemistry results (well known, e.g., in benzene and its simple derivatives[8]). If a barrier in S_0 has been passed after return to S_0 but before thermal equilibration, a "hot-ground-state" photochemical reaction results (uncommon in dense media).

Establishment of thermal equilibrium in S_1 or T_1 can be followed by processes such as a radiative or radiationless transition to S_0, intersystem crossing between S_1 and T_1, or thermally activated escape into another region of the same hypersurface (perhaps in a direction which opens up when diffusion

---

*For a discussion and leading references see Ref. 13.

brings a reaction partner). Sooner or later another minimum or region of $S_0 - S_1$ surface touching or near touching ("funnel")[4,8] is reached, until eventually a radiative or a radiationless return to the S_0 state occurs (or until a minimum in a triplet ground state is reached).

At the simplest level of investigation, in which problems of chemical dynamics are ignored, it then appears that the minimum amount of information required from theory would be prediction of the geometries at which minima in the S_1 and T_1 hypersurfaces occur, and prediction of barriers in these hypersurfaces which may hinder access to such minima if the initial excitation is not sufficiently energetic, or if "excess" energy of the photon is wasted as heat rather than utilized efficiently for motion toward and over the barriers. This assumes that techniques normally used for thermal reactivity can be used to follow the fate of the molecule after it has reached the S_0 surface as long as we know just at what geometry it has reached it. Stated in an oversimplified manner, which disregards the possibility of hot-ground-state reactions, the chemical nature of the possible products is determined by the location of minima and funnels in S_1 and T_1, along with the shape of S_0 underneath. Which of many possible products are actually formed is determined by the slopes and barriers in S_1 and T_1 in a way which generally appears impossible to study in detail without consideration of molecular dynamics. In many specific instances, it may, of course, be possible to avoid explicit use of molecular dynamics, for instance, when reaction rate is determined by escape from "thermal equilibrium" in a minimum in S_1 or T_1 over a barrier to another minimum in S_1 or T_1 from which rapid return to S_0 follows. Simple concepts such as activation energy and entropy of the escape then may be quite adequate.

Since complete searches of the S_1 and T_1 hypersurfaces are usually prohibitively expensive, a typical study of a photochemical reaction will concentrate on locating the minima and funnels through which return to S_0 occurs and then explore their relative accessibility, i.e., search for barriers and slopes.

Minima in S_1 and T_1 seem to occur at two types of geometries [4,5,8]: First, "spectroscopic minima" near the equilibrium geometries of ground-state molecules (e.g., the ordinary excited states of benzene or formaldehyde); second, at "biradicaloid" geometries, i.e., those at which the molecule has, in simple MO description, two approximately nonbonding orbitals containing only two electrons in the ground state (e.g., twisted ethylene, carbene, stretched H_2 molecule). Return to S_0 via minima of the first kind is of limited interest since, unless a hot-ground-state reaction follows, as is quite rare in condensed phase, it typically leads the molecule to the starting point (a photophysical process). On the other hand, minima at biradicaloid geometries, which are rarely accessible to direct experimental observation, are likely to return the molecule to some unusual, thermally quite inaccessible region of S_0 (which is usually quite high in energy at just these geometries), often near a top of a barrier in S_0, so that thermal equilibration need not only lead back to the starting species, but also to a different minimum in S_0, i.e., to a photochemical product.

Thus, photochemical investigations put a premium on the ability of the theoretical method to produce reliable results at biradicaloid geometries and along paths from the initial geometry to such biradicaloid geometries. As we shall see in the following, T_1 can be handled at biradicaloid geometries in the usual manner (e.g., open-shell SCF). On the other hand, S_1 and S_0 pose difficulties, since simple SCF-type descriptions are not appropriate. This is easily seen when one considers that at least three singlet configurations, $\psi_1\psi_2$, ψ_1^2, and ψ_2^2, can be of comparable low energy (ψ_1 and ψ_2 are nonbonding orbitals of the molecule).[4,14] Thus, configuration interaction including multiply excited configurations, or an equivalent procedure, is mandatory.

3. Semiempirical Methods

In the design of currently popular semiempirical methods, attention was focused on calculations for molecules at geometries at which bond lengths and angles are near their equilibrium values in the ground state. At these geometries, the closed-shell SCF approximation is usually quite adequate for the ground state, and this also simplifies excited-state calculations. Exploration of ground-state reactivity requires calculations at geometries which deviate from the equilibrium ones, but a closed-shell SCF description of the ground state is often still adequate, since thermal reaction paths tend to avoid the highly energetic biradicaloid geometries. In some instances, even the lowest energy path has to cross a region of biradicaloid geometries somewhere between reactant and product (the reaction is forbidden in the Woodward–Hoffmann sense[15]). The resulting difficulties in calculation of singlet states have been discussed in the chapter on thermal reactivity (p. 1). As just pointed out, in photochemical reactions the return to the ground state tends to occur at a biradicaloid geometry, so that almost all photochemical reaction paths proceed via such a geometry, and difficulties with theories which use the closed-shell SCF approximation for the ground state as their starting point are bound to occur as a general rule whenever an attempt is made to map potential energy hypersurfaces along the reaction path. For detailed studies of photochemical mechanisms this is a serious consideration, and, as we shall see, most common semiempirical procedures are not suitable for detailed studies of excited singlet reaction paths.

An additional problem for reactions of any multiplicity is presented by the need for a systematic geometry optimization during the search for a minimum-energy path. It is sobering to realize that a relatively sparse grid of only ten points in each dimension of the nuclear configuration space of a four-atomic molecule would require calculations at one million different geometries and, moreover, that only limited "intuition" is available for excited-state hypersurfaces.

Often, however, interesting insight about relative probabilities of several possible photochemical processes in a given molecule, or about one process in a series of molecules, can already result from inspection of the ordering and slopes of the excited-state hypersurfaces in the vicinity of the reactant geometry, since this will often produce information about possible barriers on the way. For this type of study, many more existing semiempirical procedures will be found suitable. Often, no calculation will be needed at all and simple correlation diagrams will provide the answer. Also, in relatively rare cases of photochemical reactions, mostly proton transfer processes, an equilibrium is established in the excited state, and valuable information can be obtained from mere knowledge of the energy difference between the two excited species in equilibrium.

3.1. Model Hamiltonians

Among the criteria which can be used to classify the various semiempirical models in use, two seem of particular significance for photochemical applications. These are the way in which electron–electron repulsion is handled and the way in which overlap among orbitals of the AO basis set is handled.

3.1.1. Hückel Theory

In this section we deal with models in which electron repulsion is not treated explicitly. Such models cannot distinguish between singlets and triplets and can only use the virtual orbital approximation to describe excited states. Consequently, they are incapable of properly describing the order and nature of many excited states already at the reactant geometry, e.g., singlet states of aromatic molecules[9] and of polyenes,[16] and also cannot distinguish the several low-lying singlets at biradicaloid geometries. They thus certainly appear unsuitable for studies of singlet photochemical reactions. It is well known from studies by more advanced methods, however, that neglect of CI is usually less serious for the lowest triplet state (e.g., in aromatics, in polyenes, and at biradicaloid geometries). The hypersurfaces obtained by methods of the Hückel type are therefore more likely to run roughly parallel to the correct triplet surfaces than to the correct excited singlet surfaces. In this sense, Hückel-type calculations of photochemical processes should probably be viewed as calculations for triplets, although the method itself does not really distinguish between the different multiplicities. Of course, absolute values of excitation energies, heights of barriers, etc., will generally not be in good numerical agreement with experiment.

Even within the framework of virtual orbital description of excited states, without any CI, there is an additional argument in favor of this interpretation of hypersurfaces obtained from Hückel-type calculations.[17] When electron repulsion terms are added to orbital energy differences in order to make the theory capable of distinguishing singlets and triplets, only one such term appears in the expression for the triplet excitation energy, namely the Coulomb integral J_{ij}. The expression for the singlet excitation energy also contains the exchange integral K_{ij}. Now, there is at least some hope for the Coulomb integral to remain reasonably constant as a molecule changes shape, and even within a class of molecules of similar size, since it represents the repulsion of two unit charges somehow distributed around the molecule. Then, one can pretend that the excitation energy calculated in the Hückel scheme is actually equal to $\varepsilon_j - \varepsilon_i - J_{ij}$, except perhaps for a constant. On the other hand, the exchange integral represents the self-repulsion energy of an overlap charge and its size again depends on the distribution of this charge around the molecule, but now also on its total size, which is a sensitive function of the degree of spatial overlap of orbitals involved. Thus, the error introduced into the results by lack of explicit treatment of electron repulsion terms is more likely to be serious for excited singlet than for excited triplet states.

Because of their great simplicity, Hückel-type schemes can easily handle systems with quite a few electrons, and there would seem to be little point in limiting attention to any fewer than all valence electrons in any computations of photochemical pathways. Further, it is possible to include overlap in the calculation properly with relatively little cost. The importance of the inclusion of overlap in considerations of photochemical pathways is well recognized.[18,71] In particular, it provides for proper behavior of two interacting orbitals: The out-of-phase antibonding combination is destabilized more than the in-phase bonding combination is stabilized. This will provide, for instance, the correct shape for the rotational barrier of excited ethylene, i.e., a minimum at 90° twist. As is well known,* methods which neglect overlap can also achieve this shape, namely by exaggerating the effects of hyperconjugation, but this is likely to be detrimental elsewhere.

Today's standard Hückel-type method is the extended Hückel method[20] exploited extensively by Hoffmann and others for studies of ground-state reactivity. It has been used occasionally for photochemical reactions and geometries of excited species as well.[21,22] Reparametrized and iterative forms of this simple method exist (e.g., Refs. 23), but these have not been applied to photochemical problems as far as the author is aware.

In summary, Hückel-type methods appear unsuitable for studies of photochemical reactions in the singlet state, but they can be used to derive qualitative ideas about general trends in triplet reactions, particularly in a series of similar

---

*See, for instance, the CNDO/2 study of ethylene in Ref. 19.

molecules. For an example of a nice and cautious application of this type, see Refs. 22.

3.1.2. Simple ZDO Methods

If one is willing to give up proper treatment of overlap, electron repulsion can be introduced into the semiempirical scheme in a simple fashion. It then becomes impractical to attempt to solve the model exactly for any more than about six electrons. One can either use one of the approximate methods of solution discussed in Section 3.2, and possibly adjust the semiempirical parameters accordingly, or treat explicitly only a fraction of the valence electrons present (or do both).

In all semiempirical methods, some of the electrons present in the molecule are not given an explicit treatment (except at most through a pseudopotential*), namely, inner-shell electrons (all-valence-electron methods). In the next step, one might omit $2s$ orbitals from the basis set on account of their relatively low energies, but this has apparently not been tried. The simplest procedures in use limit themselves to only a small fraction of $2p$ orbitals, usually at most one on a carbon or other heavy atom. All of the orbitals are typically of π symmetry in either the reactant or the product, or both. For instance, only the orbitals shown in the following formulas would be considered in such a study of *cis–trans* isomerization of butadiene, or of its electrocyclic closure to cyclobutene:

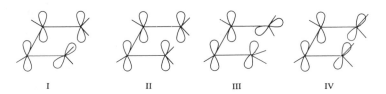

It is obvious that methods of this type have a severely limited range of applicability; however, it is interesting to note that some of the most important theoretical notions concerning pericyclic photochemical reactions have resulted from such a very crude model, for the butadiene–cyclobutene conversion, solved exactly.[25] Almost identical results were later obtained in a thorough *ab initio* study.[26]

In the present section, we discuss models in which overlap and differential overlap between basis set AOs are formally neglected. These are referred to as, simple zero-differential-overlap (ZDO) or neglect-of-differential-overlap (NDO) methods. It is well known[27] that these models can be reinterpreted as really working with Löwdin-orthogonalized[28] AO basis sets. Methods which

---

*For references to recent studies of core–valence separation see Ref. 24.

recognize the nonorthogonality of the AO basis set formally and proceed to orthogonalize the basis set explicitly will be discussed in Section 3.1.3, along with other, more sophisticated treatments of overlap.

a. PPP and CNDO. The model which formally neglects all electron repulsion integrals involving differential overlap (i.e., involving charge distributions represented by a product of two different orbitals) is referred to as PPP[9] (Pariser–Parr–Pople) if π electrons only are treated, and CNDO[10,29] (complete neglect of differential overlap) if all valence electrons are treated. A great number of variants of each procedure exist.

The PPP scheme has been used since the 1950s for calculations of electronic spectra of π-electron chromophores and has achieved a considerable degree of success.[9] It would be hopeless to attempt to list all of the parametrization schemes proposed; we shall only note that it takes a fairly complicated scheme to reproduce simultaneously various properties of the ground and excited states.[30] The schemes of most interest for photochemical studies are those designed to reproduce excitation energies. When only monoexcited CI among singlets is performed, it is usually considered best to use the Mataga–Nishimoto formula[31] for two-center electron repulsion integrals (the Dewar–Ohno–Klopman formula[32] is also sometimes used but, in the author's experience, it gives inferior results for spectral properties). When a more exact solution of the model is attempted, using multiply excited singlet configurations in the CI procedure, a less steep functional dependence of the two-center repulsion integrals on distance must be used* (for several such parametrizations, see Refs. 25, 34, 35, and 102). For triplet states, somewhat different parameters appear to be best.[36]

Several studies of photochemical processes using the PPP method and treating explicitly only a part of the valence electrons have been published[25,37] and some of them (in particular, Ref. 25) played an important role in the development of present-day notions about photochemical mechanisms. Of course, the usefulness of the PPP model for photochemical applications is limited to processes involving primarily π electrons. It provides a quite reasonable guide for state energy diagrams for large molecules needed for construction of correlation diagrams and aids in the construction of the latter (e.g., Ref. 38). It can also give indications of the shape of potential energy hypersurfaces along the reaction path if solved reasonably exactly, i.e., for singlet states with inclusion of multiply excited configurations in the CI procedure, but difficulties connected with neglect of overlap and hyperconjugation in twisted π systems should be kept in mind. In this type of application, it should typically be followed by a more elaborate study unless the size of the molecule precludes it. Finally, the PPP model can be quite adequate for calculations of reversible photochemical reactions of π-electron systems (acid–

---

*For a discussion of the reasons for this see Ref. 33.

base equilibria in excited states), and even some aromatic photosubstitutions (e.g., hydrolysis of trifluoromethylphenols[39]) in which knowledge of differences of total energies is presumed to be sufficient. Use of the ordinary static reactivity indices such as π-electron densities instead of appropriate representations of the structure of intermediates can be dangerous, particularly for comparison of states of different multiplicities.[40]

Unlike the PPP model, the original SCF-MO versions of the CNDO model[29] (CNDO/1, using only ionization potentials for derivation of atomic parameters; CNDO/2, using also electron affinities) were parametrized so as to reproduce *ab initio* calculations at the SCF level.[10] Many reparametrizations have been proposed, some adjusted to reproduce experimental results.[41] For photochemical applications, the CNDO/S version[42] is of particular interest, since it was parametrized to reproduce excited-state energies, using limited monoexcited CI. Other modifications of the CNDO/2 method for spectral purposes exist.[43] The CNDO/S version itself has been modified and improved,[44,45] in particular by use of the Mataga–Nishimoto formula[31] for electron repulsion integrals. Although it is well known[46] that the results of the ordinary CNDO/2 model are affected quite drastically when multiply excited configurations are introduced, to the author's knowledge, only one attempt seems to have been made to develop a parametrization suitable for more complete solutions of the model by inclusion of multiply excited configurations. In this work,[47] the penetration term was adjusted in order to correct imbalance in description of small rings.

Several photochemical studies using the CNDO model have been published.[19,48] The increased cost of CNDO calculations compared to PPP calculations is, one hopes, compensated by several advantages. First, reactions of systems other than approximately planar unsaturated molecules can be studied. In calculations for π-electron molecules, results of the two methods are quite similar. A little less experience is available with CNDO, particularly for larger molecules, and lack of a suitable parametrization for extensive CI presently prevents proper description of excited states for which doubly excited configurations are important (polyenes, biradicaloid hydrocarbons). Second, incorporation of heteroatoms is less ambiguous in CNDO, and states such as $n\pi^*$ are calculated along with $\pi\pi^*$ states. Thus, for the types of discussions in which only results for initial and for final geometries are needed, as for discussion of total energies or construction of correlation diagrams, the two methods are comparable for unsaturated hydrocarbons, but CNDO becomes superior as heteroatoms are introduced, except possibly for singlet states of biradicaloid systems (this limitation could probably be removed by a suitable reparametrization such as that of Ref. 47). For applications involving an exploration of the reaction path at nonplanar geometries, both methods suffer from neglect of overlap, but at first sight CNDO ought to be considerably better, since it handles all valence electrons on the same footing. Unfortunately, it appears that even those versions of CNDO that were specifically designed to reproduce ground-state properties are unreliable for a study of conformations, although they are often qualitatively correct.[49-52] Interestingly, it appears that conformations predicted using the same CNDO Hamiltonian and the PCILO approximation to the solution of the Schrödinger equation are more reliable.[51] However, both ordinary CNDO/2 and the PCILO version grossly overestimate the stability of small rings (for numerical examples, see Ref. 47). These deficiencies in the descriptions of the ground state make one fear far worse deficiencies in the excited states, so that it would appear ill-advised to take the details of the calculated hypersurfaces too seriously. In this respect, the use of the less pretentious PPP approach may actually be less misleading.

One's enthusiasm about the ability of the CNDO model to handle nonplanar systems is further dampened by the fact that the model is known to greatly overestimate hyperconjugative interactions.[53] This is an error in the opposite direction from that committed by PPP-type calculations for a few selected "π electrons" in nonplanar systems in which the background framework of classical sigma bonds is not allowed to conjugate at all with the "π system," even where symmetry would permit it (e.g., in twisted ethylene).

In conclusion, it is the author's opinion that the main value of PPP and CNDO calculations of photochemical paths lies in the discovery of new qualitative concepts, to be confirmed later by more expensive methods, rather than in furnishing numbers concerning energetic barriers and the like.

b. INDO. On the next higher level of the ZDO hierarchy, one keeps one-center exchange integrals. The resulting approximation is called intermediate neglect of differential overlap (INDO). The original INDO/1 and INDO/2 models[10,54] are quite similar to the corresponding CNDO models; they have also been subject to reparametrizations. The best known of these are Dewar's MINDO/1 to MINDO/3, designed to reproduce ground-state experimental data rather than *ab initio* SCF results,[55] but others have also been proposed,[56] in particular, several schemes aimed at reproducing excitation energies.[57,58] The main advantage of the method is the proper treatment of singlet and triplet terms resulting from the same configuration; unlike the CNDO method, it does properly reproduce the singlet–triplet splitting in $n\pi^*$ states of heterocycles. The available "spectroscopic" version of the method proposed by Zerner (INDO/S) achieves considerably better accuracy than CNDO/S in calculation of excitation energies of planar heterocycles on which it has been tested.[57,59] Both for this reason and because of the improved treatment of singlet–triplet splittings, it appears desirable to use the only slightly more expensive INDO/S method instead of the CNDO/S method in photochemical applications. Unfortunately, little is known about parametrization suitable for calculations which include multiply excited singlet configurations in the CI procedure, and thus INDO/S is at present probably not appropriate for construction of potential hypersurfaces for singlet states along paths leading through biradicaloid geometries. Like CNDO, INDO and MINDO/2 are known to overestimate hyperconjugative interactions,[53,60] and appear to be unreliable for conformational studies.[50] MINDO/2 combined with limited CI appears to place $\sigma \rightarrow \pi^*$ levels much too low in energy.[61] Less is known at the moment about the reliability of MINDO/3 in these respects.

To the author's knowledge, no photochemical studies by the INDO method have been published. Some interesting applications of MINDO/3 to calculation of triplet surfaces have been published recently.[62] Again, it seems unlikely that much more than qualitative concepts and general guidance can be extracted from calculations at the INDO level.

A recently proposed DRINDO[63] (dipoles retained INDO) scheme, which is based on approximation of repulsion integrals through multipole–multipole interactions, stands between INDO and NDDO in complexity and, presumably, in performance, but no applications have appeared so far. An IRDO (intermediate retention of differential overlap) method has been proposed.[64] It goes beyond INDO in that two-center overlap charges are kept for bonded centers.

c. NDDO. A model neglecting all diatomic differential overlap but retaining all terms involving one-center overlap distributions is referred to as NDDO (neglect of diatomic differential overlap).[10,29,65] The method has been implemented several times[66] but not tested extensively. It appears to hold some promise (for a skeptical view, see Ref. 67), but is also noticeably more expansive than CNDO and INDO. No spectroscopic parametrization has been reported, and the author is not aware of any applications to photochemical problems.

3.1.3. Explicit Treatment of Overlap

Since there are convincing arguments in favor of treating overlap properly in calculations on photochemical problems,[18] methods of this section are of considerable potential interest. Unfortunately, their complexity is such that extensive application to exploration of potential energy hypersurfaces of large molecules appears prohibitively expensive at present, particularly since no sophisticated automated search and optimization procedures have been developed. It is regrettable that most of the authors developing new approximate methods do not automatically think in terms of such searches (e.g., calculation of energy derivatives with respect to nuclear displacement[55,68]), since they would thus make their methods much more valuable for investigations of the real substance of chemistry, i.e., chemical (and photochemical) reactivity.

a. NNDO. One way to avoid the cost problems associated with methods which treat overlap explicitly is to cut down the number of electrons considered. In π-electron approximation, the distinction between neighbor and nonneighbor atoms is given by the underlying sigma framework, and it is then possible to formulate the NNDO (neglect of nonneighbor diatomic overlap) model.[18,69] This was developed by Baird and collaborators specifically for photochemical problems, with a parameterization suitable for triplet states. Mulliken approximation[70] was used to evaluate multicenter integrals. Singlet states, with their intrinsic difficulties at biradicaloid geometries, have not been examined. This appears to be an excellent and reliable method for triplet reactions of those systems to which the $\sigma-\pi$ separation can be reasonably applied, such as olefin *cis–trans* isomerization, although one still wonders about the effects of the total neglect of hyperconjugation. Clearly, however, the method is not applicable generally. A similar approach has been taken by Warshel and Karplus,[71] who performed calculations for singlet states, but did not explore actual photochemical pathways. Their results provide the first detailed understanding of the origin of vibronic fine structure in large unsaturated chromophores.

b. Hückel–Mulliken Methods. A relatively cheap possibility for an all–valence–electron model consists in keeping overlap between all orbitals, evaluating multicenter integrals by the Mulliken approximation,[70] and doing a reasonable amount of CI, but cutting down the labor by doing less than SCF for the ground state; using the extended Hückel method or one of its iterative versions to obtain the starting MOs.[72] This method is bound to give almost as poor total energies as the ordinary Hückel methods, since the ground-state energy is only obtained as twice the sum of orbital energies, but should produce much more accurate spectroscopic properties than the latter, since it distinguishes singlets and triplets and allows properly for configuration interaction. It appears suitable for qualitative purposes.

c. More Sophisticated Methods. The NDDO approximation is much less severe when applied to a Löwdin-orthogonalized[28] basis set, and methods have been developed along this line.[73] They are relatively complex, and no applications to spectroscopic or photochemical problems have been reported.

Among other methods which explicitly consider overlap densities at a higher level than NDDO and which have been parametrized to reproduce *ab initio* results, one should mention the early PDDO (Projection of diatomic differential overlap)[74] and more accurate LEDO (limited expansion of diatomic overlap)[75] methods (for their comparative evaluation, see Ref. 76), and the method proposed by Nicholson.[67] The method outlined by Body[77] appears to be much less accurate. All of these methods are fairly expensive to execute, PDDO and LEDO being only slightly faster than the variational *ab initio* STO-3G method. More recently, they have been superseded by the faster methods PRDDO (partial retention of diatomic differential overlap over an orthogonalized basis set, believed to be similar to LEDO in accuracy and only slightly inferior to the STO–3G method)[78] and the AAMOM method proposed by Zerner,[79] somewhat less accurate but faster than PRDDO. For molecules of intermediate size, PRDDO is estimated to be about four to six times and Zerner's method about twice slower than INDO; both reproduce *ab initio* SCF results much better than any of the NDO methods, but they still execute much faster than STO-3G (PRDDO by a factor of 10–30). Other methods of this category have been proposed recently (NEVE,[80] SAMO[81]). Reports about improvement of NDO methods by providing for a better balance of repulsive and attractive forces have also appeared.[82] In summary, there is nowadays an essentially continuous transition between "approximate *ab initio*" and "exact *ab initio*" methods.

The methods discussed in this section have all concentrated on calculation of ground-state properties. It would be interesting to use the same or similar model Hamiltonians for calculation of excited states. In particular, the relatively rapid method proposed by Zerner[79] appears promising for applications to reaction surfaces in a semiquantitative fashion. So far, however, no attempts to apply any of these methods to excited states and photochemistry have been reported, and reliability cannot be estimated.

3.2. Solving the Models

Most of the models described in the preceding section are too complicated to permit exact solutions. Indeed, for molecules with more than about six or seven electrons, none of the semiempirical Hamiltonians which consider electron repulsion explicitly can be solved exactly in practice, i.e., by a procedure equivalent to full CI. The price one pays for use of reasonably realistic model Hamiltonians, then, is uncertainty as to the origin of errors which are inevitably present: They could be due to approximations made when solving the model, or they may be inherent to the model.

The ordinary choice of parameters is either such as to reproduce experimental results, in our case usually electronic absorption spectra, and sometimes relative energies of starting materials and products, for a given level of approximation in solving the model (true "semiempirical" methods), or such as to reproduce *ab initio* calculations performed at the same level of approximation as that used for the model ("approximate" methods). In the latter case, it is hoped that the exact solution of the model would then presumably resemble the exact *ab initio* solution in the minimum basis set. In the following discussion, we shall at first assume that the model has been parametrized in such a manner that its exact solution optimally reproduces experimental data or exact *ab initio* results, and subsequently we shall inquire whether the shortcomings of a less-than-exact solution of the model can be made up for by semiempirical parameter adjustment.

3.2.1. SCF Methods

In these methods, the many-electron wave function of the excited state has the form of one Slater determinant (or of a simple linear combination of very few of these), and the form of the MOs from which the determinants are constructed is optimized. In the simplest approach, the so-called virtual orbital method (VO), the orbitals are the canonical Hartree–Fock orbitals obtained by optimizing the energy of the closed–shell ground state. This usually ensures a reasonable description of the ground state, but the virtual orbitals obtained are not well suited for the description of excited states, since they describe an electron moving in the field of $2N$ electrons in completely filled lowest N orbitals, as in an anion formed by electron attachment (cf. Koopmans' theorem), instead of the field of $2N-1$ electrons which the excited electron really sees. This problem is particularly striking in *ab initio* calculations with a flexible basis. In minimum-basis-set calculations, the space of virtual orbitals is very small and they are no more diffuse than those occupied in the ground state, so that the problem is not nearly as severe. Still, there is no particular reason why the virtual orbitals should be well adapted for the description of excited states by a small number of Slater determinants. Orthogonal transformations within the set or virtual orbitals (IVO, improved virtual orbitals), or the set of occupied orbitals, or a combination of two such transformations, one on each set of orbitals, can be used to improve the description of the excited state of interest. In photochemistry, this would typically be the lowest excited state. Methods of this type can be viewed as economical substitutes for extensive configuration interaction between singly excited configurations, and in the limit give the same result as full monoexcited CI* (in the one semiempirical test of which the author is aware, the performance of an IVO procedure is rather disappointing[84]). To our knowledge, schemes of this type have not been used so far for computation of potential energy hypersurfaces for excited states of photochemical interest by semiempirical methods.

A serious difficulty for any method based on an initial computation of a closed-shell SCF ground state is the inappropriateness of such a ground-state description at biradicaloid geometries, i.e., just at those of great interest. An alternative approach is construction of molecular orbitals optimized for the particular excited state in question, i.e., general open-shell SCF (for numerous references and application to CNDO and MINDO Hamiltonians, see Ref. 85). In these calculations, it is necessary to ensure orthogonality of the excited-state wave function to those of all lower states during the SCF iterations. This is not trivial to do for excited singlet states which have the same symmetry as some of

---

*For leading references see Ref. 83.

the lower states, as would be almost always the case for all excited states if one were to map a sizeable region of the nuclear configuration space of just about any organic molecule (at most points in this space the molecular symmetry is very low), and is normally not done in semiempirical methods. It is dangerous to use orbitals derived by one of the approximate open-shell SCF schemes that do not distinguish between singlets and triplets, since the ionic nature of the excited singlets at biradicaloid geometries is then likely to be exaggerated (cf. the results for trimethylenemethane[86]).

The proper SCF-type method for singlet states at biradicaloid geometries is the MC SCF (multiconfigurational SCF) procedure. For a long list of references to the method, see Ref. 87; for applications to semiempirical Hamiltonians, see Refs. 88 and 89. An open-shell SCF procedure[90] which gives good predictions of "ordinary" excited-state geometries at the INDO level forms the basis of recently developed MC SCF method for ground-state reactions.[91] A discussion of the use of MC SCF for excited states is given in Ref. 92. Ideally, the MC SCF procedure should be followed by CI (see below).

Since orthogonality of triplet states to all singlet states is ensured by their spin functions, and since there usually is only one low-lying triplet configuration, even at biradicaloid geometries, single-determinant open-shell SCF calculations on the lowest triplet state are relatively easy (for a recent review of *ab initio* calculations, see Ref. 11). The procedure of Roothaan,[93] the less rigorous procedure of Pople and Nesbet,[94] and Dewar's half-electron method,[95] which replaces an electron in a singly occupied orbital by "half-electrons" of opposite spins and is very simple computationally, are all in use. The performance of Roothaan's and Dewar's procedure on an *ab initio* level for triplets has been recently compared with that of the simple virtual orbital procedure.[96] The half-electron method gave much better results than the use of virtual orbitals from a ground-state SCF calculation, but its results still differed significantly from those of the Roothaan method. In semiempirical calculations, the half-electron method has the great advantage that no reparametrization ought to be required—the same parameter values can be used for the singlet ground state and for the lowest triplet.[95]

Of other methods of the SCF type, we shall only note the existence of methods for direct calculation of excitation energies (for a review see Ref. 97), which are attractive for calculation of spectral properties. In their simplest form (ordinary random phase approximation) they would suffer from instability of the closed-shell SCF solution at biradicaloid geometries[98]; the higher forms are more complex and have not been used in semiempirical calculations.

To summarize, closed-shell SCF techniques do not provide a sound basis for consideration of photochemical paths. For triplet reactions, the open-shell SCF methods[93-95] are probably quite adequate. For singlet reactions, it is the author's belief that MC SCF is by far the most hopeful of SCF-type methods. Still, it is very likely that additional consideration of CI is quite essential.

3.2.2. CI Methods

Extensive CI based on SCF MOs is commonly used in *ab initio* calculations on excited-state properties. A recently proposed perturbation approximation appears particularly appealing for large molecules.[99] Difficulties again arise for singlets at biradicaloid geometries, at which the closed-shell SCF procedure is not appropriate and the starting MOs are then hard to select. For triplets, Roothaan's version of open-shell SCF has been used with the CNDO/S Hamiltonian in spectral calculations.[100] It is possible to use orbitals obtained from a triplet SCF calculation to do CI on singlets, but difficulties must be expected.[86] The recently proposed[101] use of an "intermediate" Hamiltonian appears preferable. Ideally, the CI should be based on orbitals obtained by an MC SCF calculation. Neither of these more sophisticated SCF-CI approaches has been used in semiempirical methods for excited singlets so far.

The ordinary versions of existing semiempirical procedures that use configuration interaction, in particular PPP,[9] CNDO/S,[42,44] and INDO/S,[57] have been parametrized to reproduce experimental excitation energies when singly excited but no more highly excited configurations are used in the CI procedure. Since low-energy singlet states with large weights from doubly excited configurations have been recently observed experimentally in π-electron molecules[16,102] and since there is no doubt that such configurations are essential for description of low-lying singlet states at biradicaloid geometries, these methods need to be reparametrized for some suitable larger extent of CI before use in photochemical calculations. This is, of course, much easier for π-electron methods because of the strongly reduced number of electrons considered. Indeed, such a parametrization was performed by van der Lugt and Oosterhoff for the full CI level[25] and by Warshel and Karplus,[71] as well as Downing,[34] for a level intermediate between SCI and SDCI. Although the effect of large-scale CI on CNDO-INDO results has been investigated in some detail,[46] and some progress made in readjustment of parameters,[46] a generally applicable suitable set of parameters has not been developed.

3.2.3. Other Methods

CI calculations for excited states can be based on interaction between configurations constructed from localized orbitals or atomic orbitals. In semiempirical work, the former has been tested (PCILO, exciton model),[48b,103,104] but valence-bond type calculations are uncommon. The only case known to the author is the work on van der Lugt and Oosterhoff, which can be also viewed as a full CI calculation.[25] It is interesting to note that valence CI based on a minimum basis set of AOs is a very successful procedure in *ab initio* calculations on diatomics[105a]* and that an extremely simple

---

*For leading references to more recent use see Ref. 105b.

semiempirical VB method[106] DIM (diatomic-in-molecules) is highly valued by authors interested in potential surfaces of very small molecules.[107] The reason for the lack of popularity of methods such as these in semiempirical calculations on larger molecules is the rapid increase in the number of possible AO configurations as the number of atoms gets larger, and the difficulty in selecting a reasonable number of them.

Finally, we need to address the question whether an exact solution of the model Hamiltonian is really called for and whether the effects of correlation should not be absorbed into semiempirical parameters rather than attacked by elaborate tools attempting to approach full CI. Then, one would use a well-defined cheap approximation in the solution of the model. This goal has been approached for ground states and triplets of molecules at ordinary geometries by the MINDO/3 scheme of Dewar,[55] using the closed-shell and half-electron open-shell SCF approximations, respectively. It can only be justified empirically, by comparison with experiment, and it may be a little too early to tell at the moment. The initial report is very encouraging. However, it is obvious that one must go beyond the SCF level for singlets at biradicaloid geometries,[4,14,55,108] which are essential for any consideration of photochemical reactivities. This opens a Pandora's box: According to Dewar,[55,108] a 2×2 CI is sufficient for the ground state, while other arguments[4,14,109] suggest 3×3 as an absolute minimum for excited singlet states, and still others indicate that much larger CI is necessary to even obtain the correct order of the lowest two excited states at biradicaloid geometries encountered in pericyclic reactions.[5,110] A specific instance has been considered[111] to require a 15×15 CI for proper description of dissociation limits; in others, even the spectra of the starting materials cannot be described properly without fairly large CI including doubly excited configurations.[16] Thus, while schemes hoping to reproduce experimental data for large molecules, for which model Hamiltonians cannot be solved exactly, undoubtedly will continue to absorb much of the correlation energy into semiempirical parameters, it is the author's belief that a simple compromise has not yet been found with the need to include a reasonable amount of CI for description of low-lying singlets.

4. Examples of Application

The number of photochemical reactions for which theoretical insight has been derived from semiempirical calculations is relatively limited. The two processes which seem to have attracted by far most of the attention are *cis–trans* isomerization about double bonds and electrocyclic ring closure in unsaturated compounds.

It is not our intention to discuss all of the various photochemical reactions which have been studied. Instead, we shall briefly mention some of the work which has been done on what appears to be one of the easiest problems in organic photochemistry, namely phototautomerism, and on what appears to be one of the hardest tackled so far, namely electrocyclic ring closure and opening.

4.1. Phototautomerism

Excitation of molecules with acidic hydrogen or basic sites usually results in a change of the acidity and basicity of the various sites, so that a new acido–basic equilibrium is often established after excitation.[112] Many molecules contain both an acidic hydrogen and a basic site, for instance, the laser dye, 4-methylumbelliferone (V):

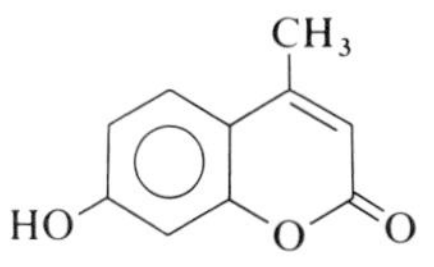

In some instances, excitation increases the acidity of the acidic hydrogen and the basicity of the basic site so much that the stable form in the excited state is a tautomer different from the one which predominates in the ground state. In the case of V, the stable excited from is VI*. Since equilibrium between V* and VI* in the excited state is established under suitable conditions in an adiabatic photochemical process without loss of excitation, a simple theoretical treatment of the process is possible. The free energy difference for the process $V^* \rightleftarrows VI^{**}$ is assumed to differ from that for the process $V \rightleftarrows VI$ by an energy term only, and the energies are estimated using one of the semiempirical methods. In the published work on V, the extended Hückel method gave an adequate answer.[114] Similar studies were performed by other authors, and the PPP method is usually found to be quite adequate.[115] For a discussion of the use of reactivity indices for this type of problem, see Ref. 40.

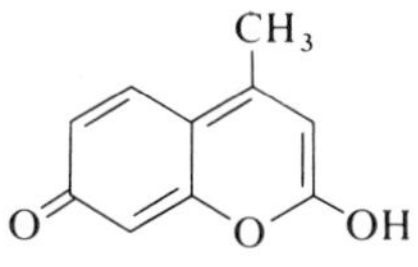

Although this level of treatment of the problem appears somewhat trivial, an ability to predict phototautomerism is of great practical significance in development of laser dyes with wide tunability range and of photostable UV-protection agents.

4.2. Electrocyclic Reactions

One of the best known singlet photochemical processes subject to Woodward–Hoffmann rules is electrocyclic ring opening and closure,[1] exemplified by the disrotatory butadiene ⇌ cyclobutene conversion:

VII   VIII

Other ring sizes are possible. The processes are stereospecific (disrotatory or conrotatory, depending on the ring size), and the bond breaking and bond making are therefore believed to proceed simultaneously, with partial retention of bonding throughout a cyclic array during the whole photochemical reaction process. This does not guarantee the absence of intermediates in the reaction, but shows that any such intermediates must retain cyclic bonding.*

The reactions are ordinarily diabatic, i.e., excited state of the product is not reached (this is often excluded by energetic considerations alone[117]). No direct experimental information is available on the processes occurring between the initial excitation and the appearance of ground-state product; this is, of course, typical of most organic photochemical reactions.

First insight into the origin of the stereospecificity of electrocyclic processes was obtained from Woodward–Hoffmann rules, which predict a barrier in the excited state for the conrotatory ring closure in VII, but not for the disrotatory closure. This is perhaps best seen from correlation diagrams such as those of Fig. 1 (for a more detailed review, see Refs. 1 and 8). Zimmerman[118] used simple arguments on a one-electron theory level to show that the photochemically allowed pericyclic path contains a point at which the excited state becomes degenerate with the ground state, providing an easy return to the ground-state surface and a reasonable mechanism for the photochemical process, in addition to ground-state forbiddenness.

This is about as far as one can go with simple qualitative arguments. Note that both arguments[1,118] would predict identical behavior for excited singlet and triplet, at variance with experimental data (triplet VII does not cyclize to VIII).

The next step was taken by Feler,[21a] who performed an extended Hückel calculation on the system VII ⇌ VIII also explored nuclear motion corresponding to *cis–trans* isomerization. This essentially confirmed the qualitative arguments of Zimmerman[118] concerning the importance of the touching of the excited with the ground state. Distinction between triplet and singlet behavior was first obtained in the work of van der Lugt and Oosterhoff,[25] which provided a basis for the present-day ideas about pericyclic processes in general. These authors performed a complete VB (full CI) calculation in the PPP approximation, considering only the four orbitals and electrons "involved" in

---

*In the case of the related excited singlet 2 + 2 cycloaddition and cycloreversion processes (e.g., 2 ethylenes ⇌ cyclobutane) there is indirect evidence for an intermediate.[116]

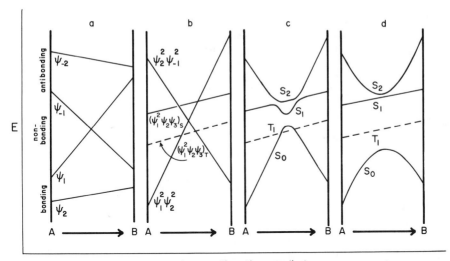

nuclear configuration coordinate

Fig. 1. (a) Orbital, (b) configuration, and (c, d) state correlation diagrams for a typical ground-state Woodward–Hoffmann forbidden pericyclic reaction path, for instance, A = VII, B = VIII (disrotatory). MOs are labeled ψ_2, ψ_1, ψ_{-1}, ψ_{-2}; singlet configurations and states are shown as solid lines, triplets as dashed lines. Reproduced by permission from Ref. 8.

the reaction (cf. Formulas II–IV). The results, reproduced schematically in Fig. 1, showed a barrier in the excited state along the conrotatory path, and details of the behavior of several low-lying states along the disrotatory path. First of all, it showed the well-known optically allowed, singly excited singlet state of butadiene, described simply as due to excitation from the highest occupied to the lowest free MO of VII, and the corresponding triplet. These states correlated directly with the corresponding states of VIII and showed no minima (cf. Fig. 1). In addition, an optically forbidden, "doubly excited" state was predicted at slightly higher energy in VII (and has since been observed in other polyenes[16]). This is the "doubly excited" state known from the correlation diagrams of Woodward–Hoffmann type.[1] It attempts to correlate with the ground state of the product VIII, while the ground state of the starting material VII attempts to correlate with a similar doubly excited state of the product VIII. The result of their avoided crossing is a barrier in the ground state and a well in the excited state (Fig. 1). According to van der Lugt and Oosterhoff, this well penetrates deep below the optically allowed, singly excited state in the region of the avoided crossing (biradicaloid geometry), so that it then represents the lowest excited singlet, as indicated schematically in part c of Fig. 1. The detailed proposed mechanism then is as follows[25]: excitation into the singly excited state of VII, motion along the disrotatory path over a small barrier to a point of crossing with the "doubly excited" state, reaching the minimum (an intermediate with cyclic bonding and possibly long enough lifetime to be observable), internal conversion to the ground state, and finally vibrational equilibration to give VII or VIII. The triplet state lacks any analogous minimum at biradicaloid geometry (Fig. 1) and the molecule in triplet state follows other paths (*cis–trans* isomerization, dimerization). All of these results can be understood on the basis of general principles and do not depend on the details of the model used, except that one may wonder about the order of excited states at the biradicaloid geometry. If the well in the "doubly excited" state were above the singly excited state, as shown in part d of Fig. 1, it would never be reached and the singlet would behave just like the triplet. It is not obvious why at these geometries the "doubly excited" state should lie below the "singly excited" one. Indeed, simple 3×3 CI would predict just the opposite ordering.[4,14] The opposite ordering was also obtained in a CNDO–PCILO study.[48b]

Subsequently, more complicated cyclobutenes were found in which correlation diagrams of the Woodward–Hoffmann type predict a barrier separating the starting geometry from the biradicaloid well; indeed, experimentally, the reactions do not proceed unless high-energy photons are used for excitation.[6,38,119] This fits nicely into the theory.

Van der Lugt and Oosterhoff's results were rederived qualitatively by Kikuchi[61] using the MINDO/2 method. Quantitatively, the all-valence-electron results were less satisfactory, showing excitations involving σ electrons at much too low energies, but the qualitative picture was the same. Essentially the same results as those of van der Lugt and Oosterhoff were also obtained in a very recent *ab initio* study by Grimbert *et al.*[26] Insight into the origin of the state ordering at the biradicaloid geometry of the presumed intermediate was obtained from analysis of results for bonding in the isoelectronic H_4 model species in VB terms[5,110] and a simple explanation has been offered[110] for the observation[120] that some cyclobutene ring openings proceed in an adiabatic fashion (if the reaction is very highly exothermic, the product side of the correlation diagram is pulled down, the well at biradicaloid geometry tilts and "spills out its contents").

This interplay of the various available theoretical models provides the best illustration known to the author of the kind of qualitative insight which semiempirical theory has provided for organic photochemical reactions.

In addition to suggesting a detailed mechanism for the electrocyclic transformation, theory also should provide guidance as to which of several possible photochemically allowed electrocyclic paths will be taken in reality, i.e., which biradicaloid minimum (funnel) will be reached preferentially. This is a difficult problem and in principle requires quite accurate calculations of hypersurfaces. Qualitative results can be obtained simply, however, by considerations of the initial slope. Perturbation theory arguments can be used to derive this from initial charge distributions and bond orders.

For example, extended Hückel charge distributions have been used to rationalize the preferential formation of IX if Z is electron-donating and of X if Z is electron-withdrawing.[121]

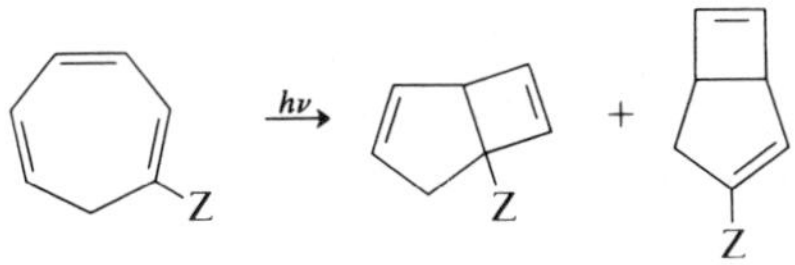

Similar investigations of substituent effects by the extended Hückel method were published for the case of six-membered ring systems.[21b] A result of great practical importance for synthetic chemistry of polycyclic hydrocarbons of the helicene type was the empirical finding[122] that the direction of cyclization of stilbene analogs, in which a new six-membered ring is formed, is correlated with a sum of ground-state free valence indices on the atoms involved in the ring closure. A more satisfactory formulation of this useful rule of thumb has been found recently: The relative ease of cyclization is related to the change in Mulliken overlap population upon excitation as calculated by the extended Hückel method.[123]

5. Summary and Outlook

It is very difficult to elucidate the detailed course of photochemical reactions by experimental means, even in this era of picosecond spectroscopy. Even for such well-studied processes as *cis–trans* isomerization of stilbene, there is no unanimous agreement among experimentalists. Insight provided by reliable calculations would be of great help to further development of the field. Unfortunately, as we have seen above, it is also very difficult to perform reliable calculations of the requisite energy hypersurfaces for molecules of the size required. It is often relatively easy to calculate an equilibrium geometry in the lowest excited singlet or triplet state if there is one near the ground-state equilibrium geometry, but following the reaction path over possibly present barriers to the biradicaloid geometry at which return to the ground state occurs is quite another matter, and just determining the correct ordering of states at the latter geometry is a nontrivial task. It is sometimes stated that knowledge of the lowest excited hypersurface of each multiplicity would be adequate for a detailed understanding of a photochemical process, and this is probably often true. However, in order to be sure of the nature of the lowest excited state, it appears imperative to calculate well several of them, since relatively minor changes in a calculation can change their order. Also, effects involving higher excited states such as wavelength dependence in photochemistry of large molecules are just beginning to be of considerable experimental interest. Finally, it must be remembered that calculation of the hypersurfaces represents only half of the overall problem, probably the easier half. Dynamics of the motion on and jumps between surfaces, including those of different multiplicity (intersystem crossing), will probably turn out to be even harder.

Clearly, at the present time and in the foreseeable future, the value of theory to organic photochemistry does not and will not lie so much in providing numbers such as barrier heights and other energy differences, but in providing novel qualitative concepts and general guidance. In the author's opinion, this is equally true of semiempirical and *ab initio* calculations. Much of the present understanding of organic photochemistry has come from such simple models as extended Hückel and PPP and there is probably more to be learned from them. Relatively few of the basic concepts originated in studies of CNDO and INDO types, but this could change in the future. Much has already come from *ab initio* work.

To make a guess for the future, the author would expect continued emphasis on the quest for understanding of the nature of chemical bonding in excited states, to be extracted from three main lines of inquiry. First, from calculations by simple semiempirical methods such as extended Hückel, PPP, CNDO, INDO, and probably soon NDDO, for representative molecules at a relatively large number of geometries along suspected reaction paths, preferably with automated geometry optimization. It appears desirable to use several methods simultaneously in such applications, since each has specific strong points: extended Hückel, inclusion of overlap; PPP, cutting down the number of electrons to be considered, thus making full or nearly full CI feasible and avoiding the dilemmas of SCF methods; INDO, treatment of electron repulsion, handling all valence electrons on equal footing (CNDO is only a little cheaper and offers little advantage over INDO). Rapid progress at the INDO level is likely for triplet states, for which open-shell SCF is probably adequate. Serious difficulties are faced in the case of singlet states; these will perhaps be resolved by approaches of the MC SCF type.

The second line of work is less detailed geometrical search using variational *ab initio* methods on molecules of interest, and, perhaps soon, also approximate *ab initio* methods of the PRDDO class, if the latter are reformulated to include excited states. Again, triplets are relatively easy (except when their relative energy with respect to the singlets is important), and singlets are hard.

A third line is work on simple model molecules in the hope that the principles of bonding in simple models are the same as those in actual molecules of interest. Just as an understanding of the nature of the single bond in ground-state H_2 promoted understanding of single bonds in general, it can be hoped that understanding of the excited states of H_2, H_4, H_6, etc., at a variety of geometries, relatively cheaply obtained by a fairly reliable *ab initio* method, will promote understanding of bonding in excited states of collections of one, two, three, etc., bonds. This type of work is pursued in the author's laboratory.[5,110]

In the spirit of traditional physical organic chemistry, each of these lines of inquiry will probably be complemented by an investigation of minor structural effects such as substitution. For this purpose, calculations limited to the initial molecular geometry may sometimes be sufficient.

Finally, it is to be hoped that some novel and unpredicted breakthroughs will occur: perhaps a practical version of the VB method, conceivably along DIM lines?

ACKNOWLEDGMENTS

Much of the work on this chapter was accomplished during the author's stay at the Chemistry Department, Aarhus University, Denmark. It is a pleasure to thank Prof. Jan Linderberg and Dr. Erik Thulstrup for their warm hospitality. Permission to reproduce Fig. 1 from *Topics Curr. Chem.* **46**, 141 (1974) is gratefully acknowledged.

References

1. R. B. Woodward and R. Hoffmann, *The Conservation of Orbital Symmetry*, Verlag Chemie, Weinheim, Germany, and Academic Press, New York (1970).
2. R. C. Dougherty, *J. Am. Chem. Soc.* **93**, 7187 (1971); M. J. S. Dewar and R. C. Dougherty, *The PMO Theory of Organic Chemistry*, Plenum Press, New York (1975), Chapter 6.
3. H. E. Zimmerman, *Acc. Chem. Res.* **4**, 272 (1971); N. D. Epiotis, *Angew. Chem. Int. Ed. Engl.* **13**, 751 (1974).
4. J. Michl, *Mol. Photochem.* **4**, 257 (1972).
5. J. Michl, *Pure Appl. Chem.* **41**, 507 (1975).
6. J. Michl, *Mol. Photochem.* **4**, 243, 287 (1972); in *Chemical Reactivity and Reaction Paths* (G. Klopman, ed.), Wiley, New York (1974), Chapter 8.
7. W. C. Herndon, *Topics Curr. Chem.* **46**, 141 (1974).
8. J. Michl, *Topics Curr. Chem.* **46**, 1 (1974).
9. R. G. Parr, *The Quantum Theory of Molecular Electronic Structure*, Benjamin, New York (1964); H. Suzuki, *Electronic Absorption Spectra and Geometry of Organic Molecules*, Academic Press, New York (1967); L. Salem, *The Molecular Orbital Theory of Conjugated Systems*, Benjamin, New York (1966).
10. J. A. Pople and D. L. Beveridge, *Approximate Molecular Orbital Theory*, McGraw-Hill, New York (1970).
11a. A. Devaquet, *Topics Curr. Chem.* **54**, 1 (1975).
11b. W. G. Dauben, L. Salem, and N. J. Turro, *Acc. Chem. Res.* **8**, 41 (1975).
12. W. Rhodes, *Chem. Phys.* **4**, 259 (1974); see also J. Jortner and S. Mukamel, in *The World of Quantum Chemistry. Proceedings of the First International Congress of Quantum Chemistry held at Menton, France, July 4–10, 1973* (R. Daudel and B. Pullman eds.), D. Reidel, Dordrecht, Holland (1974), pp. 145–209; G. W. Robinson and C. A. Langhoff, *Chem. Phys.* **5**, 1 (1974).
13. K. G. Kay, *J. Chem. Phys.* **61**, 5205 (1974).

14. L. Salem and C. Rowland, *Angew. Chem. Int. Ed. Engl.* **11**, 92 (1972).

15. R. Hoffmann and R. B. Woodward, *Acc. Chem. Res.* **1**, 17 (1968).

16. B. Hudson and B. Kohler, *Ann. Rev. Phys. Chem.* **25**, 437 (1974); see also R. M. Gavin, Jr. and S. A. Rice, *J. Chem. Phys.* **60**, 3231 (1974); J. Čizek, J. Paldus, and I. Hubač, *Int. J. Quant. Chem.* **8**, 951 (1974).

17. J. Michl and E. W. Thulstrup, *Tetrahedron*, **32**, 205 (1976).

18. N. C. Baird, *Mol. Phys.* **18**, 39 (1970); *J. Chem. Soc. D. Chem. Commun*, **1970**, 199.

19. A. J. Lorquet, *J. Phys. Chem.* **74**, 895 (1970).

20. M. Wolfsberg and L. Helmholz, *J. Chem. Phys.* **20**, 837 (1952); R. Hoffmann and W. N. Lipscomb, *J. Chem. Phys.* **37**, 2872 (1962); L. L. Lohr, Jr., and W. N. Lipscomb, *J. Chem. Phys.* **38**, 1607 (1963); R. Hoffmann, *J. Chem. Phys.* **39**, 1397 (1963).

21. (a) G. Feler, *Theor. Chim. Acta* **12**, 412 (1968); (b) K. A. Muszkat and W. Schmidt, *Helv. Chim. Acta* **54**, 1195 (1971); W. Schmidt, *Helv. Chim. Acta* **54**, 862 (1971); (c) R. Hoffmann, *Tetrahedron* **22**, 521, 539 (1966); R. Hoffmann and R. A. Olofson, *J. Am. Chem. Soc.* **88**, 943 (1966); G. W. Van Dine and R. Hoffmann, *J. Am. Chem. Soc.* **90**, 3227 (1968); S.-i. Kita and K.-i. Fukui, *Bull. Chem. Soc. Japan* **42**, 66 (1969); H. E. Zimmerman, R. W. Binkley, R. S. Givens, G. L. Grunewald, and M. A. Sherwin, *J. Am. Chem. Soc.* **91**, 3316 (1969); R. Hoffmann, *J. Am. Chem. Soc.* **90**, 1475 (1968).

22. W. D. Stohrer, G. Wiech, and G. Quinkert, *Angew. Chem. Int. Ed. Engl.* **13**, 200 (1974); W. D. Stohrer, P. Jacobs, K. H. Kaiser, G. Wiech, and G. Quinkert, *Topics Curr. Chem.* **46**, 181 (1974).

23. J. L. Chenot, *Theor. Chim. Acta* **28**, 201 (1973); B. L. Kalman, *J. Chem. Phys.* **59**, 5184 (1973); P. Coffey and J. R. Van Wazer, *J. Chem. Phys.* **61**, 85 (1974); K. Sakamoto and Y. J. I'Haya, *Bull. Chem. Soc. Japan* **44**, 1201 (1971); A. B. Anderson, *J. Chem. Phys.* **62**, 1187 (1975); J. Spanget–Larsen, *J. El. Spectrosc. Rel. Phenomena* **2**, 33 (1973); J. Linderberg and Y. Öhrn, *Propagators in Quantum Chemistry*, Academic Press, New York (1973), Chapter 9.

24. P. Coffey, C. S. Ewig, and J. R. Van Wazer, *J. Am. Chem. Soc.* **97**, 1656 (1975).

25. W. Th. A. M. van der Lugt and L. J. Oosterhoff, *J. Am. Chem. Soc.* **91**, 6042 (1969).

26. D. Grimbert, G. Segal, and A. Devaquet, *J. Am. Chem. Soc.* **97**, 6629 (1975).

27. I. Fischer-Hjalmars, *J. Chem. Phys.* **42**, 1962 (1965).

28. P.-O. Löwdin, *J. Chem. Phys.* **18**, 365 (1950).

29. J. A. Pople, D. P. Santry, and G. A. Segal, *J. Chem. Phys.* **43**, S129 (1965).

30. J. Pancíř, I. Matoušek, and R. Zahradník, *Coll. Czech. Chem. Commun.* **38**, 3039 (1973).

31. N. Mataga and K. Nishimoto, *Z. Phys. Chem. (N. F.)* **13**, 140 (1957).

32. M. J. S. Dewar and N. L. Hojvat (Sabelli), *J. Chem. Phys.* **34**, 1232 (1961); *Proc. Roy. Soc. Ser. A.* **264**, 431 (1961); K. Ohno, *Theor. Chim. Acta* **2**, 219 (1964); G. Klopman, *J. Am. Chem. Soc.* **86**, 4550 (1964).

33. J. Koutecký, *J. Chem. Phys.* **47**, 1501 (1967).

34. J. W. Downing, Ph.D. Dissertation, University of Utah, (1974).

35. J. Karwowski, *Chem. Phys. Lett.* **18**, 47 (1973).

36. J. Pancíř and R. Zahradník, *J. Phys. Chem*, **77**, 107 (1973).

37. K. Inuzuka and R. S. Becker, *Bull. Chem. Soc. Japan* **44**, 3323 (1971); P. Borrell and H. H. Greenwood, *Proc. Roy. Soc. Lond. A* **298**, 453 (1967); K. Inuzuka and R. S. Becker, *Bull. Chem. Soc. Japan* **45**, 1557 (1972); R. S. Becker, K. Inuzuka, J. King, and D. E. Balke, *J. Am. Chem. Soc.* **93**, 43 (1971); R. S. Becker, K. Inuzuka, and J. King, *J. Chem. Phys.* **52**, 5164 (1970).

38. J. Michl, *J. Am. Chem. Soc.* **93**, 523 (1971).

39. P. Seiler and J. Wirz, *Helv. Chim. Acta* **55**, 2693 (1972).

40. R. Constanciel, O. Chalvet, and J.-C. Rayez, *Theor. Chim. Acta* **37**, 305 (1975).

41. J. M. Sichel and M. A. Whitehead, *Theor. Chim. Acta* **7**, 32 (1967); **11**, 220, 239, 254, 263 (1968); K. B. Wiberg, *J. Am. Chem. Soc.* **90**, 59 (1968); H. Fischer and H. Kollmar, *Theor. Chim. Acta* **13**, 213 (1969); R. D. Brown and F. R. Burden, *Theor. Chim. Acta* **12**, 95 (1968); F. Hirota and S. Nagakura, *Bull. Chem. Soc. Japan* **43**, 1642 (1970); A. Rauk, J. D. Andose, W. G. Frick, R. Tang, and K. Mislow, *J. Am. Chem. Soc.* **93**, 6507 (1971); R. J. Boyd and M. A. Whitehead, *J. Chem. Soc., Dalton Trans.* **1972**, 73, 79, 81; K. Ohno, T. Hirooka,

Y. Harada, and H. Inokuchi, *Bull. Chem. Soc. Japan* **46**, 2353 (1973); A. DasGupta and S. Huzinaga, *Theor. Chim. Acta* **35**, 329 (1974).

42. J. Del Bene and H. H. Jaffé, *J. Chem. Phys.* **48**, 1807, 4050 (1968); **49**, 1221 (1968).
43. A. J. Lorquet and J. C. Lorquet, *J. Chem. Phys.* **49**, 4955 (1968); P. A. Clark and J. L. Ragle, *J. Chem. Phys.* **46**, 4235 (1967).
44. R. L. Ellis, G. Kuehnlenz, and H. H. Jaffé, *Theor. Chim. Acta* **26**, 131 (1972).
45. A. Tajiri, N. Ohmichi, and T. Nakajima, *Bull. Chem. Soc. Japan* **44**, 2347 (1971); R. E. Linder, H. Weiler-Feilchenfeld, G. Barth, E. Bunnenberg, and C. Djerassi, *Theor. Chim. Acta* **36**, 135 (1974).
46. C. Giessner-Prettre and A. Pullman, *Theor. Chim. Acta* **13**, 265 (1969); **17**, 120 (1970); **18**, 14 (1970); **20**, 378 (1971).
47. E. M. Evleth and G. Feler, *Chem. Phys. Lett.* **22**, 499 (1973).
48. (a) G. M. White, A. J. Yarwood, and D. P. Santry, *Chem. Phys. Lett.* **13**, 501 (1972); J. C. Leclerc, J. A. Horsley, and J. C. Lorquet, *Chem. Phys.* **4**, 337 (1974); F. Momicchioli, G. R. Corradini, M. C. Bruni, and I. Baraldi, *J. Chem. Soc. Faraday II* **71**, 215 (1975); F. Momicchioli, M. C. Bruni, I. Baraldi, and G. R. Corradini, *J. Chem. Soc. Faraday II* **70**, 1325 (1974); A. Dargelos, D. Liotard, and M. Chaillet, *Theor. Chim. Acta* **38**, 79 (1975); (b) J. Langlet and J.-P. Malrieu, *J. Am. Chem. Soc.* **94**, 7254 (1972).
49. J. Weber and R. Gerdil, *Helv. Chim Acta* **56**, 1565 (1973).
50. L. L. Combs and M. Holloman, *J. Phys. Chem.* **79**, 512 (1975).
51. D. Perahia and A. Pullman, *Chem. Phys. Lett.* **19**, 73 (1973).
52. D. A. Condirston and D. C. Moule, *Theor. Chim. Acta.* **29**, 133 (1973).
53. E. Heilbronner and A. Schmelzer, *Helv. Chim. Acta* **58**, 936 (1975).
54. J. A. Pople, D. L. Beveridge, and P. A. Dobosh, *J. Chem. Phys.* **47**, 2026 (1967); R. N. Dixon, *Mol. Phys.* **12**, 83 (1967).
55. R. C. Bingham, M. J. S. Dewar, and D. H. Lo, *J. Am. Chem. Soc.* **97**, 1285 (1975).
56. F. A. Van-Catledge, *J. Phys. Chem.* **78**, 763 (1974); H. P. Figeys, P. Geerlings, and C. Van Alsenoy, *Bull. Soc. Chim. Belg.* **84**, 145 (1975).
57. J. Ridley and M. Zerner, *Theor. Chim. Acta* **32**, 111 (1973).
58. Z.-i. Yoshida and T. Kobayashi, *J. Chem. Phys.* **58**, 334 (1973); S.-Y. Chen and R. M. Hedges, *Theor. Chim. Acta* **31**, 275 (1973); F. A. Van-Catledge, *J. Am. Chem. Soc.* **93**, 4365 (1971); **95**, 1173 (1973); H. Kato, H. Konishi, and T. Yonezawa, *Bull. Chem. Soc. Japan* **40**, 1017 (1967); H. Kato, H. Konishi, H. Yamabe, and T. Yonezawa, *Bull. Chem. Soc. Japan* **40**, 2761 (1967); K. Yamaguchi and T. Fueno, *Bull. Chem. Soc. Japan* **44**, 43 (1971).
59. J. E. Ridley and M. C. Zerner, *J. Mol. Spectrosc.* **50**, 457 (1974).
60. G. Ahlgren, *Tet. Letters* **1974**, 989.
61. O. Kikuchi, *Bull. Chem. Soc. Japan* **47**, 1551 (1974).
62. M. J. S. Dewar and S. Kirschner, *J. Am. Chem. Soc.* **96**, 7578 (1974); M. J. S. Dewar, S. Kirschner, and H. W. Kollmar, *J. Am. Chem. Soc.* **96**, 7579 (1974); R. S. Case, M. J. S. Dewar, S. Kirschner, R. Pettit, and W. Slegeir, *J. Am. Chem. Soc.* **96**, 7581 (1974).
63. B. Voigt, *Theor. Chim. Acta* **31**, 289 (1973).
64. R. G. Jesaitis and A. Streitwieser, Jr., *Theor. Chim. Acta* **17**, 165 (1970).
65. J. A. Pople, D. L. Beveridge, and N. S. Ostlund, *Int. J. Quant. Chem.* **1S**, 293 (1967); see also J. J. Kaufman, *J. Chem. Phys.* **43**, S152 (1965); J. P. Dahl, *Acta Chem. Scand.* **21**, 1244 (1967).
66. R. Sustmann, J. E. Williams, M. J. S. Dewar, L. C. Allen, and P. von R. Schleyer, *J. Am. Chem. Soc.* **91**, 5350 (1969); R. B. Davidson, W. L. Jorgensen, and L. C. Allen, *J. Am. Chem. Soc.* **92**, 749 (1970); P. Birner, H.-J. Köhler, and C. Weiss, *Chem. Phys. Lett.* **27**, 347 (1974).
67. B. J. Nicholson, *Adv. Chem. Phys.* **18**, 249 (1970).
68. J. W. McIver, Jr., and A. Komornicki, *J. Am. Chem. Soc.* **94**, 2625 (1972); J. Pancíř, *Theor. Chim. Acta* **29**, 21 (1973); P. Pulay and F. Török, *Mol. Phys.* **25**, 1153 (1973).
69. N. C. Baird and R. M. West, *J. Am. Chem. Soc.* **93**, 4427 (1971); N. C. Baird, *J. Am. Chem. Soc.* **94**, 4941 (1972); N. C. Baird and R. M. West, *Mol. Photochem.* **5**, 209 (1973).
70. R. S. Mulliken, *J. Chim. Phys. Physicochim. Biol.* **46**, 497, 675 (1949).

71. A. Warshel and M. Karplus, *J. Am. Chem. Soc.* **94**, 5612 (1972); A. Warshel, *J. Chem. Phys.* **62**, 214 (1975).
72. A. Trautwein and F. E. Harris, *Theor. Chim. Acta* **30**, 45 (1973).
73. D. B. Cook, P. C. Hollis, and R. McWeeny, *Mol. Phys.* **13**, 553 (1967); R. D. Brown, F. R. Burden and G. R. Williams, *Theor. Chim. Acta* **18**, 98 (1970); R. D. Brown, F. R. Burden, G. R. Williams, and L. F. Phillips, *Theor. Chim. Acta* **21**, 205 (1971).
74. M. D. Newton, N. S. Ostlund, and J. A. Pople, *J. Chem. Phys.* **49**, 5192 (1968); M. D. Newton, *J. Chem. Phys.* **51**, 3917 (1969); M. D. Newton, W. A. Lathan, W. J. Hehre, and J. A. Pople, *J. Chem. Phys.* **51**, 3927 (1969).
75. F. P. Billingsley II and J. E. Bloor, *J. Chem. Phys.* **55**, 5178 (1971).
76. A. Okniński, *J. Chem. Phys.* **60**, 4098 (1974).
77. R. G. Body, *Theor. Chim. Acta* **18**, 107 (1970).
78. T. A. Halgren and W. N. Lipscomb, *J. Chem. Phys.* **58**, 1569 (1973).
79. M. Zerner, *J. Chem. Phys.* **62**, 2788 (1975).
80. J. B. Peel, *Aust. J. Chem.* **27**, 923 (1974).
81. J. E. Eilers and D. R. Whitman, *J. Am. Chem. Soc.* **95**, 2067 (1973); B. J. Duke, J. E. Eilers, and B. O'Leary, *J. Chem. Soc. Faraday II* **70**, 386 (1974).
82. J. M. Herbelin and F. E. Harris, *J. Am. Chem. Soc.* **93**, 2565 (1971); P. G. Burton, *Chem. Phys.* **6**, 419 (1974); P. G. Burton and R. D. Brown, *Chem. Phys.* **4**, 220 (1974); P. G. Burton, *Chem. Phys.* **4**, 226 (1974).
83. T. L. Gilbert, *J. Chem. Phys.* **60**, 1789 (1974); S. Iwata and K. Morokuma, *Theor. Chim. Acta* **33**, 285 (1974); R. Carbó, *Int. J. Quant. Chem.* **8**, 423 (1974); K. Hirao, *J. Chem. Phys.* **61**, 3247 (1974).
84. D. Mukherjee, *Int. J. Quant. Chem.* **8**, 247 (1974).
85. R. Caballol, R. Gallifa, J. M. Riera, and R. Carbó, *Int. J. Quant. Chem..* **8**, 373 (1974).
86. W. T. Borden, *J. Am. Chem. Soc.* **97**, 2906 (1975).
87. A. Gołebiewski and E. Nowak-Broclawik, *Mol. Phys.* **26**, 989 (1973).
88. C. W. Eaker and J. Hinze, *J. Am. Chem. Soc.* **96**, 4084 (1974).
89. K. Jug, *Theor. Chim. Acta* **30**, 231 (1973).
90. G. A. Segal, *J. Chem. Phys.* **53**, 360 (1970).
91. O. Kikuchi and K. Aoki, *Bull. Chem. Soc. Japan* **47**, 2915 (1974).
92. G. Das, *J. Chem. Phys.* **58**, 5104 (1973).
93. C. C. J. Roothaan, *Rev. Mod. Phys.* **32**, 179 (1960).
94. J. A. Pople and R. K. Nesbet, *J. Chem. Phys.* **22**, 571 (1954).
95. M. J. S. Dewar, J. A. Hashmall, and C. G. Venier, *J. Am. Chem. Soc.* **90**, 1953 (1968); M. J. S. Dewar and N. Trinajstic, *J. Chem. Soc. A* **1971**, 1220.
96. N. C. Baird and R. F. Barr, *Theor. Chim. Acta* **36**, 125 (1974).
97. P. Jørgensen, *Ann. Rev. Phys. Chem.* **26**, 359 (1975).
98. C. A. Coulson and I. Fischer, *Phil. Mag.* **40**, 386 (1949); V. Bonačić and J. Koutecký, *J. Chem. Phys.* **56**, 4563 (1972); A. R. Gregory, *J. Chem. Phys.* **60**, 1680 (1974).
99. B. Huron, J. P. Malrieu, and P. Rancurel, *J. Chem. Phys.* **58**, 5745 (1973).
100. H. M. Chang, H. H. Jaffé, and C. A. Masmanidis, *J. Phys. Chem.* **79**, 1109 (1975).
101. L. Salem, C. Leforestier, G. Segal, and R. Wetmore, *J. Am. Chem. Soc.* **97**, 479 (1975).
102. J. Downing, V. Dvořák, J. Kolc, A. Manzara, and J. Michl, *Chem. Phys. Lett.* **17**, 70 (1972); J. Kolc, J. W. Downing, A. P. Manzara, and J. Michl, *J. Am. Chem. Soc.*, **98**, 930 (1976).
103. A. Denis, J. Langlet, and J. P. Malrieu, *Theor. Chim. Acta* **29**, 117 (1973), J. Langlet and J. P. Malrieu, *Theor. Chim. Acta* **30**, 59 (1973).
104. J. Langlet and J. P. Malrieu, *Theor. Chim. Acta* **33**, 307 (1974).
105a. H. F. Schaefer III and F. E. Harris, *J. Chem. Phys.*, **48**, 4946 (1968).
105b. H. F. Schaefer III, *The Electronic Structure of Atoms and Molecules*, Addison-Wesley, Reading, Mass. (1972), p. 202; E. W. Thulstrup and A. Andersen, *J. Phys. B.* **8**, 965 (1975).
106. F. O. Ellison, *J. Am. Chem. Soc.* **85**, 3540 (1963); F. O. Ellison, N. T. Huff, and J. C. Patel, *J. Am. Chem. Soc.* **85**, 3544 (1963).
107. J. C. Tully, *J. Chem. Phys.* **58**, 1396 (1973); C. W. Bauschlicher, Jr., S. V. O'Neil, R. K. Preston, H. F. Schaefer III, and C. F. Bender, *J. Chem. Phys.* **59**, 1286 (1973).

108. R. C. Bingham and M. J. S. Dewar, *J. Am. Chem. Soc.* **94**, 9107 (1972).
109. W. J. Hehre, L. Salem and M. R. Willcott, *J. Am. Chem. Soc.* **96**, 4328 (1974).
110. W. Gerhartz, R. D. Poshusta, and J. Michl, *J. Am. Chem. Soc.* **98**, 6427 (1976).
111. G. A. Segal, *J. Am. Chem. Soc.* **96**, 7892 (1974).
112. E. Vander Donckt, *Prog. Reaction Kinetics* **5**, 273 (1970).
113. A. M. Trozzolo, A. Dienes, and C. V. Shank, *J. Am. Chem. Soc.* **96**, 4699 (1974).
114. J. R. Huber, M. Nakashima, and J. A. Sousa, *J. Phys. Chem.* **77**, 860 (1973).
115. R. Daudel, *Advan. Quant. Chem.* **5**, 1 (1970); J. Bertrán, O. Chalvet, and R. Daudel, *Theor. Chim. Acta* **14**, 1 (1969); V. I. Pechenaya and V. I. Danilov, *Chem. Phys. Lett.* **11**, 539 (1971).
116. G. Kaupp, *Liebigs Ann. Chem.* **1973**, 844.
117. W. G. Dauben, in *Reactivity of the Photoexcited Organic Molecule*, (Proc. 13th Conf. Chem., Univ. Brussels, 1965), Wiley–Interscience, New York (1967), p. 171.
118. H. E. Zimmerman, *J. Am. Chem. Soc.* **88**, 1566 (1966).
119. J. M. Labrum, J. Kolc. and J. Michl, *J. Am. Chem. Soc.* **96**, 2636 (1974).
120. N. C. Yang, R. V. Carr, E. Li, J. K. McVey, and S. A. Rice, *J. Am. Chem. Soc.* **96**, 2297 (1974); R. V. Carr, B. Kim, J. K. McVey, N. C. Yang, W. Gerhartz, and J. Michl, *Chem. Phys. Lett.* **39**, 57 (1976).
121. L. Libit, *Mol. Photochem.* **5**, 327 (1973).
122. M. Scholz, M. Mühlstädt, and F. Dietz, *Tet. Lett.* **1967**, 665; W. H. Laarhoven, Th. J. H. M. Cuppen, and R. J. F. Nivard, *Tetrahedron* **26**, 4865 (1970); T. Sato and T. Morita, *Bull. Chem. Soc. Japan* **45**, 1548 (1972).
123. K. A. Muszkat and S. Sharafi-Ozeri, *Chem. Phys. Lett.* **20**, 397 (1973).

4

Approximate Methods for the Electronic Structures of Inorganic Complexes

C. J. Ballhausen

1. Inorganic Complexes Contrasted to Organic Molecules

A few years ago a chapter on approximate methods of inorganic complexes would have been tantamount to a paper on crystal field theory. This theory is specific to inorganic d^n and f^n complexes and hence no overlap with organic molecules was encountered. This is, however, no longer the case. Modern calculational and experimental methods have made it possible to probe deep into the electronic structures of all molecules and ions, thereby breaking down what was formerly a natural division between organic and inorganic theories. All the methods and problems met with in this volume can therefore be of importance when dealing with inorganic molecules and ions.

In the following I shall concentrate on the energetic aspects of the electronic structures of inorganic complexes containing transition metal elements. The discussion will deal with the properties of the ground and excited states. Due to the enormous material found in the literature a selection will have to be made. I have therefore elected to discuss representative cases in order to show what can be done. Likewise I have concentrated the discussion on those theories and results that at the moment of writing seem to me to be the most interesting.

The methods of treating the electronic structures of inorganic complexes are strongly influenced by the fact that these molecules or ions have a "natural

C. J. Ballhausen • Department of Physical Chemistry, H. C. Ørsted Institutet, Københavns Universitet, Copenhagen, Denmark

center," a metal atom or ion. In contrast to organic molecules of corresponding complexity, say benzene, the octahedral or tetrahedral complex has a transition metal atom placed at a center of high symmetry. Hence the high degeneracy of the metal orbitals, so characteristic of central symmetry, is only partially removed. Since the degenerate orbitals may be only partially filled with electrons, the electronic term problem is nearly as complicated as that encountered in the theory of atomic structure. A further distinction between inorganic and organic systems is that spin–orbit coupling often plays a great role in inorganic complexes.

The presence of the characteristic metal atom made it natural to focus attention on its electrons. Very early this gave rise to a completely localized theory of electronic motions for complexes, namely the crystal field theory,[1] The surroundings of the metal ion, the so-called ligands, play, however, a more than static role. It was consequently realized by Van Vleck[2] that a delocalized description offered many advantages. Crystal field theory is nowadays defined as an electronic theory of inorganic complexes which neglects the overlap of metal orbitals with ligand orbitals. The corresponding theory with overlap not neglected is termed ligand field theory. However, the two schemes of approximation each have their particular merits. Depending on the nature of the problem, it may be appropriate to focus attention on the metal and/or the ligand electronic structures. It serves no purpose dogmatically to rule out crystal field theory as "unphysical."

Experimentally, the complexes are often studied in the solid state. Hence interactions may occur, but I shall here adopt the tight-binding approach, treating each molecule or ion as an entity in its own right. The historical development will not govern the discussion; for that aspect I refer to the book *Symmetry in Chemical Theory*,[3] which is a reprint volume containing a selection of early papers. As starting point I shall take the Hartree–Fock equations for the electronic motions in molecules. In retrospect this seems the natural point of departure for reaching an understanding of the various approximation schemes used when dealing with the electronic structures of complexes.

2. The Orbitals

Approximate methods for calculating electronic structures of molecules distinguish between a full-electron treatment and a valence-electron treatment. In this chapter we shall solely be concerned with the latter. We define a core consisting of the metal and ligand nuclei and the nonvalence electrons of the metal and ligands. The core electrons are assumed to reside in the same atomic orbitals as in the free ions. The electrons that take part in the bonding of the ligands to the metal ion are the valence electrons.

The effect of the core is incorporated into an "effective" Hamiltonian for the valence electrons. As shown by Lykos and Parr,[4] such a description is only practical if all valence orbitals are kept orthogonal to all core orbitals. This so-called "strong orthogonality condition" is very important and should always be fulfilled.

The valence molecular orbitals of the complexes will be approximated by linear combinations of atomic orbitals. We shall primarily be interested in complexes of the transition series where the atomic nd shells are being filled up. The valence atomic orbitals for the transition metal are taken to be nd, $(n+1)s$, and $(n+1)p$. In the first transition series $n = 3$(Sc to Zn). The ligand valence atomic orbitals are usually the $2s$ and $2p$ orbitals.

By χ_M we designate a metal atomic orbital and by χ_λ a ligand atomic orbital. The LCAO MOs for the valence electrons are then of the type

$$\psi_\gamma = \alpha\chi_{M\gamma} + \beta \sum_{i=1}^{L} c_{i\gamma}\chi_{\lambda i} \tag{1}$$

The γ indices refer to the symmetry type of the molecular orbital. The coefficients $c_{i\gamma}$ in front of the L ligand orbitals are for simplicity assumed to be symmetry determined. α and β are variational parameters to be determined by a variational procedure of the Hartree–Fock type.

The $c_{i\gamma}$ coefficients are in most cases simple to determine. Consider an octahedral molecule, where the coordinate systems for the metal and ligands are placed as in Fig. 1. We can without loss of generality take all the χ's to be real functions. By using standard group theory, or simply by matching the overlap of the metal orbital with the ligand functions (Figs. 2 and 3), we can easily construct the linear combinations of Table 1 by inspection. Notice that

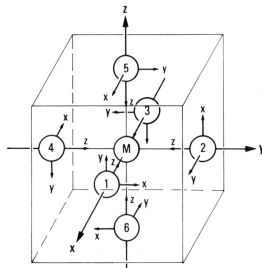

Fig. 1. Coordinate systems for an octahedral molecule.

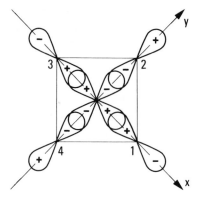

Fig. 2. The matching of σ ligand orbitals with $d_{x^2-y^2}$.

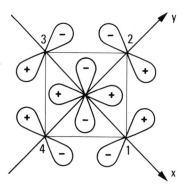

Fig. 3. The matching of π ligand orbitals with d_{xy}.

Table 1. Metal and Ligand Orbitals for the Molecular Orbitals of an Octahedral Complex

Represen-tation	Metal orbital	Ligand orbitals	
		σ	π
a_{1g}	s	$(1/\sqrt{6})(\sigma_1+\sigma_2+\sigma_3+\sigma_4+\sigma_5+\sigma_6)$	—
e_g	$d_{x^2-y^2}$	$\frac{1}{2}(\sigma_1-\sigma_2+\sigma_3-\sigma_4)$	—
	d_{z^2}	$(1/2\sqrt{3})(2\sigma_5+2\sigma_6-\sigma_1-\sigma_2-\sigma_3-\sigma_4)$	—
t_{1u}	p_x	$(1/\sqrt{2})(\sigma_1-\sigma_3)$	$\frac{1}{2}(p_{y2}+p_{x5}-p_{x4}-p_{y6})$
	p_y	$(1/\sqrt{2})(\sigma_2-\sigma_4)$	$\frac{1}{2}(p_{x1}+p_{y5}-p_{y3}-p_{x6})$
	p_z	$(1/\sqrt{2})(\sigma_5-\sigma_6)$	$\frac{1}{2}(p_{y1}+p_{x2}-p_{x3}-p_{y4})$
t_{2g}	d_{xz}	—	$\frac{1}{2}(p_{y1}+p_{x5}+p_{x3}+p_{y6})$
	d_{yx}		$\frac{1}{2}(p_{x2}+p_{y5}+p_{y4}+p_{x6})$
	d_{xy}		$\frac{1}{2}(p_{x1}+p_{y2}+p_{y3}+p_{x4})$
t_{1g}	—	—	$\frac{1}{2}(p_{y1}-p_{x5}+p_{x3}-p_{y6})$
			$\frac{1}{2}(p_{x2}-p_{y5}+p_{y4}-p_{x6})$
			$\frac{1}{2}(p_{x1}-p_{y2}+p_{y3}-p_{x4})$
t_{2u}	—	—	$\frac{1}{2}(p_{y2}-p_{x5}-p_{x4}+p_{y6})$
			$\frac{1}{2}(p_{x1}-p_{y5}-p_{y3}+p_{x6})$
			$\frac{1}{2}(p_{y1}-p_{x2}-p_{x3}+p_{y4})$

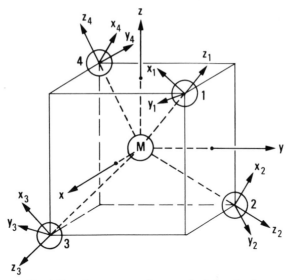

Fig. 4. Coordinate systems for a tetrahedral molecule.

overlap between the various ligand functions has been neglected in the normalization constant of the ligand combinations. If desired that can of course be easily incorporated. The neglect is, however, consistent with the approximation schemes which will be presented. In Fig. 4 and Table 2 we give the corresponding numbering of coordinate systems and ligand orbital combinations for regular tetrahedral complexes.

In cases of lower symmetry than O_h, say D_{4h}, corresponding to "tetragonal symmetry," the linear combinations of ligand orbitals given in Tables 1 and 2 are easily modified by the inclusion of some additional variational parameters.

Table 2. Basis Functions for T_d Molecules

Representation	Metal	Ligand orbitals
a_1	s	$\frac{1}{2}(s_1+s_2+s_3+s_4), \frac{1}{2}(p_{z1}+p_{z2}+p_{z3}+p_{z4})$
e	d_{z^2}	$\frac{1}{2}(p_{x1}-p_{x2}-p_{x3}+p_{x4})$
	$d_{x^2-y^2}$	$\frac{1}{2}(p_{y1}-p_{y2}-p_{y3}+p_{y4})$
t_2	p_x, d_{yz}	$\frac{1}{2}(p_{z1}-p_{z2}-p_{z3}-p_{z4}), \frac{1}{2}(s_1-s_2+s_3-s_4),$
		$\frac{1}{4}[p_{x1}+p_{x2}-p_{x3}-p_{x4}+\sqrt{3}(-p_{y1}-p_{y2}+p_{y3}+p_{y4})]$
	p_y, d_{xz}	$\frac{1}{2}(p_{z1}+p_{z2}-p_{z3}-p_{z4}), \frac{1}{2}(s_1+s_2-s_3-s_4),$
		$\frac{1}{4}[p_{x1}-p_{x2}+p_{x3}-p_{x4}+\sqrt{3}(p_{y1}-p_{y2}+p_{y3}-p_{y4})]$
	p_z, d_{xy}	$\frac{1}{2}(p_{z1}-p_{z2}-p_{z3}+p_{z4}), \frac{1}{2}(s_1-s_2-s_3+s_4),$
		$-\frac{1}{2}(p_{x1}+p_{x2}+p_{x3}+p_{x4})$
t_1	—	$\frac{1}{4}[\sqrt{3}(p_{x1}+p_{x2}-p_{x3}-p_{x4})+p_{y1}+p_{y2}-p_{y3}-p_{y4}]$
		$-\frac{1}{4}[\sqrt{3}(p_{x1}-p_{x2}+p_{x3}-p_{x4})-p_{y1}+p_{y2}-p_{y3}+p_{y4}]$
		$-\frac{1}{2}(p_{y1}+p_{y2}+p_{y3}+p_{y4})$

Other geometries of interest are square-planar and trigonal pipyramidal. The necessary LCAO MO combinations can again be written down by inspection.

3. The Ligand Field and the Crystal Field Methods

The state energies W_n of a system having a core and N valence electrons are found as solutions to

$$\mathcal{H}_{val}\Psi(1,\ldots,N) = W_n\Psi(1,\ldots,N) \tag{2}$$

with

$$\mathcal{H}_{val} = \sum_{j=1}^{N} \hat{h}_{core}(j) + \sum_{i<j}^{N} 1/r_{ij} \tag{3}$$

and

$$\hat{h}_{core}(1) = -\frac{1}{2}\nabla_1^2 - \sum_{\eta}\frac{Z_\eta}{r_{\eta 1}} + \sum_{\substack{core\\orbitals}} (2\hat{J}_k - \hat{K}_k) \tag{4}$$

The three terms in $\hat{h}_{core}$ are the kinetic energy, the potential energy due to the η nuclei, and the potential stemming from all the filled electronic inner shells of the metal and ligands. $\hat{J}_j$ and $\hat{K}_j$ are the usual Coulomb and exchange operators. By definition

$$\hat{J}_j\psi_i(1) = \int \psi_j(2)\psi_j(2)\frac{1}{r_{12}}dv_2\,\psi_i(1) \tag{5}$$

$$\hat{K}_j\psi_i(1) = \int \psi_j(2)\psi_i(2)\frac{1}{r_{12}}dv_2\,\psi_j(1) \tag{6}$$

The self-consistent field operator for the orbitals is for a system possessing closed shells of valence orbitals

$$\hat{F} = \hat{h}_{core}(1) + \sum_{j=1}^{N} (2\hat{J}_j - \hat{K}_j) \tag{7}$$

where we have to solve

$$\hat{F}\psi_i = w_i\psi_i \tag{8}$$

In order to illustrate the transition from the ligand field model to the crystal field model, we shall consider an octahedral complex of O_h symmetry. The five d orbitals span representation of t_{2g} and e_g symmetry, and suitable ligand orbital combinations can be made up to go together with $e_g(\sigma)$ and with $t_{2g}(\pi)$. The ligand valence orbitals are filled with electrons. Hence the metal valence electrons will occupy the antibonding $t_{2g}^*(\pi)$ and $e_g^*(\sigma)$ orbitals.

We now take a single-determinantal closed-shell state function

$$|(\text{core})(\psi_{e_g})^4(\psi_{t_{2g}})^6|$$

where

$$\psi_\gamma = \alpha\chi_{M\gamma} + \beta\chi_L \tag{9}$$

where $\chi_L = \sum_i^L c_{i\gamma}\chi_{\lambda i}$ and $\chi_{M\gamma}$ is a metal orbital.

Using the set of equations (8), a secular equation is obtained for each symmetry orbital

$$\begin{vmatrix} F_{MM} - w_\gamma & F_{ML} - w_\gamma G_{ML} \\ F_{ML} - w_\gamma G_{ML} & F_{LL} - w_\gamma \end{vmatrix} = 0 \tag{10}$$

where F_{MM}, etc., are the elements of the Hartree–Fock operator (7) and G_{ML} is the overlap integral $\langle\chi_{M\gamma}|\chi_L\rangle$. Most inorganic complexes will have

$$F_{MM} - F_{LL} \gg |F_{ML} - F_{LL}G_{ML}| \tag{11}$$

The low "bonding root" is then to good approximation given by

$$w_\gamma = F_{LL} - \frac{(F_{ML} - F_{MM}G_{ML})^2}{F_{MM} - F_{LL}} \tag{12}$$

and the high "antibonding" root by

$$w_\gamma^* = F_{MM} + \frac{(F_{ML} - F_{LL}G_{ML})^2}{F_{MM} - F_{LL}} \tag{13}$$

Solving for α and β in (9) using (11) leads to molecular orbitals correct to G_{ML}^2:

$$\psi_\gamma = \chi_L \tag{14}$$

and

$$\psi_\gamma^* = (1 - G_{ML}^2)^{-1/2}(\chi_{M\gamma} - G_{ML}\chi_L) \tag{15}$$

Notice that the antibonding orbital is a metal orbital, which is Schmidt-orthogonalized on the "pure" bonding ligand function.

We assume now that ψ_γ and ψ_γ^* are reasonable approximations to the Hartree–Fock orbitals for the complex. Hence

$$\hat{F}\psi_\gamma^* = w_\gamma^*\psi_\gamma^* \tag{16}$$

and

$$\hat{F}\psi_\gamma = w_\gamma\psi_\gamma \tag{17}$$

Substituting (15) in (16) and making use of (17), we obtain

$$\hat{F}\chi_{M\gamma} + (w_\gamma^* - w_\gamma)G_{ML}\sum_{i=1}^{L} c_{i\gamma}\chi_{\lambda i} = w_\gamma^*\chi_{M\gamma} \tag{18}$$

Defining a so-called pseudopotential $\hat{U}$ by

$$\hat{U}\chi_{M\gamma}(1) = (w_\gamma^* - w_\gamma)\int \chi_{M\gamma}^*(2)\sum_{i=1}^{L} c_{i\gamma}\chi_{\lambda i}(2)\,dv_2\sum_{j=1}^{L} c_{j\gamma}\chi_{\lambda j}(1) \tag{19}$$

we can rewrite (18) as

$$(\hat{F} + \hat{U})\chi_{M\gamma} = w_\gamma^*\chi_{M\gamma} \tag{20}$$

The definition of $\hat{U}$ is seen to run parallel to the definition of an exchange operator.

From (18) we get

$$\langle\chi_{M\gamma}|\hat{F} + (w_\gamma^* - w_\gamma)G_{ML}^2|\chi_{M\gamma}\rangle = w_\gamma^* \tag{21}$$

The form of this equation implies that we can write w_γ^* as a sum of two terms—one independent of G_γ and one dependent upon G_γ^2.

In ligand field theory a parameter $10Dq$ is defined in octahedral symmetry by a difference in core matrix elements,

$$10Dq = \bar{h}(e_g^*) - \bar{h}(t_{2g}^*) \tag{22}$$

The core potential here includes the ten electrons located in $\psi(t_{2g})$ and $\psi(e_g)$. Making use of (15), we get

$$\langle e_g^*|\hat{h}_{core}|e_g^*\rangle = \frac{1}{1 - G_e^2}[\langle d_{e_g}|\hat{h}_{core}|d_{e_g}\rangle + G_e^2\langle\chi_L|\hat{h}_{core}|\chi_L\rangle - 2G_e\langle\chi_L|\hat{h}_{core}|d_{e_g}\rangle] \tag{23}$$

Expanding the d orbitals on the ligand orbitals, we have, to first order in the overlap, $d_e \simeq G_e\chi_L$. Hence we notice that (23) does not depend in first order upon G_e. Performing the same expansion for $\langle t_{2g}^*|\hat{h}_{core}|t_{2g}^*\rangle$, we see that (22) can be written as a power series in G_γ:

$$10Dq = \bar{h}(d_{e_g}) - \bar{h}(d_{t_{2g}}) + G_{e_g}^2 \cdot \tilde{w}_e - G_{t_{2g}}^2 \cdot \tilde{w}_t \tag{24}$$

Notice that the core parameters of crystal field theory differs from the w_γ^* values of self-consistent field theory. The crystal field theory $\bar{h}$(core) splittings therefore cannot be assumed to reflect self-consistent field energy differences. This point is often overlooked when comparing SCF calculations with parameters derived from crystal field theory. The level order of the d and f antibonding orbitals can, however, with fair success be surmised from the size of the group overlap integral G_{ML}.

The actual calculation of the energies in some Hartree–Fock scheme can proceed in various ways. Assuming the molecular orbitals to be of the type $\psi_i = \sum_\mu c_{\mu i}\chi_\mu$, we find that the Harteee–Fock–Roothaan equation for a closed-shell system takes the form[6]

$$\sum_\nu (F_{\mu\nu} - w_i S_{\mu\nu})c_{\nu i} = 0 \tag{25}$$

where the elements of the Hartree–Fock operator $\hat{F}$ are

$$F_{\mu\nu} = (\mu|\hat{h}_{core}|\nu) + \sum_{\lambda,\sigma} P_{\lambda\sigma}[(\mu\nu|\lambda\sigma) - \tfrac{1}{2}(\mu\lambda|\nu\sigma)] \tag{26}$$

Here

$$(\mu|\hat{h}_{core}|\nu = \int \chi_{\mu}(1)\hat{h}_{core}(1)\chi_{\nu}(1) \, dv_1 \tag{27}$$

$$(\mu\nu|\lambda\sigma) = \iint \chi_{\mu}(1)\chi_{\nu}(1)\frac{1}{r_{12}}\chi_{\lambda}(2)\chi_{\sigma}(2) \, dv_1 \, dv_2 \tag{28}$$

$$S_{\mu\nu} = \int \chi_{\mu}(1)\chi_{\nu}(1) \, dv_1 \tag{29}$$

and $P_{\lambda\sigma}$, the so-called charge and bond order matrix, is given by

$$P_{\lambda\sigma} = 2 \sum_{i}^{occ} c_{\lambda i}c_{\sigma i} \tag{30}$$

In the calculation of the electronic structures of complexes it is very common to find that the ground state of the complex has one or more open shells. In these cases it is usually possible to write down a single Slater determinant which transforms in an orbitally correct way and which is an eigenstate of $\hat{S}_Z$. However, in general it is not an eigenfunction of $\hat{S}^2$, but is "contaminated" with other spin states. A spin doublet $(S_Z = \tfrac{1}{2})$ will, e.g., contain quartet $(S_Z = \tfrac{1}{2})$ and higher states. In order to keep the contamination low, a spin projection[7] can be carried out after each iteration in the Hartree–Fock equations to remove the contaminating parts.

The self-consistent field for unfilled shells assumes two completely independent sets of molecular orbitals for the α and β spin orbitals. In the so-called unrestricted molecular orbital approach the optimum values of the coefficients in the LCAO MOs are found by varying the ψ^{α} and ψ^{β} orbitals independently. Two coupled equations are to be solved:

$$\sum_{\nu} (F_{\mu\nu}^{\alpha} - \omega_i^{\alpha}S_{\mu\nu})c_{\nu i}^{\alpha} = 0 \tag{31}$$

$$\sum_{\nu} (F_{\mu\nu}^{\beta} - \omega_i^{\beta}S_{\mu\nu})c_{\nu i}^{\beta} = 0 \tag{32}$$

The two Hartree–Fock operators are

$$F_{\mu\nu}^{\alpha} = (\mu|\hat{h}_{core}|\nu) + \sum_{\lambda,\sigma} [P_{\lambda\sigma}(\mu\nu|\lambda\sigma) - P_{\lambda\sigma}^{\alpha}(\mu\sigma|\lambda\nu)] \tag{33}$$

$$F_{\mu\nu}^{\beta} = (\mu|\hat{h}_{core}|\nu) + \sum_{\lambda,\sigma} [P_{\lambda\sigma}(\mu\nu|\lambda\sigma) - P_{\lambda\sigma}^{\beta}(\mu\sigma|\lambda\sigma)] \tag{34}$$

where $P_{\mu\nu}^{\alpha}$ and $P_{\mu\nu}^{\beta}$ are density matrices defined by

$$P_{\mu\nu}^{\alpha} = \sum_{i}^{p} c_{\mu i}^{\alpha}c_{\nu i}^{\alpha} \quad \text{and} \quad P_{\mu\nu}^{\beta} = \sum_{i}^{q} c_{\mu i}^{\beta}c_{\nu i}^{\beta} \tag{35}$$

with

$$P_{\mu\nu} = P_{\mu\nu}^{\alpha} + P_{\mu\nu}^{\beta} \tag{36}$$

So far Eqs. (25)–(36) have been exact inside the framework spanned by the LCAO MO Hartree–Fock–Roothaan method. However, the number of integrals to be treated is prohibitively large, and a number of simplifying assumptions are introduced in order to make the calculations tractable. Basically these approximations go back to the zero-differential-overlap (ZDO) method. In this method, the full equations (25) or (31) and (32) are taken as a starting point, but the following approximations are introduced:

1. All overlap integrals are neglected.
2. $(r|\hat{h}_{core}|s) = 0$ unless χ_r and χ_s belong to the same center or are "bonded."
3. $(\mu\nu|\lambda\sigma) = (\mu\mu|\lambda\lambda)\delta_{\mu\nu}\delta_{\lambda\sigma}$.

Notice that the problem of evaluating the difficult three- and four-center, two-electron integrals is avoided. Naturally these drastic approximations lead to a great simplification of the secular equation. The various Hartree–Fock approximation schemes which have been used for inorganic complexes, viz. CNDO (complete neglect of differential overlap) and INDO (intermediate neglect of differential overlap), differ now solely in the extent to which the approximation (3) is invoked.

As shown by Pople *et al.*,[8] the CNDO method will only give results independent of the choice of atomic coordinate systems provided all atomic orbitals are taken to behave like s orbitals in the integral evaluations. The additional approximation of giving $(\mu\mu|\lambda\lambda) = \gamma_{AB}$ one value per atom pair is therefore introduced. γ_{AB} is seen to be an average electrostatic repulsion between any electron on A and any electron on B.

In the CNDO approximation all one-center, two-electron exchange integrals are put equal to zero. The splittings of the spin multiplets arising from the electronic configurations of an open shell are therefore not correctly reflected in the calculations. This shortcoming is amended in INDO, where the approximation scheme retains the one-center exchange integrals. Since most transition metal complexes are conspicuous for their variety of states all originating from one electronic configuration, INDO is the "natural" approximation scheme to use here.

We shall not inquire into the validity of the above approximations to the Hartree–Fock–Roothaan equations (26), (33), and (34), that has been dealt with in other places in this volume. Rather, taking the above approximation schemes at their face value, it is possible to proceed in two basically different ways: One may obtain the numerical values of the matrix elements alone by computation.[9] Such evaluations can be carried out using available standard programs, such as, e.g., the DIATOM programs.[10] Alternatively, one may introduce semiempirical features. In that case numerical values for the one-center integrals are obtained from an analysis of atomic spectral data for the various atoms in the molecule. Two-center core integrals are obtained from molecular spectral data. Finally, two-center, two-electron integrals are obtained by interpolation techniques. For long distances R between the centers, $(\mu\mu|\lambda\lambda) \simeq 1/R$, and for short distances, $(\mu\mu|\lambda\lambda)$ goes into one-center atomic integrals.

Seen from a theoretical point of view, the first method is to be preferred. The balance of the calculation can be held at a uniform level, and no "fudge factors" are introduced. Furthermore, semiempirical calculations on inorganic complexes have the disadvantage that the number of parameters one has to use is very large. The ambiguities are therefore many, and the results are consequently not very reliable. The temptation to use a heavily scaled approach is, however, great since the results often conform to expectations.

4. Koopmans' Theorem

The one-electron orbital energies w_i which come out of the solutions to the Hartree–Fock equations are given by

$$w_i = (\psi_i|\hat{h}_{core}|\psi_i) + \sum_{j=1}^{N} (2J_{ij} - K_{ij}) \tag{37}$$

for a closed electronic system. w_i is seen to be the energy of an electron in ψ_i interacting with the core and with the other $N-1$ valence electrons. As first proven by Koopmans, provided that there is no reorganization of the other $N-1$ valence electrons upon ionization, $-w_i$ is found to be equal to the ionization potential of an electron in ψ_i. Koopmans' theorem has been very widely used to correlate the measured ionization potentials with the calculated SCF energies w_i. However, careful calculations on a number of organometallic complexes have shown [11,12] that no water-tight conclusions can be drawn from the photoelectron spectrum regarding the sequence of molecular orbitals in the ground state.

The reason for the breakdown of Koopmans' theorem is that different extents of electronic rearrangement occur upon ionization, depending upon the nature of the orbital involved in the ionization process. For a primarily ligand orbital there is little electronic rearrangement upon ionization. The ionization potential taken as the energy difference between the molecule and the ion is close to the calculated orbital energy. However, for an orbital that has both metal- and ligand-orbital character there is a marked rearrangement upon ionization. The amount of ligand-orbital character in the molecular orbitals changes drastically by going to the ion. The "crystal field orbitals" of ferrocene, $Fe(Cp)_2$, become, e.g., nearly pure $3d$ orbitals when going to $Fe(Cp)_2^+$. The computed ionization energy may therefore differ from the orbital energy by as much as 6 eV.

We must conclude that Koopmans' theorem is of limited value when dealing with inorganic complexes. The orbitals in which we are primarily interested are just the ones whose composition are the most susceptible to ionization. The failure of Koopmans' theorem therefore cannot be taken as a proof that a simple "aufbau" principle is inoperative for the electronic structure of the ground state.

5. Spin–Orbit Coupling

Due to the large spin–orbit coupling constant associated with the metal center, and in some cases with the ligands, spin–orbit coupling plays quite an important role for inorganic complexes. Restricting ourselves to the interaction between the spin and orbital motions of the same electron, we may write for the spin–orbit perturbation

$$\mathcal{H}^{(1)} = \sum_j^N \frac{1}{2m^2c^2} (\text{grad } \mathcal{V} \times \mathbf{p}_j) \cdot \mathbf{s}_j \tag{38}$$

Here, $\mathcal{V}$ is the potential experienced by electron j, m is the mass of the electron, c is the velocity of light, $\mathbf{p}_j$ is the linear momentum, and $\mathbf{s}_j$ is the spin momentum operator. The summation runs over all the valence electrons in the molecule. The spin–orbit coupling term assumes importance by virtue of its ability to split the spin multiplets and to "mix" the spin states that differ by $\Delta S = 1$. This feature breaks down the validity of the spin quantum number S.

The formal group-theoretical treatment classifies the states simultaneously after their orbital and spin transformation properties, making use of the "double groups."[1] By virtue of the fact that the spin–orbit coupling operator transforms as the totally symmetric representation in the double group of the molecule, it can only couple states together possessing the same double-group state designation.

With η nuclei in the molecule it is the usual procedure to approximate $\mathcal{V}$ in (38) as a sum of η spherical potentials located on the η centers,

$$\mathcal{V} = \mathcal{V}_1 + \mathcal{V}_2 + \cdots + \mathcal{V}_\eta \tag{39}$$

This approximation turns (38) into

$$\mathcal{H}^{(1)} = \sum_\eta \sum_j \frac{1}{2m^2c^2} \left(\frac{1}{r_j} \frac{\partial \mathcal{V}_\eta}{\partial r_j} \right) \mathbf{l}_j \cdot \mathbf{s}_j$$

or, collecting the radial dependence into functions $\zeta_\eta(r)$ characteristic of each atom in the molecule,

$$\mathcal{H}^{(1)} = \sum_\eta \sum_j \zeta_\eta(r) \mathbf{l}_j \cdot \mathbf{s}_j \tag{40}$$

However, it must be remembered that the approximation of Eq. (39) is quite drastic since it completely neglects the charge distribution due to the formation of the molecule. The basis of the approximation is that grad $\mathcal{V}$ most changes close to the nuclei. However, with all overlap terms thrown away one can only expect such a treatment to give an order of magnitude for the molecular spin–orbit coupling. Minor deviations from expected "atomic" values of the coupling constants are therefore not significant.

Consider now an octahedral or tetrahedral complex. The orbital angular momentum operator transforms under T_{1g} of the group O_h and under T_1 in T_d. It follows from group theory [5] that the multiplet structure is expected to be pronounced only for T_1 or T_2 states.

The three components of a T_1 state (respectively, a T_2 state) can be characterized[5] as Φ_1, Φ_0, and Φ_{-1}, in direct analogy with an atomic P state, where for $M_L = 1, 0, -1$ we write P_1, P_0, and P_{-1}. We find that we can take linear combinations of the basis functions which span T_1 and T_2, such that

$$\hat{L}_z \Phi_{M_T} = \alpha \hbar M_T \Phi_{M_T}, \qquad M_T = 1, 0, -1 \tag{41}$$

where for instance for a $^3T_{1g}(t_{2g})^2$ state $-3/2 \leq \alpha \leq -1$. The step-up and step-down operators $\hat{L}_x \pm i L_y$ act upon the T_1 and T_2 components as on a P state, only weighted with α.

Defining the total angular momentum $\mathbf{J} = \mathbf{L} + \mathbf{S}$, we have

$$\mathbf{L} \cdot \mathbf{S} = \tfrac{1}{2}(\hat{J}^2 - \hat{L}^s - \hat{S}^s) \tag{42}$$

The energies in the spin multiplet $^{2S+1}T_1$ or $^{2S+1}T_2$ may then be written

$$W_{J,S} = \tfrac{1}{2}\lambda\alpha[J(J+1) - 1 \cdot 2 - S(S+1)] \tag{43}$$

Consider a $^4T_1$ state. J can be $5/2$, $3/2$, or $1/2$. Hence the $^4T_1$ is split into three levels: one sixfold degenerate $W_{5/2,3/2} = (3/2)\lambda\alpha$, one fourfold degenerate $W_{3/2,3/2} = -\lambda\alpha$, and one twofold degenerate $W_{1/2,3/2} = -(5/2)\lambda\alpha$. The spin–orbit components of the $^4T_1$ should therefore occur with energy separations in the ratio $5:3$.

Suppose now that we have diagonalized the molecular orbitals under the spin–orbit operator, or, in other words, have found a set of molecular orbitals which transform like the representations of the double group. The total electronic wave function is a determinant and since the spin–orbit coupling term is a one-electron operator, we get

$$W_{J,S} = \sum_{t=1}^{N} \langle \psi_t | \mathcal{H}^{(1)} | \psi_t \rangle \tag{44}$$

The summation over t runs over all the occupied double-group molecular orbitals ψ_t.

The summation in (44) over the filled sets of double-group orbitals, with parentage in an irreducible representation of the simple group, yield zero. Only "open shells" will contribute to $W_{J,S}$.

In an inorganic complex where the molecular orbitals are approximated by

$$\psi_\gamma = \alpha \chi_{M\gamma} + \beta \sum_i^L c_{i\gamma} \chi_{\lambda i}$$

all the low-lying, open-shell "crystal field states" are derived from orbitals having $\alpha \approx 1$. The spin–orbit splittings of the "crystal field states" will therefore be dominated by ζ(metal). In the higher lying, so-called charge-transfer states an electron has either been moved from an orbital primarily of ligand character ($\beta \approx 1$) to an orbital primarily of metal character ($\alpha \approx 1$) or the reverse. Hence a contribution to the spin–orbit splitting of the ensuing state arises from an open shell of ligand orbitals. If one (or more) of the ligands are heavy atoms or ions with a high value of ζ(ligand), the contribution to the spin–orbit splitting from that particular ligand may be dominant. Provided ζ(ligand) $\gg \zeta$(metal), the charge transfer bands can therefore show spin–orbit splittings characteristic of the ligands.

This effect was observed by Bird and Day[13] in the charge transfer spectrum of $CoBr_4^{2-}$. The ground state of the tetrahedral complex is $\cdots (t_1)^6 (e)^4 (t_2)^4 \, {}^4A_2$. Exciting a $t_1(\pi)$ electron from an orbital located on the ligands to the partly filled $t_2(3d)$ shell, the electronic configuration $\cdots (t_1)^5 (e)^4 (t_2)^4$ gives rise, among others, to one ${}^4T_1$ state. Transitions ${}^4A_2 \to {}^4T_1$ are orbitally allowed, and the three spin–orbit components were observed with a total spread of 3500 cm^{-1}.

6. Nonempirical CNDO and INDO Methods

In these schemes all numerical values of the nonneglected matrix elements in Eqs. (27)–(29) in the Hartree–Fock–Roothaan equations are obtained by computation alone. A basis set of orthogonalized atomic orbitals is chosen. The core integrals (27) are evaluated using the DIATOM programs.[10] The basic approximation used in the evaluation of the two-center (a and b) electron integrals (28) lies in the treatment of the two-center charge distributions $\chi_a(1)\chi_b(1)$. We write

$$\chi_a(1)\chi_b(1) = \mu \chi_a(1)\chi_a(1) + \nu \chi_b(1)\chi_b(1) \tag{45}$$

where in order to preserve charge we must have

$$S_{ab} = \mu + \nu \tag{46}$$

For $\mu = \nu = \frac{1}{2}S_{ab}$, Eq. (45) gives us the Mulliken approximation, appropriate to the use of Löwdin symmetrically orthogonalized orbitals.[9] This is a natural choice of orbitals when the center of gravity for the charge distribution $\chi_a(1)\chi_b(1)$ is located about halfway between the centers a and b. This is the case when χ_a and χ_b are two ligand orbitals or if χ_b is a ligand orbital and χ_a a metal $3d$ orbital. If, however, χ_a is a metal $4s$ or $4p$ orbital and χ_b is a ligand orbital, the charge distribution has its maximum close to the ligand. In this case we take $\mu = 0$ and $\nu = S_{ab}$ appropriate to the use of a Schmidt orthogonalization of χ_a and χ_b.

Nonempirical CNDO and INDO calculations have been performed on a number of closed-shell tetrahedral complexes, such as MnO_4^- and $TiCl_4$. Unrestricted CNDO calculations have been performed on VCl_4 and MnO_4^{2-}. We shall briefly describe the results.

For the permanganate ion Dahl and Johansen[14] carried out a systematic search for an optimal basis set of atomic orbitals. Such a set should result in a lower molecular energy than any other basis set of the same size. In their procedure they varied the manganese atomic orbitals of the type $3d$, $4s$, and $4p$ but kept the oxygen orbitals fixed. They concluded that one should use $3d$ atomic orbitals corresponding to the atomic configurations $3d(d^7)$ or $3d(d^6)$ for both e and t_2 representations. For the $Mn(4s)$ and $Mn(4p)$, orbitals corresponding to the atomic configurations $4s(d^5s^2)$ and $4p(d^5p^2)$ gave the best results.

The dependence of the molecular orbital energies upon the bond distance Mn–O was also looked into. The indications this energy variation give concerning the band widths of the electronic transitions were used in their band assignments. At the equilibrium point the calculated energies of the two highest filled orbitals, using the metal orbitals $3d(d^4)$, $4s(d^5s^2)$, and $4p(d^5p^2)$, were -12.368 eV for $w(t_1)$ and -12.979 eV for $w(1e)$. A comparison with an all-electron SCF LCAO MO calculation[15] which used Gaussian-type functions and a large basis set with near-Hartree–Fock accuracy shows, however, large discrepancies. In this last calculation one gets -7.619 eV for $w(t_1)$.

In $TiCl_4$ both metal and ligand valence orbitals were extensively varied in the search for the best atomic orbitals.[16] It was found possible to optimize the $Ti(4s)$ and $Ti(4p)$ orbitals. For the $Ti(3d)$ and $Cl(4s, 4p)$, atomic functions which are slightly expanded over those of the free atoms had to be assumed. As expected, the ground state for $TiCl_4$ was $\cdots(t_1)^6 \, {}^1A_1$. However, the first virtual orbitals were found to be $3a_1$ and $4t_2$, followed by $2e$. This level ordering is of course not what one would expect from a crystal field model, where the order should be $2e$ followed by $4t_2$ and $3a_1$.

Utilizing a somewhat more polarized set of orbitals than Becker and Dahl,[16] an INDO version of the CNDO scheme used on $TiCl_4$ restores, however, the traditional ligand field scheme for the virtual orbitals of tetrahedral complexes.[17] The filled orbitals and the ground state are the same in the two methods. The calculations show therefore the great importance that the choice of basis sets has for the order of virtual orbitals. A comparison of the found and calculated ionization potentials gave reasonable agreement for the t_1 level. In view of our previous remarks concerning Koopmans' theorem, this is to be expected since the t_1 orbital is a pure ligand orbital.

Direct comparisons of the results of nonempirical INDO approximation scheme with *ab initio* calculations on $Ni(CO)_4$, $Ni(CN)_4^{2-}$, $Cr(CO)_6$, and $Fe(C_5H_5)_2$ have been carried out by Ziegler.[18] Koopmans' theorem is found to be reasonably well obeyed for $Ni(CN)_4^{2-}$ in contrast to the findings of Demuynck and Veillard.[19] It is suggested that the reason for the improved agreement with experiment lies in the use of a $3d$ polarization Slater orbital as contrasted to a single Gaussian orbital. This is, however, a moot point.

Copeland and Ballhausen[20] performed an unrestricted nonempirical CNDO MO calculation on VCl_4. In contrast to the expectations for a d^1 tetrahedral system, where crystal field theory demands a ^2E ground state, they found that the lone electron in VCl_4 would occupy the $3a_1$ orbital. A slight relaxation of the valence energy minimization criterion was found necessary to ensure that the d^1 systems of MnO_4^{2-} and CrO_4^{3-} have the experimentally required ^2E ground state.[21]

Using an open-shell method in which the basis set expansion for an orbital is performed in such a way that it is kept orthogonal to all the remaining occupied orbitals, Truax *et al.*[22] also looked at VCl_4. This calculational scheme avoids the explicit use of Lagrangian multipliers, and gave the "traditional" ^2E ground state.

The Hartree–Fock–Slater Xα Method

The presence of an exchange operator in the Hartree–Fock equations makes the number of integrals which have to be treated large and the numerical evaluation difficult. Therefore it would be nice if it was possible to replace the $\hat{K}$ operator with another operator which would incorporate the exchange effect but would be easier to handle.

Let us assume that an exchange term may be found as the potential at the point x arising from a spherical distribution of uniform charge density holding one electron. The distribution is centered at x and has an electronic density equal to $\rho(x)$. The exchange potential $-U(x)$ is then, with R the radius of the spherical distribution,

$$-U(x) = 4\pi \int_0^R \frac{1}{r}\rho(x)r^2 \, dr = 2\pi\rho(x)R^2 \tag{47}$$

Since a sphere of radius R contains one electron, we have also

$$(4\pi/3)R^3\rho(x) = 1 \tag{48}$$

Eliminating R from (47) and (48), we find

$$U(x) = -[(9\pi/2)\rho(x)]^{1/3} \tag{49}$$

Introducing a constant α, we can use a more flexible exchange potential

$$U_{X\alpha}(x) = -3\alpha[(3/4\pi)\rho(x)]^{1/3} \tag{50}$$

In applying this method to a molecule,[23] the effective Hamiltonian becomes

$$\mathcal{H} = -\tfrac{1}{2}\nabla^2 + \mathcal{V}_{nuc} + \mathcal{V}_{Coul} + U_{X\alpha} \tag{51}$$

where the first three terms are the conventional Hartree–Fock kinetic, nuclear attraction, and Coulomb terms, respectively. One-electron eigenfunctions are then obtained by using a LCAO basis set, and a secular equation is constructed. The method is semiempirical in the sense that it contains the variable parameter α. It has been found[23] that $\alpha = 1$ gives good results.

$TiCl_4$ and VCl_4 have been treated using the above method.[23,24] For $TiCl_4$ the ground state was the expected $\cdots(2e)^4(9t_2)^6(2t_1)^6\,{}^1A_1$, while for VCl_4 the "crystal field" ground state $\cdots(2e)^4(9t_2)^6(2t_1)^6(3e)^1\,{}^2E$ was found. In both cases an α value of about 0.85 yielded ionization potentials in reasonable agreement with the photoionization data.

Other tetrahedral molecules which have been treated by this method[25] include FeO_4^{2-}, RuO_4, RuO_4^-, RuO_4^{2-}, and OsO_4. The d^0, d^1, and d^2 systems yield the expected ground states of ${}^1A_1$, ${}^2E$, and ${}^3A_2$, respectively. However unrestricted Hartree–Fock calculations on the ground-state orbital energies were not performed.

7. Semiempirical CNDO and INDO Methods

We shall now turn to the approximation schemes where the various matrix elements in the Hartree–Fock equations are evaluated using experimental data. A discussion of work performed before 1968 can be found in a review paper by Dahl and Ballhausen.[9]

In 1970 Brown and Roby[26,27] developed a new semiempirical approximation scheme for inorganic complexes. They proposed a procedure for the calculation of the matrix elements of the core Hamiltonian, and introduced an empirical formula for the necessary scaling, thus allowing indirectly for electron correlation and deviations from Hartree–Fock atomic orbitals. Based on this approach an unrestricted Hartree–Fock method was developed[28] in order to account for the transferred spin densities found on fluorine in octahedral complexes of type MF_6^{n-}, M = Mn, Cr, Fe, and Ni. It was found that neither the metal $4s$ or $4p$ orbitals were strongly partitioned in the bonding M–F. This result is at variance with most other calculations, semiempirical or all-electron.

A revised INDO method has been proposed by van der Lugt.[29] After the orbital calculation a configuration interaction was carried out. The reliability of the calculated results was judged by a comparison between the calculated and observed charge transfer spectra of, e.g., $CuCl_4^{2-}$.

CNDO methods have been used by Clack and co-workers[30,31] to treat CrO_4^{2-} and the fluorine spin densities of MF_6^{n-} ions, M ranging from V^{2+} to Ni^{2+}. Relying heavily on atomic input data, it was possible to obtain reasonable agreement between the calculated and measured transferred spin densities. The method has also been used to study, e.g., $Ni(CO)_4$.[32]

The above methods have been extended to an INDO scheme,[33] including all one-center, two-electron exchange integrals. Unfortunately, as it stands, the method is not invariant to the rotation of axes. It is, however, claimed that this deficiency is not too serious. Using an unrestricted open-shell Hartree–Fock method, the minimized wavefunctions are not eigenfunctions of $\hat{S}^2$. The expectation values of $\hat{S}^2$ do not, however, deviate much from the expected values for pure spin configurations. The reported calculations therefore do not include a spin projection.

Using the INDO method, it is possible to study the crossing of states having different spin multiplicities.[34,35] For instance, octahedral complexes having a d^6 electronic configuration may have either a high-spin ($S = 2$) or a low-spin ($S = 0$) ground state. A calculation of the total molecular energy of, say, CoF_6^{3-} reveals that at the equilibrium bond distance the high-spin state $^5T_{2g}$ is the ground state. Conversely, for $Co(H_2O)_6^{3+}$ the low-spin state of $^1A_{1g}$ is stable by 2.1 eV over $^5T_{2g}$ at the equilibrium point. In accord with the old proposal of van Santen and van Wieringen,[36] it was found that larger equilibrium distances M–L correspond to an increase of the occupancy number of electrons in the e_g shell. States with different spin multiplicities but arising from the same electronic configuration $(t_{2g})^n(e_g)^m$ were found to have the same shape and the same equilibrium bond distance. This is borne out by the well-known fact that transitions between states having the same electronic configuration produce structured bands.

8. The Excited States

The best procedure for calculating the energies of the excited states is to perform an independent Hartree–Fock minimization of each state. However, the usual procedure is to use the virtual orbitals emerging from the solutions of the ground-state Hartree–Fock equations. Provided the ground-state configuration is a closed shell, the excitation of an electron from an occupied orbital ψ_i to a virtual orbital ψ_k will give rise to a spin singlet and a spin triplet state with excitation energies[9]

$$W(^1\Psi(i \to k)) - W_0 = w_k - w_i - J_{ik} + 2K_{ik} \qquad (52)$$

$$W(^3\Psi(i \to k)) - W_0 = w_k - w_i - J_{ik} \qquad (53)$$

As is well known, one obtains much better values for the energies of the excited states than given by (52) and (53) by performing a configuration interaction. Indeed, from calculations on conjugated π systems we know that extensive configuration interaction is required in order to place the excited states in the right order. Nevertheless, most investigators of the excited states of inorganic complexes have used (52) and (53) in order to assess the position of these states. In the CNDO approximation, K_{ik} of Eq. (52) is put equal to zero. In INDO, where the one-center exchange integrals are retained, a separation of the spin multiplets will, however, occur.

Extensive calculations of the excited states have been carried out on the almost "ionic" complexes CrF_6^{3-} and NiF_6^{4-}. The professed purpose of these investigations was to calculate the crystal field parameter $10Dq$. What the investigators did was, however, to calculate the energy separation of the ground state and first excited state. Equating $10Dq$ to state energy differences would mean that the definition of the parameter would depend upon the actual mode of calculation.[9]

Using a systematic set of approximations in an open-shell-type calculation of the Roothaan type, Offenhartz[37,38] obtained a separation of 6300 cm^{-1} for the $^3A_{2g} \rightarrow {}^3T_{2g}$ transition of NiF_6^{4-}. The experimental value is 7300 cm^{-1}. Even though this report does not deal with *ab initio* calculations, it is pertinent here to point to the work of Richardson and co-workers.[39] Using very few numerical approximations, open- and closed-shell spin-restricted Hartree–Fock–Roothaan equations were set up for the various states of the $(t_{2g})^n(e_g)^m$ configurations. Where appropriate, configuration interaction was included. For NiF_6^{4-} the transition $^3A_{2g} \rightarrow {}^3T_{2g}$ yielded 7126 cm^{-1}, and for CrF_6^{3-} the energy difference $^4A_{2g} \rightarrow {}^4T_{2g}$ was calculated to be 15,900 cm^{-1}. The experimental value is 16,100 cm^{-1}.

Attempts have been made to calculate the position of the excited states for d^n complexes by heavily scaled theories. However, I think it fair to say that knowing now as we do that the best theoretical calculational methods[39] can indeed accurately predict the positions of the excited states for d^n systems, we can rest assured that the basic theory is well understood. Hence we need not use computer time and effort to try to reproduce what simple crystal field diagrams can tell us in seconds (see Section 9).

Turning now to the many complexes that do not possess d or f electrons, the situation is different. Whereas nearly all excited d^n states have been identified using crystal field diagrams, the identification of the so-called charge transfer states has been made by a mixture of molecular orbital calculations and experiments. Especially useful experiments are polarization[40] and magnetic circular dichroism (MCD) measurements.[41] A simple calculation or just a hunch predicts a certain behavior for an excited state, and experiments confirm or rule out the assignment.

Consider as an example the tetrahedral MnO_4^- complex. Extensive [42,9,14] approximate calculations had indicated that the proposed assignment[43] for the 18,000-cm^{-1} allowed transition $(t_1)^6 {}^1A_1 \rightarrow (t_1)^5(e)^1 {}^1T_2$ was a strong possibility. The t_1 orbital, being made up solely of oxygen π orbitals (Table 2), is completely symmetry determined. The sign and the magnitude of the magnetic circular dichroism of the transition $^1A_1 \rightarrow {}^1T_2$ are determined[41] by the value of the matrix element $\langle {}^1A_1|\hat{L}_z|{}^1T_2\rangle$. Since $\hat{L}_z$ has no matrix elements within an e manifold, the calculation of the magnetic circular dichroism should be quite reliable as to the sign and order of magnitude. The measurements of Schatz et $al.$[44] showed indeed that one may assign the 18,000-cm^{-1} band to $^1A_1(t_1)^6 \rightarrow$ $^1T_2(t_1)^5(e)^1$.

The electronic configuration $(t_1)^5(e)^1$ should, however, also give rise to a $^1T_1$ state together with two triplet states, $^3T_1$ and $^3T_2$. A sharp band system found at about 14,500 cm^{-1} was assigned[45] to the orbitally forbidden transition $^1A_1 \rightarrow {}^1T_1$. No splitting or broadening of the zero-phonon line could be detected in a Zeeman experiment.[46] Hence it is an excited spin singlet state.

McGlynn and co-workers,[47,48] using linearly polarized light, have observed splitting of both the transitions $^1A_1 \rightarrow {}^1T_1$ and $^1A_1 \rightarrow {}^1T_2$ in the spectrum of MnO_4^- dissolved in $Ba(ClO_4)_2$. The site group of MnO_4^- in this lattice is only C_3, and a splitting of the states $^1T_1$ and $^1T_2$ into $^1A_1$ and ^1E is expected. Transition from $^1A_1$ to $^1A_1$ is allowed with the electric vector parallel to the C_3 axis and from $^1A_1$ to ^1E with the electric vector perpendicular to the C_3 axis. The transition whose perpendicular component has a zero-phonon line at 14,247 cm^{-1} showed a splitting of 152 cm^{-1}. The splitting and the polarization behavior thus prove the $^1A_1$ to $^1T_1$ assignment in the parent tetrahedral molecule.

Dissolving MnO_4^- in $LiClO_4 \cdot 3H_2O$ leads to a site group C_{3v} of MnO_4^-. Here $^1T_1$ splits into ^1E and $^1A_2$. Transition from the ground state is allowed to ^1E with the electric vector perpendicular to the C_3 axis. On applying pressure to the crystal, the "forbidden line" $^1A_1 \rightarrow {}^1A_2$ shows up,[49] and the "allowed line" $^1A_1 \rightarrow {}^1E$ splits in two. The polarization behavior of the three lines is exactly as would be expected for a forbidden electric dipole $^1A_1 \rightarrow {}^1T_1$ transition in the parent tetrahedral point group.

The calculation of Dahl and Johansen[14] showed that the transitions $t_1 \rightarrow 2e$, $3t_2 \rightarrow e$, and $3t_2 \rightarrow 3a$ allow the formation of excited states which are stable with respect to symmetric distortions, and with an Mn–O distance close to the one found in the ground state. These three transitions are therefore expected to lead to absorption bands possessing pronounced vibrational structure. Using (52) with $K_{ik} = 0$, the first two of these transitions are found to be placed at nearly the same energy, while the third one is placed at much higher energy. The 18,000-cm^{-1} band has already been identified with a transition $^1A_1 \rightarrow {}^1T_2$ of mostly $t_1 \rightarrow 2e$ character. It follows that the band system at 33,000 cm^{-1}, which shows a very regular vibrational structure, must be identified with a $^1A_1 \rightarrow {}^1T_2$ transition of mostly $3t_2 \rightarrow 2e$ character. However, the degree of interaction between the states is of course not known.

The singlet states $^1T_1$ and $^1T_2$, both primarily arising from a $(t_1)^5(e)^1$ configuration, are observed to be separated by some 4000 cm^{-1}. INDO calculations[46] then indicate that the two triplet states $^3T_1$ and $^3T_2$, arising from the same electronic configuration, should be placed on the red side of the $^1T_1$ state by some 1000 cm^{-1}. However, no absorption has been seen in the region of interest.

9. The Crystal Field Theory

Crystal field theory deals with the behavior of electrons located in antibonding orbitals of the type of Eq. (15):

$$\psi_\gamma^* = (1 - G_{ML}^2)^{-1/2}(\chi_{M\gamma} - G_{ML}\chi_L) \tag{54}$$

For small values of G_{ML} the orbitals in which the electrons responsible for the low-lying transitions move are therefore nearly pure d or f orbitals.

In the evaluation of the state energies using the Hamiltonian (3), the core integrals $\bar{h}(\gamma)$ are treated as parameters to be determined for the individual complexes by spectroscopic investigations.[5] Using orbitals of the type (54), we expand the two-electron integrals. We have, for instance,

$$K_{xz,yz}^{\text{complex}} = (1 - G_\pi^2)^{-2}\left[\iint d_{xz}(1) \, d_{yz}(1)\frac{1}{r_{12}} d_{xz}(2) \, d_{yz}(2) \, dv_1 \, dv_2\right.$$

$$\left. -4G_\pi \iint d_{xz}(1) \, d_{yz}(1)\frac{1}{r_{12}} d_{xz}(2)\chi_{Lyz}(2) \, dv_1 \, dv_2 + \cdots\right]$$

Expanding the ligand orbital on the metal set, we have, to first order in G_γ,

$$\chi_{Lyz} \approx G_\pi d_{yz} + \cdots$$

Correct to second order in G_π we get, therefore,

$$K_{xz,yz}^{\text{complex}} \approx (1 - 2G_\pi^2)K_{xz,yz}^{\text{atomic}} \tag{55}$$

Similarly

$$J_{xz,x^2-y^2}^{\text{complex}} \approx (1 - G_\pi^2 - G_\sigma^2)J_{xz,x^2-y^2}^{\text{atomic}} \tag{56}$$

A small reduction of the molecular integrals over those of corresponding atomic ones is therefore expected.

The atomic J and K integrals are evaluated in terms of the Condon–Shortley–Slater F_n integrals.[50] The ones for d orbitals are given in Table 3. In crystal field theory the F_n integrals are treated as slowly varying parameters, having approximately the values known from atomic spectroscopy.

Fifteen different J integrals, ten different K integrals, and nine other two-electron integrals can be met with in a $(\psi_{xz}, \psi_{yz}, \psi_{xy}, \psi_{x^2-y^2}, \psi_{z^2})$ set of

Table 3. Coulomb Integrals J and Exchange Integrals K of the nd Set t_{2g}
(xz, yz, xy) and $e_g(x^2 - y^2, z^2)$ According to Ref. 5

$J(z^2, z^2) = J(x^2 - y^2, x^2 - y^2) = J(xy, xy) = J(xz, xz) = J(yz, yz) = F_0 + 4F_2 + 36F_4$

$J(x^2 - y^2, xz) = J(x^2 - y^2, yz) = J(xy, yz) = J(xy, xz) = J(xz, yz) = F_0 - 2F_2 - 4F_4$

$J(z^2, xz) = J(z^2, yz) = F_0 + 2F_2 - 24F_4$

$J(z^2, xy) = J(z^2, x^2 - y^2) = F_0 - 4F_2 + 6F_4$

$J(x^2 - y^2, xy) = F_0 + 4F_2 - 34F_4$

$K(xy, yz) = K(xy, xz) = K(xz, yz) = K(x^2 - y^2, xz) = K(x^2 - y^2, yz) = 3F_2 + 20F_4$

$K(z^2, x^2 - y^2) = K(z^2, xy) = 4F_2 + 15F_4$

$K(z^2, xz) = K(z^2, yz) = F_2 + 30F_4$

$K(x^2 - y^2, xy) = 35F_4$

molecular orbitals. Symmetry of the molecule will, however, impose certain restrictions on the independence of these 34 integrals. In octahedral (O_h) symmetry we encounter, for instance, only ten independent two-electron integrals.

The configuration $|\psi_{xz}^\alpha \psi_{yz}^\alpha \psi_{xy}^\alpha|$ gives rise in O_h symmetry to a $^4A_{2g}$ state, and a component of an excited $^4T_{2g}$ state is given by $|\psi_{xz}^\alpha \psi_{yz}^\alpha \psi_{x^2-y^2}^\alpha|$. We find

$$W(^4T_{2g}) - W(^4A_{2g}) = 10Dq + 2J_{xz,x^2-y^2} - 2J_{yz,xz} - 2K_{xz,x^2-y^2} + 2K_{yz,xz} \qquad (57)$$

where $10Dq$ is defined in Eq. (22).

Taking G_π and G_σ equal to zero, we find by use of (55), (56), and Table 3

$$2J_{xz,x^2-y^2} - 2J_{yz,xz} + 2K_{yz,xz} - 2K_{xz,x^2-y^2} = 0$$

The assumption that the two-electron integrals can be evaluated using pure atomic orbitals is basic to the crystal field model. We find therefore that (57) in crystal field theory is changed into

$$W(^4T_{2g}) - W(^4A_{2g}) = 10Dq \qquad (58)$$

Notice that only in crystal field theory is it possible to identify a measured state energy difference with the core parameter $10Dq$.

Crystal field theory offers the possibility of calculating the so-called spin-pairing energies very easily.[5] The most detailed studies have been carried out by König and Kremer.[51,52] With spin–orbit coupling thrown in for good measure, a fairly accurate picture of the possible ground-state multiplets of d^4, d^5, d^6, and d^7 systems was obtained. Using the phenomenological parameters of crystal field theory, we must conclude that this aspect of the theory is extremely well understood.

Most important, and where crystal field theory has celebrated its greatest triumphs, is its ability to identify excited states.

Consider as an example the ground state of a $(3d)^6$ diamagnetic octahedral Co^{3+} complex. The electronic configuration is $(t_{2g}^*)^6$, corresponding to the wave function

$$\Psi(^1A_{1g}) = |\psi_{xz}^\alpha \psi_{xz}^\beta \psi_{yz}^\alpha \psi_{yz}^\beta \psi_{xy}^\alpha \psi_{xy}^\beta| \tag{59}$$

Exciting an electron will lead to the configuration $(t_{2g}^*)^5(e_g^*)^1$, producing $^1T_{1g}$ and $^1T_{2g}$ states together with the corresponding spin triplets.

Using a descent in symmetry[5] to construct the excited states, we get for one component each of $^1T_{1g}$ and $^1T_{2g}$

$$\Psi(^1T_{1g}) = (1/\sqrt{2})(|\psi_{xz}^\alpha \psi_{xz}^\beta \psi_{yz}^\alpha \psi_{yz}^\beta \psi_{yz}^\alpha \psi_{x^2-y^2}^\beta| - |\psi_{xz}^\alpha \psi_{xz}^\beta \psi_{yz}^\alpha \psi_{yz}^\beta \psi_{xy}^\beta \psi_{x^2-y^2}^\alpha|) \tag{60}$$

$$\Psi(^1T_{2g}) = (1/\sqrt{2})(|\psi_{xz}^\alpha \psi_{xz}^\beta \psi_{yz}^\alpha \psi_{yz}^\beta \psi_{xy}^\alpha \psi_{z^2}^\beta| - |\psi_{xz}^\alpha \psi_{xz}^\beta \psi_{yz}^\alpha \psi_{yz}^\beta \psi_{xy}^\beta \psi_{z^2}^\alpha|) \tag{61}$$

In the molecular orbital scheme we find easily, with $10\,Dq$ being given in Eq. (22),

$$W(^1T_{1g}) - W(^1A_{1g}) = 10Dq + 4J_{yz,x^2-y^2} - 4J_{yz,xy} + 2K_{xz,xy}$$
$$- 2K_{xz,x^2-y^2} + J_{xy,x^2-y^2} - J_{xy,xy} + K_{xy,x^2-y^2} \tag{62}$$

and

$$W(^1T_{2g}) - W(^1A_{1g}) = 10Dq + 4J_{yz,z^2} - 4J_{yz,xy} + 2K_{xz,xy}$$
$$- 2K_{xz,z^2} + J_{xy,z^2} - J_{xy,xy} + K_{xy,z^2} \tag{63}$$

Making the transition to the crystal field model by setting G_σ and G_π equal to zero and evaluating the J and K integrals by means of Table 3, we obtain

$$W(^1T_{1g}) - W(^1A_{1g}) = 10Dq - 35F_4 \tag{64}$$

$$W(^1T_{2g}) - W(^1A_{1g}) = 10Dq + 16F_2 - 115F_4 \tag{65}$$

For complexes in the first transition series orders of magnitude for F_2 and F_4 are $F_2 \approx 10F_4 \approx 1000\,cm^{-1}$. For $Co(NH_3)_6^{+3}$ the first spin-allowed band is found at 21,400 cm^{-1}. Hence from Eq. (64) we get $10Dq \simeq 25,000\,cm^{-1}$. This value of $10Dq$ predicts the $^1A_{1g} \rightarrow {}^1T_{2g}$ band to be placed at 29,500 cm^{-1} (experimental result is 29,600 cm^{-1}). The identification of the $^1A_{1g} \rightarrow {}^1T_{1g}$ band is provided by "splitting" the band by looking at a lower molecular symmetry, viz. investigating *trans*-$[Co(en)_2Cl_2]^+$, where en = ethylenediamine. The effective symmetry of this compound is D_{4h}, and $^1T_{1g}$ splits into two states of symmetry $^1A_{2g}$ and ^1E_g. The vibronic selection rules for the transitions, taken in conjunction with the polarized spectrum, prove the assignment.[53]

Equations (64) and (65) have been derived using the so-called strong crystal field approximation. Here the crystal field splitting parameter $10Dq$ is much larger than the electronic repulsion terms. As opposed to this, the weak field approximation scheme uses as starting point the atomic states of the metal ion. The wave functions and energies of these states are well known.[50] Using perturbation theory, the crystal field state energies are then easily calculated.[54–60] The ensuing energy level diagrams with spin–orbit coupling included are unsurpassed in all molecular theory in their power of dealing with the position of the excited states. I shall give some examples.

One of the best investigated $3d^3$ complexes is $Cr(en)_3Cl_3$. The Cr^{3+} ion is here surrounded by six nitrogen atoms, with an $O_h(D_3)$ symmetry. The D_3 classification is appropriate when dealing with the natural optical activity of the complex and for detailed energetic considerations.

The crystal field matrices for a d^3 system in trigonal symmetry have been given by Macfarlane.[61] Figure 5 shows part of the full crystal field energy diagram for d^3 in octahedral symmetry. The spectrum of $Cr(en)_3^{3+}$ has been

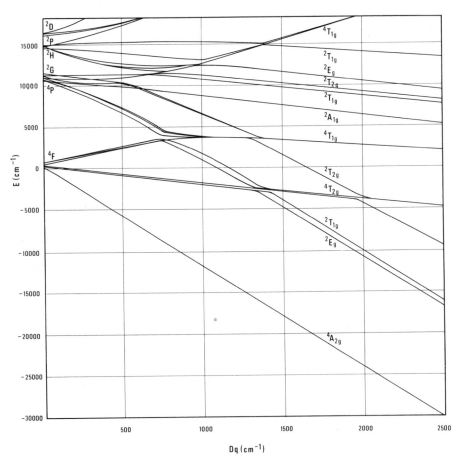

Fig. 5. Part of the "complete" crystal field energy diagram[60] for a d^3 configuration in O_h. $F_2 = 1100$ cm^{-1}, $F_4 = 80$ cm^{-1}, and $\xi_{3d} = 175$ cm^{-1}.

given by McCarthy and Vala.[62] The first spin-allowed transition shows up with a maximum of 22,000 cm^{-1}. According to Eq. (58), we have therefore $Dq = 2200$ cm^{-1}. As can be seen from the energy level diagram in Fig. 5, around this value of Dq the spin-forbidden transition $^4A_{2g} \to {}^2T_{2g}$ interacts strongly with the $^4A_{2g} \to {}^4T_{2g}$ transition. This shows up in the spectrum,[62] presumably as a so-called antiresonance.[63] The second spin-allowed band is for the value of Dq expected at 29,000 cm^{-1}. It is found at 28,800 cm^{-1}. The polarization behavior of the bands confirms the assignments; as group theory demands, in D_3 the two split components of $^4T_{2g}$ are seen with the electric vector polarized parallel (π) and perpendicular (σ) to the threefold axis of the complex. On the other hand, the $^4A_{2g} \to {}^4T_{1g}$ transition is mostly seen in σ polarization, again in conformity with the theoretical predictions. The $^4A_{2g} \to {}^4T_{2g}$, $^4T_{1g}$ transitions can further be distinguished on the basis of their optical activities.[64] The rotational strengths are grouped into the two sets, depending upon whether the respective transitions are magnetically allowed or forbidden. The $^4A_{2g} \to {}^4T_{2g}$ band should on this ground show stronger optical activity than $^4A_{2g} \to {}^4T_{1g}$. Again measurements confirm the above assignments.[65]

The transitions $^4A_{2g} \to {}^2E_g$, $^2T_{1g}$ should be located at about 14,500 cm^{-1} (Fig. 5). They are indeed also seen here as very detailed absorption lines. Their sharpness allows a vibrational analysis to be carried out,[62] and Zeeman studies[66,67] can confirm the assignments.

The states of octahedral $Ni(H_2O)_6^{2+}$ complexes follow the d^8 diagram given in Fig. 6. The first spin-allowed absorption band is seen at 8500 cm^{-1} and is identified with the transition $^3A_{2g} \to {}^3T_{2g}$. Crystal field theory gives[5] $10Dq = W(^3T_{2g}) - W(^3A_{2g})$. Hence $Dq = 850$ cm^{-1}. In conformity with the Cr^{3+} complexes, the $^3A_{2g} \to {}^3T_{2g}$ band should show stronger optical activity than the next spin-allowed transition $^3A_{2g} \to {}^3T_{1g}$ (I). This is indeed observed[68] in optically active crystals of α-[Ni(H_2O)_6]SO_4.

The second absorption region observed around 15,000 cm^{-1} exhibits an interesting fine structure. Figure 6 reveals that at the value of $Dq \simeq 850$ cm^{-1} a ^1E_g level tries to cross over the $^3T_{1g}$ (I) manifold. The fine structure of the transition $^3A_{2g} \to {}^3T_{1g}$ (I), ^1E_g has been analyzed by Pryce et al.[69] and Solomon and Ballhausen.[70] The $^3T_{1g}$ (I) multiplet is split by the spin–orbit coupling into four octahedral double-group components $\Gamma_1, \Gamma_3, \Gamma_4, \Gamma_5$. The ^1E state has the double group symmetry Γ_3. From Fig. 6 it is seen how the attempted crossing of the two Γ_3 states manifests itself. The energy splittings among the vibronic origins of the five spin–orbit levels is, however, found[70] to differ from those calculated by crystal field theory. Couplings to the vibrations allow the Γ_1, Γ_4, and Γ_5 origins to shift some 1100 cm^{-1} to lower energy. On the other hand, the two Γ_3 levels have been effectively uncoupled by the large spin–orbit coupling.

The crystal field method is completely general in the sense that the formalism can be used whenever we have a complex where the d and f shells

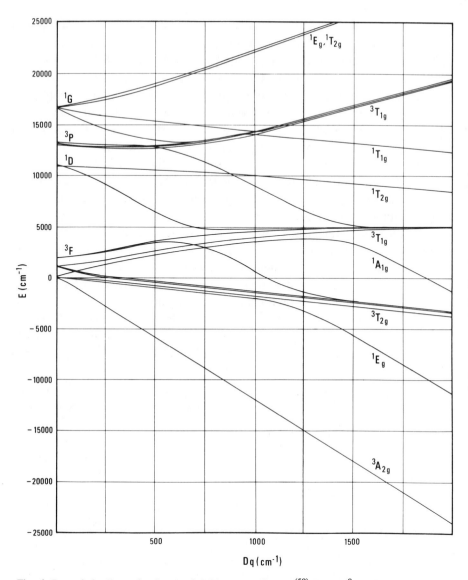

Fig. 6. Part of the "complete" crystal field energy diagram[59] for a d^8 configuration in O_h. $F_2 = 1120$ cm^{-1}, $F_4 = 90$ cm^{-1}, and $\xi_{3d} = 550$ cm^{-1}.

are partly filled. However, moving from $3d$ *to* $4d$ *to* $5d$ shells causes the group overlap integrals G_{ML} to increase. Hence the orbital splittings get larger and the correlative effects smaller. At the same time the spin–orbit coupling assumes more importance. We move therefore from a situation where the orbital splittings are comparable to the multiplet splittings, but large compared to the spin–orbit coupling, to a situation where spin–orbit coupling and core splittings are of the same order of magnitude and large compared to the correlative interactions between the electrons. At the same time the charge transfer transitions move down in energy, thereby often blotting out the excited crystal field states. Due to these complications, the application of the crystal field theory meets with increasing difficulties as we move from $3d$ complexes to $4d$ and $5d$.

A beautiful example of spectral assignments for complexes having $5d$ electrons is found in the series ReF_6 $(5d)^1$, OsF_6 $(5d)^2$, IrF_6 $(5d)^3$, and PtF_6 $(5d)^4$. Looking at the configurations $(t_{2g})^n$, $n = 1-4$, Moffitt et al.,[71] using two parameters values $\xi_{5d} = 3400$ cm^{-1} and $G = 3F_2 + 20F_4 = 2400$ cm^{-1}, succeeded in accounting for the lowest 13 band systems of the four hexafluorides. Similarly, the low absorption bands of the series TcF_6 $(4d)^1$, RuF_6 $(4d)^2$, RhF_6 $(4d)^3$, and PdF_6 $(4d)^4$ can be accounted for[72] using $\xi_{4d} = 1280$ cm^{-1} and $G = 3000$ cm^{-1}.

Turning our attention to the complexes of the rare earths, where the $4f$ shell is being filled up, a very small value of G_{ML} is found. The $4f$ electrons are shielded from their ligands by the filled $5s$ and $5p$ shells. Hence the crystal field core integrals are orders of magnitude less than both the multiplet splittings and the spin–orbit coupling constant. Observations of the excited electronic states are therefore effectively giving us the atomic line spectra of the ions.

A completely different situation is encountered for octahedrally complexed actinides, where the $5f$ shell is being filled. Here the spin–orbit coupling parameter and the crystalline field core parameters are of comparable magnitude. The level diagrams for octahedrally coordinated $(5f)^n$ systems are therefore very complicated. On the other hand, tetrahedral crystal fields provide little alteration of the states found in the actinide complexes.[73]

In O_h the seven f orbitals span a_{2u}, t_{1u}, and t_{2u} representations. The energy matrices in this representation have been given by Eisenstein and Pryce.[74] We place the core energies at $\bar{h}(t_{2u}) = 0$, $\bar{h}(a_{2u}) = -\Delta$, and $\bar{h}(t_{1u}) = \Theta$. The level order for, say, UCl_6 $(5f)^1$ is then $W(\Gamma_7) < W(\Gamma_8) < W(\Gamma_8') < W(\Gamma_6)$. The absorption spectrum can be fitted with $\xi_{5f} = 1940$ cm^{-1}, $\Delta = 1940$ cm^{-1}, and $\Theta = 3710$ cm^{-1}. One should, however, be aware of the fact that the values of Δ and Θ as evaluated from absorption spectra[73] are extremely sensitive to small variations of ξ_{5f}.

The electronic structures of the linear complexes UO_2^{2+} $(5f)^0$, NpO_2^{2+} $(5f)^1$, and PuO_2^{2+} $(5f)^2$ have been treated by Eisenstein and Pryce.[75,76] Using a pure crystal field argument, they considered the strong linear field to repel the $5f$ electrons. The lowest state of NpO_2^{2+} should therefore be the one in which the electron is furthest away from the O–Np–O axis. This is a $\varphi_u(e_{3u})$ level. On the other hand, assuming that the $5f$ orbitals partake in the bonding, Brint and McCaffery[77] argued that the first virtual orbital in UO_2^{2+} is a $\delta_u(e_{2u})$ level. The ground state of UO_2^{2+} is $^1\Sigma_g^+$. The low-energy bands are, after Brint and McCaffery, due to transitions to the triplet states $^3\Pi_g$ and $^3\Delta_g$. Even though MCD measurements are invoked in the support of these assignments, this system still appears to be a very puzzling one.

10. Extended Hückel Theory. Angular Overlap Model

Semiempirical molecular orbital theories have been developed which use different criteria for the estimation of the molecular orbitals than a minimization of the total energy in a Hartree–Fock scheme. Taking as a starting point the secular equation $|H_{ij} - WG_{ij}| = 0$, we can relate the H_{ij} integrals, by using various assumptions and approximations, to atomic ionization potentials. These potentials are steeply varying functions of the atomic charge and electronic configuration. In the extended Hückel method, a set of H_{ij} values is chosen corresponding to a certain electronic population of the atoms in question. The secular equation is solved and a set of molecular orbitals is obtained. The orbitals are now analyzed using a charge partitioning and it is investigated whether the calculated and assumed charge distributions agree. If not, a new set of H_{ij} values corresponding to the calculated charge distribution is chosen and the procedure is repeated, until a self-consistent atomic charge and electronic configuration is obtained. Intuitively appealing as such a procedure is, it is nevertheless of dubious theoretical value.

The background for the interest in Hückel-like methods[78–83] was that calculations aiming at even very approximate solutions to the Hartree–Fock equations for medium and large molecules and ions were completely out of question before 1965. However, the coming of the big electronic computers made it possible to solve the molecular Hartree–Fock–Roothaan equations, and the need for Hückel methods disappeared. The primary justification for the extended Hückel methods was their simplicity. Attempts to improve the procedure should therefore not be so elaborate[84] that the labor involved is comparable to doing the same calculation in a Hartree–Fock scheme. Even though the results are claimed to be comparable, one must retain some reservations as to the utility of a theoretically unjustified procedure.

The alleged aim of the angular overlap model is to parameterize the splittings of a metal d or f basis under the influence of a crystal field potential. In order to do so, an undefined quantity called orbital energy is introduced. As presented,[85,86] the method makes heavy use of advanced algebra. In discussing the basis of the theory we shall, however, take as our point of departure Eqs. (22)–(24).

It follows from Eq. (24) that the orbital splittings can be written as a sum of terms in the overlap integrals G_γ. Most noteworthy is the fact that no terms linear in G_γ appear. Calculations of Kleiner[87] showed that the term $\bar{h}(d_{e_g}) - \bar{h}(d_{t_{2g}})$ for $Cr(H_2O)_6^{3+}$ is -5500 cm^{-1}. This has the wrong sign compared with the experimental orbital splitting of $17{,}400 \text{ cm}^{-1}$.

The group overlap integral G_γ is defined by

$$G_\gamma = \int \chi_{M\gamma} \sum_i^L c_{i\gamma}\chi_{\lambda i} \, dv = \omega_\gamma S(l, \mu) \tag{67}$$

where ω_γ is a factor which relates the group overlap integral to a diatomic overlap integral $S(l, \mu)$. Clearly the manifestations of the molecular geometry are contained in the ω_γ factor. With ν the dimension of the representation γ, one can show[88]

$$\sum \nu_\gamma \omega_\gamma^2 = L \tag{68}$$

where L is the number of orbitals in the ligand set and the summation includes those representations that are shared between the metal and ligand set. For instance, in an octahedral molecule ML_6 the sigma ligand set and the $3d$ set have one common representation e_g, and the pi ligand set and the $3d$ set share t_{2g}. Hence $2\omega_{e_g}^2 = 6$ and $3\omega_{t_{2g}}^2 = 12$.

Defining the parameters

$$e'_\sigma = \tilde{w}_e S^2(d_\sigma, L_\sigma) \tag{69}$$

$$e'_\pi = \tilde{w}_t S^2(d_\pi, L_\pi) \tag{70}$$

we can write (24) as

$$10Dq = \bar{h}(d_{e_g}) - \bar{h}(d_{t_{2g}}) + 3e'_\sigma - 4e'_\pi \tag{71}$$

The characteristic octahedral "orbital energy difference" used in the angular overlap model[85,86] is, however, taken as

$$\Delta(O_h) = 3e'_\sigma - 4e'_\pi \tag{72}$$

retaining only the dominant part of (71). The use of the truncated expression (72) is equivalent to basing the parametrization on the properties of the potential $\hat{U}$ alone, Eq. (20).

In practice the e'_γ parameters are extracted from spectral data assuming that the J and K integrals can be evaluated taking $G_{ML} = 0$. The introduction of the e'_γ quantities of Eq. (72) thus focuses attention on the properties of the individual ligands. The parametrization process in the angular overlap model is consequently neither better nor worse than the one used in the pure ionic crystal field theory.

11. An Example, Ni(CN)₄²⁻. Conclusions

In order to illustrate the workings of the methods with which we have been dealing, we shall take a look at the electronic structures of $Ni(CN)_4^{2-}$. The molecular structure of the ion is square planar, D_{4h}. Measurements of the magnetic susceptibility show the complex to be diamagnetic. Looking at the group overlaps of the $3d$ orbitals with the ligand orbitals, the order of the core values of the five $3d$ orbitals are presumably[89] (note that the coordinate system used differs from that in Ref. 86)

$$\bar{h}(e_g \pi^*) < \bar{h}(b_{2g} \pi^*) < \bar{h}(a_{1g} \sigma^*) \ll \bar{h}(b_{1g} \sigma^*)$$

The ground state is therefore $(e_g)^4 (b_{2g})^2 (a_{1g})^2 \, {}^1A_{1g}$.

Single electronic excitations into the virtual b_{1g} orbital lead to the following singlet states having the indicated excitation energies:

$$a_{1g} \to b_{1g}, \; {}^1B_{1g}, \qquad \Delta W = \bar{h}(b_{1g}) - \bar{h}(a_{1g}) - 4F_2 - 15F_4 \qquad (73)$$

$$b_{2g} \to b_{1g}, \; {}^1A_{2g}, \qquad \Delta W = \bar{h}(b_{1g}) - \bar{h}(b_{2g}) - 35F_4 \qquad (74)$$

$$e_g \to b_{1g}, \; {}^1E_g, \qquad \Delta W = \bar{h}(b_{1g}) - \bar{h}(e_g) - 3F_2 - 20F_4 \qquad (75)$$

Consider the ${}^1A_{1g} \to {}^1A_{2g}$ transition. We notice that the orbitals involved in the transition are located in the plane of the molecule; they are the ψ_{xy}^* and $\psi_{x^2-y^2}^*$ orbitals. The situation is therefore formally exactly as found in octahedral diamagnetic d^6 complexes [compare Eq. (65)]. In $Co(CN)_6^{3-}$ the transition ${}^1A_{1g} \to {}^1T_{1g}$ is located[90] at 32,400 cm^{-1}, and the ${}^1A_{1g} \to {}^1T_{2g}$ transition is at 39,000 cm^{-1}. We expect the orbital separation $\bar{h}(b_{1g}) - \bar{h}(b_{2g})$ to be somewhat less in the Ni^{2+} system than in the Co^{3+} system, but how much less is of course unknown. Anyhow, 32,400 cm^{-1} is certainly an upper limit to the ${}^1A_{1g} \to {}^1A_{2g}$ transition in the $Ni(CN)_4^{2-}$ complex. From the position of the two $Co(CN)_6^{3-}$ bands we get that $F_2 = 900$ cm^{-1} and $F_4 = 100$ cm^{-1}, which accounts for the separation of the two bands [Eqs. (65) and (66)].

The packing of the $Ni(CN)_4^{2-}$ units in a crystal[91] shows a layer structure of the type *ABABAB* along the Ni–Ni axis (the crystallographic c axis). All A units occupy translationally equivalent sites, as do the B units, but A and B units may be rotated through an angle with respect to each other about the common Z(Ni–Ni) axis. In $BaNi(CN)_4 \cdot 4H_2O$ the A and B units are rotated 45°; in $SrNi(CN)_4 \cdot 5H_2O$ the units are not rotated. Using simple crystal theory, it is found that the b_{2g} orbital is the only metal orbital whose energy is sensitive to the rotation; the more rotated A and B are with respect to each other, the lower is $\bar{h}(b_{1g}) - \bar{h}(b_{2g})$.

Experiments[89,91] showed some spectral changes when comparing the absorption spectrum of $BaNi(CN)_4$ with that of $SrNi(CN)_4$. A band polarized with **E** perpendicular to the c axis showed up at 22,500 cm^{-1} in $BaNi(CN)_4$, which was absent in $Sr\,Ni(CN)_4$. We interpret the 22,500-cm^{-1} $BaNi(CN)_4$ band as the ${}^1A_{1g} \to {}^1A_{2g}$ transition. The transition is allowed as a magnetic dipole transition, and the expected polarization agrees with the observation. Hence for $BaNi(CN)_4$, $\bar{h}(b_{1g}) - \bar{h}(b_{2g}) \approx 26,000$ cm^{-1}, rising to perhaps 30,000 cm^{-1} in $SrNi(CN)_4 \cdot 5H_2O$.

Another band was seen[89] in all crystals at 24,000 cm^{-1}. It is polarized parallel to the c axis. Tentatively this band can be assigned ${}^1A_{1g} \to {}^1B_{1g}$, yielding $\bar{h}(b_{1g}) - \bar{h}(a_{1g}) = 29,000$ cm^{-1}. The observed polarization is interesting. Since A_{1g} and B_{1g} are even states, a mechanism giving vibronic intensity will have to be invoked.[5] With the vibrations present in the $Ni(CN)_4^{2-}$ unit the band should, however, appear both parallel and perpendicular to the c axis.

Knowing that it is primarily seen in π polarization, we see that the composition of the excited state must be ${}^1B_{1g} + \lambda({}^1A_{2u})$. The mixing of ${}^1B_{1g}$

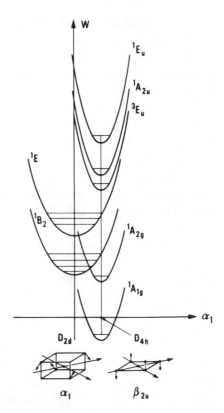

Fig. 7. Potential energy curves for $Ni(CN)_4^{2-}$. The ground state is characterized by D_{4h} symmetry. The β_{2u} vibration becomes an α_1 vibration on distorting the equilibrium geometry to D_{2d}.

with $^1A_{2u}$ must be caused by a β_{2u} vibration. This is the vibration which will take the D_{4h} symmetry of $Ni(CN)_4^{2-}$ to a D_{2d} conformation (Fig. 7). We propose that the electronic configuration that gives rise to the excited $^1B_{1g}$ state when minimized will yield a molecular equilibrium structure corresponding to a flattened tetrahedron. Such a geometry is supported by a state correlation to d^8 in a T_d crystalline field.[89] A similar deformation should occur for the ^1E state.

Looking at the polarized spectrum of $[(n-C_4H_9)_4N]_2Ni(CN)_4$ at liquid helium temperature,[92] a low-intensity band, polarized perpendicular to the c axis, is observed at 31,600 cm^{-1}. Presumably this can be identified with $^1A_{1g} \rightarrow {}^1E$. Measurements[93] of MCD also point toward the presence of a degenerate excited state around 30,000 cm^{-1}. If this is true, then $\bar{h}(e_g) - \bar{h}(b_{1g}) \approx 36,000$ cm^{-1} (Fig. 7).

Turning our attention to the charge transfer states, simple molecular orbital considerations indicated[94] that the intense, near-ultraviolet bands should be assigned as transitions from the occupied d orbitals to an $a_{2u}(\pi^*CN)$ virtual orbital. Calculating the transition energy as the difference of the energies of the two states individually minimized, Demuynck and Veillard[19] found $\cdots (a_{1g})^2(a_{2u})^0 \, {}^1A_{1g} \rightarrow (a_{1g})^1(a_{2u})^1 \, {}^1A_{2u}$ placed at 33,900 cm^{-1}. Incidentally, if the formula (52) was used, the same transition was expected at $80,330 \text{ cm}^{-1}$.

Experimentally,[92] a π-polarized band is found at $34,400 \text{ cm}^{-1}$. However, a second π-polarized band turns up at $35,840 \text{ cm}^{-1}$. The finding of two intense π-polarized bands can be understood in terms of allowed transitions to two spin–orbit mixed states arising from the mixing of $(^1A_{2u}, {}^3E_u)$. Such an assignment was indeed proposed relying on careful MCD measurements.[95] Finally, the strong σ polarization of the $36,000\text{-cm}^{-1}$ band points toward a $^1A_{1g}$–^1E_u transition. MCD observations again substantiate[95] this assignment.

There are still many facets of the Ni(CN)_4^{2-} spectrum which are less well understood. However, we believe that the above analysis has succeeded in throwing some light on the identity of some of the "crystal field" and "charge transfer" states.

In summarizing what we have learned, the most important point is probably the realization that the result of an approximate ground-state calculation is rarely of much interest. The many parameters of INDO and related methods do not add to the reliability of the result. The evaluation of the transfer of spin densities from the metal to the ligands is, as far as I can see, the most valuable information the molecular orbital methods have so far given us.

Turning to the excited states, approximate molecular orbital calculations can give us ideas to test. Indeed, the identifications of excited states is normally reached by a cross-fertilization of theory and experiments. Only the very best calculations, in which both ground and excited states are minimized independently, can be trusted, however, when it comes to numerical identifications. The crystal field level diagrams, when applicable to simple octahedral and tetrahedral molecules, are here quite unique in offering reliable identifications of both ground and excited states. They should, however, not be abused, and in particular the inclusion of low field effects seldom leads to happy results.

References

1. H. Bethe, *Ann. Phys.* **5**, 133–208 (1929).
2. J. H. Van Vleck, *J. Chem. Phys.* **3**, 807–813 (1935).
3. John P. Fackler, Jr. (ed.), *Symmetry in Chemical Theory*, Dowden, Hutchinson and Ross, Stroudsburg, Pennsylvania (1973).
4. P. G. Lykos and R. G. Parr, *J. Chem. Phys.* **24**, 1166–1173 (1956).
5. C. J. Ballhausen, *Introduction to Ligand Field Theory*, McGraw-Hill, New York (1962).
6. *J. A. Pople and D. L. Beveridge, Approximate Molecular Orbital Theory*, McGraw-Hill, New York (1970).
7. A. T. Amos, *Mol. Phys.* **5**, 91–104 (1962).
8. J. A. Pople, D. P. Santry, and G. A. Segal, *J. Chem. Phys.* **43**, S129–S135 (1965).
9. J. P. Dahl and C. J. Ballhausen, *Adv. Quant. Chem.* **4**, 170–226 (1968).
10. F. J. Corbato and A. C. Switendick, in *Methods in Computational Physics* (B. Alder, S. Fernbach, and M. Rotenburg, eds.), Academic Press, New York and London (1963), Vol. 2, pp. 155–179.

11. M.-M. Coutière, J. Demuynck, and A. Veillard, *Theor. Chim. Acta* **27**, 281–287 (1972).
12. M.-M. Rohmer, J. Demuynck, and A. Veillard, *Theor. Chim. Acta* **36**, 93–102 (1974).
13. B. D. Bird and P. Day, *J. Chem. Phys.* **49**, 392–403 (1968).
14. J. P. Dahl and H. Johansen, *Theor. Chim. Acta* **11**, 8–25 (1968).
15. H. Johansen, *Chem. Phys. Lett.* **17**, 569–573 (1972).
16. C. A. L. Becker and J. P. Dahl, *Theor. Chim. Acta* **14**, 26–38 (1969).
17. D. R. Truax, J. A. Geer, and T. Ziegler, *J. Chem. Phys.* **59**, 6662–6666 (1973).
18. T. Ziegler, *Acta Chem. Scand.* **A28**, 29–36 (1974).
19. J. Demuynck and A. Veillard, *Theor. Chim. Acta* **28**, 241–265 (1973).
20. D. A. Copeland and C. J. Ballhausen, *Theor. Chim. Acta* **20**, 317–330 (1971).
21. D. A. Copeland, *Theor. Chim. Acta* **32**, 41–47 (1973).
22. D. R. Truax, J. A. Geer, and T. Ziegler, *Theor. Chim. Acta* **33**, 299–306 (1974).
23. D. E. Ellis and T. Parameswaran, *Int. J. Quant. Chem. Symp.* **5**, 443–449 (1971).
24. T. Parameswaran and D. E. Ellis, *J. Chem. Phys.* **58**, 2088–2095 (1973).
25. A. Rauk, T. Ziegler, and D. E. Ellis, *Theor. Chim. Acta* **34**, 49–59 (1974).
26. R. D. Brown and K. R. Roby, *Theor. Chim. Acta* **16**, 175–216 (1970).
27. R. D. Brown and K. R. Roby, *Theor. Chim. Acta* **16**, 278–302 (1970).
28. R. D. Brown and P. G. Burton, *Theor. Chim. Acta.* **18**, 309–328 (1970).
29. W. Th. A. M. van der Lugt, *Int. J. Quant. Chem.* **6**, 859–880 (1972).
30. D. W. Clack, *J. Chem. Soc. Faraday II* **68**, 1672–1678 (1972).
31. D. W. Clack, N. S. Hush, and J. R. Yandle, *J. Chem. Phys.* **57**, 3503–3510 (1972).
32. A. Serafini, J.-M. Savariault, P. Cassoux, and J.-F. Labarre, *Theor. Chim. Acta* **36**, 241–247 (1975).
33. D. W. Clack, *Mol. Phys.* **27**, 1513–1519 (1974).
34. D. W. Clack and W. Smith, *Theor. Chim. Acta* **36**, 87–92 (1974).
35. D. W. Clack and W. Smith, *J. Chem. Soc. Dalton* **1974**, 2015–2020.
36. J. H. van Santen and J. S. van Wieringen, *Rec. Trav. Chim. Pays Bas* **71**, 420–430 (1952).
37. P. O'Donnell Offenhartz, *J. Am. Chem. Soc.* **91**, 5699–5704 (1969).
38. P. O'Donnell Offenhartz, *J. Am. Chem. Soc.* **92**, 2599–2602 (1970).
39. J. W. Richardson, D. M. Vaught, T. F. Soules, and R. R. Powell, *J. Chem. Phys.* **50**, 3633–3634 (1969).
40. C. J. Ballhausen, in *Spectroscopy in Inorganic Complexes* (C. N. R. Rao and J. R. Ferraro, eds.), Academic Press, New York and London (1970), Vol. 1, pp. 1–28.
41. P. J. Stephens, *Ann. Rev. Phys. Chem.* **25**, 201–232 (1974).
42. R. D. Brown, B. H. James, M. F. O'Dwyer, and K. R. Roby, *Chem. Phys. Lett.* **1**, 459–464 (1967).
43. C. J. Ballhausen and A. D. Liehr, *J. Mol. Spect.* **2**, 342–360 (1958).
44. P. N. Schatz, A. J. McCaffery, W. Suëtaka, G. N. Henning, A. B. Ritchie, and P. J. Stephens, *J. Chem. Phys.* **45**, 722–734 (1966).
45. C. J. Ballhausen, *Theor. Chim. Acta* **1**, 285–293 (1963).
46. C. J. Ballhausen, J. P. Dahl, and I. Trabjerg, *Colloques Int. Cent. Natn. Rech. Scient.* **191**, 69–72 (1970).
47. L. W. Johnson and S. P. McGlynn, *Chem. Phys. Lett.* **10**, 595–599 (1971).
48. L. W. Johnson, E. Hughes, Jr., and S. P. McGlynn, *J. Chem. Phys.* **55**, 4476–4480 (1971).
49. C. J. Ballhausen and I. Trabjerg, *Mol. Phys.* **24**, 689–693 (1972).
50. E. U. Condon and G. H. Shortley, *The Theory of Atomic Spectra*, Cambridge University Press (1935).
51. E. König and S. Kremer, *Theor. Chim. Acta* **23**, 12–20 (1971).
52. E. König and S. Kremer, *Theor. Chim. Acta* **26**, 311–320 (1972).
53. C. J. Ballhausen and W. Moffitt, *J. Inorg. Nucl. Chem.* **3**, 178–181 (1956).
54. R. Finkelstein and J. H. Van Vleck, *J. Chem. Phys.* **8**, 790–797 (1940).
55. F. E. Ilse and H. Hartmann, *Z. Naturforsc.* **6a**, 751–754 (1951).
56. Y. Tanabe and S. Sugano, *J. Phys. Soc. Japan* **9**, 753–766 (1954).
57. Y. Tanabe and S. Sugano, *J. Phys. Soc. Japan* **9**, 766–779 (1954).
58. L. E. Orgel, *J. Chem. Phys.* **23**, 1004–1014 (1955).

59. A. D. Liehr and C. J. Ballhausen, *Ann. Phys. N.Y.* **2**, 134–155 (1959).
60. A. D. Liehr, *J. Phys. Chem.* **67**, 1314–1328 (1963).
61. R. M. Macfarlane, *J. Chem. Phys.* **39**, 3118–3126 (1963).
62. P. J. McCarthy and M. T. Vala, *Mol. Phys.* **25**, 17–34 (1973).
63. J. Jortner and G. C. Morris, *J. Chem. Phys.* **51**, 3689–3691 (1969).
64. W. Moffitt, *J. Chem. Phys.* **25**, 1189–1198 (1956).
65. J.-P. Mathieu, *J. Chim. Physique* **33**, 78–98 (1936).
66. J. H. Van Vleck, *J. Chem. Phys.* **8**, 787–789 (1940).
67. H. U. Güdel, I. Trabjerg, M. Vala, and C. J. Ballhausen, *Mol. Phys.* **24**, 1227–1232 (1972).
68. L. R. Ingersoll, P. Rudnick, F. G. Slack, and N. Underwood, *Phys. Rev.* **57**, 1145–1153 (1940).
69. M. H. L. Pryce, G. Agnetta, T. Garofano, M. B. Palma-Vittorelli, and M. U. Palma, *Phil. Mag.* **10**, 477–496 (1964).
70. E. I. Solomon and C. J. Ballhausen, *Mol. Phys.* **29**, 279–299 (1975).
71. W. Moffitt, G. L. Goodman, M. Fred, and B. Weinstock, *Mol. Phys.* **2**, 109–122 (1959).
72. B. Weinstock and G. L. Goodman, *Adv. Chem. Phys.* **9**, 169–319 (1965).
73. C. J. Ballhausen, *Theor. Chim. Acta* **24**, 234–240 (1972).
74. J. C. Eisenstein and M. H. L. Pryce, *Proc. Roy. Soc. A* **255**, 181–198 (1960).
75. J. C. Eisenstein and M. H. L. Pryce, *Proc. Roy. Soc. A* **229**, 20–38 (1955).
76. J. C. Eisenstein and M. H. L. Pryce, *Proc. Roy. Soc. A* **238**, 31–45 (1956).
77. P. Brint and A. J. McCaffery, *Mol. Phys.* **25**, 311–322 (1973).
78. M. Wolfsberg and L. Helmholz, *J. Chem. Phys.* **20**, 837–843 (1952).
79. C. J. Ballhausen and H. B. Gray, *Inorg. Chem.* **1**, 111–122, (1962).
80. H. B. Gray and N. A. Beach, *J. Am. Chem. Soc.* **85**, 2922–2927 (1963).
81. H. Basch, A. Viste, and H. B. Gray, *J. Chem. Phys.* **44**, 10–19 (1966).
82. R. F. Fenske and C. C. Sweeney, *Inorg. Chem.* **3**, 1105–1113 (1964).
83. R. F. Fenske and D. D. Radtke, *Inorg. Chem.* **7**, 479–487 (1968).
84. A. Dutta-Ahmed and E. A. Boudreux, *Inorg. Chem.* **12**, 1590–1602 (1973).
85. C. E. Schäffer, in *Structure and Bonding* (J. D. Dunitz *et al.*, eds.), Springer Verlag, Berlin (1973), Vol. 14, pp. 69–110.
86. C. E. Schäffer, in *Wave Mechanics—The First Fifty Years* (W. C. Price, S. S. Chissick, and T. Ravensdale, eds.), Butterworths, London (1973), pp. 174–192.
87. W. H. Kleiner, *J. Chem. Phys.* **20**, 1784–1791 (1952).
88. C. J. Ballhausen and J. P. Dahl, *Theor. Chim. Acta* **34**, 169–174 (1974).
89. C. J. Ballhausen, N. Bjerrum, R. Dingle, K. Eriks, and C. R. Hare, *Inorg. Chem.* **4**, 514–518 (1965).
90. J. Brigando, *Bull. Soc. Chim. France* **1957**, 503–516.
91. J. P. Dahl, R. Dingle, and M. T. Vala, *Acta. Chem. Scand.* **23**, 47–55 (1969).
92. C. D. Cowman, C. J. Ballhausen, and H. B. Gray, *J. Am. Chem. Soc.* **95**, 7873–7875 (1973).
93. P. J. Stephens, A. J. McCaffery, and P. N. Schatz, *Inorg. Chem.* **7**, 1923–1925 (1968).
94. H. B. Gray and C. J. Ballhausen, *J. Am. Chem. Soc.* **85**, 260–265 (1963).
95. S. P. Piepho, P. N. Schatz, and A. J. McCaffery, *J. Am. Chem. Soc.* **91**, 5994–6001 (1969).

Approximate Molecular Orbital Theory of Nuclear and Electron Magnetic Resonance Parameters

David L. Beveridge

1. Introduction

Magnetic resonance experiments are currently one of the most important and widely available means for studying the molecular and electronic structure of molecules. Both the analysis and interpretation of magnetic resonance spectra are based on quantum mechanics, and magnetic resonance research has evolved with an especially constructive interaction between experimentalists and theorists. This chapter deals particularly with the interpretation of nuclear magnetic resonance (NMR) spectra and electron spin resonance spectra (ESR) based on quantum mechanics and particularly molecular orbital calculations.

Electron and magnetic nuclei exhibit a net spin angular momentum characterized by a vector operator $\mathbf{M}$, which obeys the general family of relationships of all angular momentum operators in quantum mechanics.[1] As with all bound quantum mechanical particles in motion, a spin system exists in one or another of discrete energy states. These spin states of the system are identified by quantum numbers M and M_z, and have eigenfunctions $|M, M_z\rangle$ such that

$$\mathbf{M}^2|M, M_z\rangle = M(M+1)|M, M_z\rangle \tag{1}$$

David L. Beveridge • Chemistry Department, Hunter College of the City University of New York, New York

and

$$\mathbf{M}_z |M, M_z\rangle = M_z |M, M_z\rangle \tag{2}$$

where M can be zero, integral, or half-integral, and M_z can take the $2M+1$ values $M, M-1, \ldots, -M$.

The spin states of a system are degenerate in the absence of an external perturbation. The presence of a magnetic field $\mathbf{H}$ resolves the degeneracy and induces a net alignment of magnetic spins in a preferred direction corresponding to one of the allowed quantum mechanical energy states E of the system, where E is given by the expression

$$E = -\boldsymbol{\mu} \cdot \mathbf{H} \tag{3}$$

where $\boldsymbol{\mu}$ is the magnetic moment representative of the spin state. The spin magnetic moment is defined as

$$\mu_M = \gamma \hbar M_z = g\beta M_z \tag{4}$$

where γ is the gyromagnetic ratio of the spinning particle, $\hbar$ is Planck's constant divided by 2π, g is the g-value characteristic of the particle, and β is the corresponding Bohr magneton.

The $2M+1$ values of M_z give rise to a manifold of spin states for the system. The presence of a secondary oscillating magnetic field of the appropriate frequency ν can induce transitions between spin states. The magnetic excitation energies consistent with the quantum mechanical selection rule $\Delta M_z = \pm 1$ can be precisely determined,

$$\Delta E = \gamma \hbar H = g\beta H = h\nu \tag{5}$$

While the analysis up to here applies to isolated spins, the magnitude of these transition energies for spin states of a molecule carries detailed information about the chemical environment of the spinning particle. This information can be used to deduce the relative position of magnetic nuclei and hence molecular structure, and also the distribution and polarizability of electrons in the vicinity of magnetic nuclei and hence electronic structure.

The analysis and interpretation of magnetic resonance spectra can conveniently be considered as two separate but of course related quantum mechanical problems. The calculation of energy levels for a molecule in a magnetic field follows from the full quantum mechanical Hamiltonian operator, including, in addition to the zeroth-order kinetic and potential energy terms, the spin and magnetic field explicitly. However, since magnetic spectral transitions arise from induced changes in the spin state of the system, magnetic resonance spectra are customarily analyzed in terms of a spin Hamiltonian. The spin

Hamiltonian arises from replacing the full Hamiltonian by an effective operator acting on purely spin space. The spin Hamiltonian includes the external magnetic field and spin operators explicitly together with certain parameters, referred to as *magnetic resonance parameters*, chosen so as to make the aforementioned replacement possible. Expressions for the magnetic energy levels of the system can be worked out in terms of the magnetic resonance parameters. Magnetic resonance spectral analysis can thus be reduced to specification of the appropriate values of the magnetic resonance parameters. The interpretation of magnetic resonance spectra is concerned with understanding the magnitudes and signs of magnetic resonance parameters in terms of the molecular electronic structure of the system. The magnetic resonance parameters relevant to this chapter are the chemical shift and nuclear spin coupling constants of nuclear magnetic resonance spectroscopy and the *g*-value shift and hyperfine constants of electron spin resonance spectroscopy.

The theoretical calculation of magnetic resonance parameters is currently a topic of central interest in theoretical chemistry. Reliable theoretical estimates of the magnitudes and signs of magnetic resonance parameters can be a valuable aid in band assignments and particularly analysis of complex spectra. The major route to spectral interpretation is via the formal quantum mechanical expressions for magnetic resonance parameters and calculations based on these expressions. The development of this area of research drew significantly from both the valence bond theory and molecular orbital theory for quantum mechanical electronic wave functions, and the history of this development has been described in several current textbooks and recent review articles. Molecular orbital theory, together with quantum mechanical perturbation theory, has emerged as the major computational approach to magnetic resonance parameters in general, particularly following the development of valence electron self-consistent field molecular orbital theory in the mid-1960s. This chapter is concerned primarily with approximate molecular orbital theory of magnetic resonance parameters and includes relevant background material, an overview of methodology, an assessment of the capabilities and limitations of current calculations, and a survey of selected applications to chemical problems.

2. Magnetic Resonance Parameters

The purpose of this section is to develop quantum mechanical expressions for the magnetic resonance parameters. To proceed we focus on a particular spin state of the system and introduce for **M** specifically the nuclear spin angular momentum vector $\boldsymbol{\mu}_N$ for nucleus N and the electron spin angular momentum vector $\boldsymbol{\mu}_e$. The energy of a molecule in an external field H, including all terms relevant to nuclear and electron magnetic resonance, is written as

$$E(\mathbf{H}, \boldsymbol{\mu}_N, \boldsymbol{\mu}_e) = E_0 - \boldsymbol{\gamma} \cdot \mathbf{H} - \tfrac{1}{2} \mathbf{H} \cdot \boldsymbol{\chi} \cdot \mathbf{H} \tag{6a}$$

$$-\sum_N \boldsymbol{\mu}_N \cdot \mathbf{H} + \sum_N \boldsymbol{\mu}_N \cdot \boldsymbol{\sigma}_N \cdot \mathbf{H}$$

$$+\tfrac{1}{2} \sum_M \sum_N \boldsymbol{\mu}_N \cdot \mathbf{K}_{MN} \cdot \boldsymbol{\mu}_N \tag{6b}$$

$$-\boldsymbol{\mu}_e \cdot \mathbf{H} + \boldsymbol{\mu}_e \cdot \boldsymbol{\Delta} \mathbf{g} \cdot \mathbf{H} + \sum_N \boldsymbol{\mu}_e \cdot \mathbf{T}_N \cdot \boldsymbol{\mu}_N \tag{6c}$$

where $\boldsymbol{\gamma}$ is the permanent magnetic moment of the system, $\boldsymbol{\chi}$ is the magnetic susceptibility tensor, $\boldsymbol{\sigma}_N$ is the magnetic shielding tensor for nucleus N, $\mathbf{K}_{MN}$ is the nuclear spin–spin coupling tensor for nuclei N and M, $\boldsymbol{\Delta}\mathbf{g}$ is the differential g-tensor, and $\mathbf{T}_N$ is the electron–nuclear hyperfine tensor. The equation fragment (6b) is relevant to nuclear magnetic resonance and (6c) is relevant to electron magnetic resonance. In an effort to maintain a parallel development of nuclear and electron magnetic resonance, Eq. (6) is formulated in terms of so-called "reduced" magnetic resonance parameters. These quantities usually require some transformation for direct comparison with experimental data.

The quantum mechanical nature of the magnetic resonance parameters may be determined by comparing the spin Hamiltonian with the energy expectation value of the complete Hamiltonian operator and matching terms bi-linear in the field and spin terms. The complete Hamiltonian operator is

$$\mathcal{H}(\mathbf{H}, \boldsymbol{\mu}_N, \boldsymbol{\mu}_e) = \mathcal{H}^0 + \mathcal{H}^{(100)} \cdot \mathbf{H} + \tfrac{1}{2} \mathbf{H} \cdot \mathcal{H}^{(200)} \cdot \mathbf{H}$$

$$+\sum_N \boldsymbol{\mu}_N \cdot \mathcal{H}_N^{(010)} + \tfrac{1}{2} \sum_N \boldsymbol{\mu}_N \cdot \mathcal{H}^{(110)} \cdot \mathbf{H}$$

$$+\tfrac{1}{2} \sum_M \sum_N \boldsymbol{\mu}_M \cdot \mathcal{H}^{(020)} \cdot \boldsymbol{\mu}_N$$

$$+\sum_i \boldsymbol{\mu}_{ei} \cdot \mathcal{H}_i^{(001)} + \sum_i \boldsymbol{\mu}_{ei} \cdot \mathcal{H}_i^{(101)} \cdot \mathbf{H}$$

$$+\sum_i \sum_N \boldsymbol{\mu}_{ei} \cdot \mathcal{H}_{iN}^{(011)} \cdot \boldsymbol{\mu}_N \tag{7}$$

where the $\mathcal{H}^{(nlm)}$ are operators representing interactions n-linear in H, l-linear in μ_N, and m-linear in μ_e. The derivations of the $\mathcal{H}^{(nlm)}$ are discussed in detail in various textbooks on magnetism and magnetic resonance.[2] The resulting expressions, with t, u, and v representing Cartesian vector and tensor components, are

$$\mathcal{H}^0 = -\sum_i \left(\frac{\hbar^2}{2m} \nabla_i^2 + \sum_N Z_N e^2 r_{iN}^{-1} \right) + \sum_{i<j} e^2 r_{ij}^{-1} \tag{8}$$

$$\mathcal{H}_t^{(100)} = \frac{e\hbar}{2mc} \sum_i L_{it} \tag{9}$$

$$\mathcal{H}_{tu}^{(200)} = \frac{e^2}{8mc^2}\sum_i (r_i^2 \delta_{tu} - r_{it}r_{iu}) \tag{10}$$

$$\mathcal{H}_{Nt}^{(010)} = \frac{e}{mc}\sum_i L_{iNt}r_{Ni}^{-3} \tag{11}$$

$$\mathcal{H}_{Ntu}^{(110)} = \frac{e^2}{2mc^2}\sum_i (\mathbf{r}_i \cdot \mathbf{r}_{iN}\,\delta_{tu} - r_{it}r_{jNu})r_{iN}^{-3} \tag{12}$$

$$\mathcal{H}_{MNtu}^{(020)} = (r_{MN}^2 \delta_{MN} - r_{MNt}r_{MNu})r_{MN}^{-5} + \frac{e^2}{2mc^2}\sum_i r_{iMt}r_{iNu}r_{iM}^{-3}r_{iN}^{-3} \tag{13}$$

$$\mathcal{H}_{it}^{(001)} = \frac{e\hbar}{2mc}\sum_N Z_N L_{Nit}r_{Ni}^{-3} \tag{14}$$

$$\mathcal{H}_{itu}^{(101)} = \frac{e^2}{4\hbar mc}\sum_N Z_N(\mathbf{r}_{iN}\cdot\mathbf{r}_i\delta_{tu} - r_{it}r_{iNu})r_{iN}^{-3} \tag{15}$$

$$\mathcal{H}_{Nitu}^{(011)} = \frac{8\pi}{3}\delta(\mathbf{r}_{iN}) + (r_{iN}^2\delta_{tu} - 3r_{iNt}r_{iNu})r_{iN}^{-5} \tag{16}$$

where c is the speed of light, m and e are electronic mass and charge, respectively, ∇_i is the gradient operator on electron i, Z_N is the nuclear charge, $\mathbf{r}_{iN}$, $\mathbf{r}_{ij}$, and $\mathbf{r}_{MN}$ are electron–nucleus, electron–electron, and nucleus–nucleus distances, respectively, and $\delta(\mathbf{r}_{iN})$ is the Dirac delta function. The symbol L refers to the orbital angular momentum operators

$$L_{it} = i\hbar(\mathbf{r}_i \times \nabla_i)_t \tag{17}$$

and

$$L_{iNt} = -i\hbar(\mathbf{r}_{iN} \times \nabla_i)_t \tag{18}$$

Also γ_N is the gyromagnetic ratio of nucleus N, g is the free electronic g-value 2.0023, and

$$\boldsymbol{\mu}_e = \sum_i \boldsymbol{\mu}_{ei} \tag{19}$$

where $\boldsymbol{\mu}_{ei}$ is the spin vector operator for electron i. Sums over lower-case latin letters are over electrons and sums over upper-case characters are over nuclei. The physical significance of each operator will be discussed as they appear in the calculation of the various magnetic resonance parameters.

An initial development of expressions for the magnetic resonance parameters σ_N and $\mathbf{K}_{MN}$ follows from a Taylor series expansion of $\Psi(\mathbf{H}, \boldsymbol{\mu}_N)$ around zero-field values of $\mathbf{H}$ and the $\boldsymbol{\mu}_N$,

$$\Psi(\mathbf{H}, \boldsymbol{\mu}_N) = \Psi^0 + \Psi^{(100)}\cdot\mathbf{H} + \sum_N \Psi_N^{(010)}\cdot\boldsymbol{\mu}_N + \cdots \tag{20}$$

where the $\Psi^{(nlm)}$ are expansion coefficients. Using Eqs. (7) and (20) to compute

$$E(\mathbf{H}, \boldsymbol{\mu}_N) = \langle\Psi(\mathbf{H}, \boldsymbol{\mu}_N)|\mathcal{H}(\mathbf{H}, \boldsymbol{\mu}_N)|\Psi(\mathbf{H}, \boldsymbol{\mu}_N)\rangle \tag{21}$$

and comparing with Eq. (6) leads to quantum mechanical expressions for the cartesian components of the nuclear magnetic resonance parameters as

$$\sigma_{Ntu} = \langle \Psi^0 | \mathcal{H}_{Ntu}^{(110)} | \Psi^0 \rangle + \langle \Psi^0 | \mathcal{H}_{Nt}^{(010)} | \Psi_u^{(100)} \rangle + \langle \Psi^0 | \mathcal{H}_t^{(100)} | \Psi_{Nu}^{(010)} \rangle + \cdots \quad (22)$$

and

$$K_{MNtu} = \langle \Psi^0 | \mathcal{H}_{MNtu}^{(020)} | \Psi^0 \rangle + \langle \Psi^0 | \mathcal{H}_{Mt}^{(010)} | \Psi_{Nu}^{(010)} \rangle + S_t^{-1} \left\langle \Psi^0 \left| \sum_i s_{it} \mathcal{H}_{Nitu}^{(011)} \right| \Psi_M^{(011)} \right\rangle + \cdots \tag{23}$$

where only the leading contributions are explicitly retained. Note that alternatively these parameters can be formally obtained as second derivatives of the energy as

$$\sigma_{Ntu} = [\partial^2 E(\mathbf{H}, \boldsymbol{\mu}_N) / \partial H_t \, \partial \mu_{Nu}]_{\mathbf{H} = \boldsymbol{\mu}_N = 0} \tag{24}$$

and

$$K_{MNtu} = [\partial^2 E(\boldsymbol{\mu}_M, \boldsymbol{\mu}_N) / \partial \mu_{Mt} \, \partial \mu_{Nu}]_{\boldsymbol{\mu}_M = \boldsymbol{\mu}_N = 0} \tag{25}$$

Equations (22) and (23) lead to the calculation of nuclear magnetic resonance parameters using sum-over-states perturbation theory, and Eqs. (24) and (25) lead to calculation of parameters using finite perturbation theory.

A parallel development of electron magnetic resonance parameters encounters certain complications, but the essential features emerge by using Eqs. (7) and (20) to develop an energy expression and selecting terms bilinear in $\mathbf{H}$ and $\boldsymbol{\mu}_e$ to provide an expression for $\Delta \mathbf{g}$ and terms bilinear in $\boldsymbol{\mu}_N$ and $\boldsymbol{\mu}_e$ for $\mathbf{T}_N$. The resulting expressions are

$$\Delta g_{tu} = S_t^{-1} \left\{ \left\langle \Psi^0 \left| \sum_i s_{it} \mathcal{H}_{itu}^{(101)} \right| \Psi^0 \right\rangle + \left\langle \Psi^0 \left| \sum_i s_{it} \mathcal{H}_{it}^{(001)} \right| \Psi_u^{(100)} \right\rangle + \cdots \right\} \tag{26}$$

and

$$T_{Ntu} = S_t^{-1} \left\{ \left\langle \Psi^0 \left| \sum_i s_{it} \mathcal{H}_{Ntu}^{(011)} \right| \Psi^0 \right\rangle + \left\langle \Psi^0 \left| \sum_i s_{it} \mathcal{H}_{it}^{(001)} \right| \Psi_{Nu}^{(010)} \right\rangle + \cdots \right\} \tag{27}$$

The rigorous development of these types of expressions has been given by McWeeney.[3] Note that alternatively we have, analogous to Eqs. (24) and (25),

$$\Delta g_{tu} = [\partial^2 E(\mathbf{H}, \boldsymbol{\mu}_e) / \partial H_t \, \partial \mu_{eu}]_{\mathbf{H} = \boldsymbol{\mu}_e = 0} \tag{28}$$

and

$$T_{Ntu} = [\partial^2 E(\boldsymbol{\mu}_N, \boldsymbol{\mu}_e) / \partial \mu_{Nt} \, \partial \mu_{eu}]_{\boldsymbol{\mu}_N = \boldsymbol{\mu}_e = 0} \tag{29}$$

The remainder of this chapter deals with the calculation of $\boldsymbol{\sigma}_N$, $\mathbf{K}_{MN}$, $\Delta \mathbf{g}$, and $\mathbf{T}_N$ using the various alternative perturbation theories and approximate molecular orbital theory. The following section gives an overview of molecular orbital theory and perturbation theory applicable to magnetic resonance parameters in general. In the final four sections each of the magnetic resonance parameters are discussed individually.

3. Molecular Quantum Mechanics[4]

Here we review briefly aspects of molecular orbital theory and perturbation theory relevant to the calculation of magnetic resonance parameters. Each of the topics discussed in this section is taken up in more detail elsewhere in this volume, and this presentation is included essentially for the introduction of terminology and notation. The topics treated here are molecular orbital theory, approximate molecular orbital methodology, and perturbation theory.

The quantum theoretical calculation of each of the magnetic resonance parameters follows from a knowledge of the electronic wave function for the system. The various electronic energy states of a 1-, 2-, ..., n-electron quantum mechanical system are characterized by an energy E and wave function Ψ such that Schrödinger's equation,

$$\mathscr{H}^0(1, 2, .., n)\Psi^0(1, 2, \ldots, n) = E^0\Psi^0(1, 2, \ldots, n) \tag{30}$$

is satisfied. The wave function Ψ^0 is an eigenfunction of $\mathscr{H}^0$ and each of the operators which commute with $\mathscr{H}^0$, particularly $\mathbf{S}^2$ and $\mathbf{S}_z$, viz.

$$\mathbf{S}^2\Psi^0 = S(S+1)\Psi^0 \tag{31}$$

$$\mathbf{S}_z\Psi^0 = S_z\Psi^0 \tag{32}$$

For a many-electron, polyatomic molecule the Schrödinger equation is normally unsolvable and molecular wave functions for practical purposes are developed in approximate form.

3.1. Molecular Orbital Theory

The Schrödinger equation is generally unsolvable due to the presence of terms in $\mathscr{H}$ involving interelectronic distances r_{ij} which prevent separation of variables. If the Schrödinger equation were separable, solutions could be developed in the form

$$\Psi^0(1, 2, \ldots, n) = \psi_1(1)\psi_2(2) \cdots \psi_m(n) \tag{33}$$

The spectroscopic evidence for the shell structure of atoms and molecules led the early quantum theorists to believe that on approximate solution of this form could be suitable for a description of the electronic structure of a many-electron system. Thus molecular orbital theory adopts the functional form of Eq. (33) as a working approximation, identifying the one-electron functions ψ_i as molecular orbitals. Self-consistent field molecular orbital theory prescribes the determination of molecular orbitals for an optimum approximation to the many-electron wave function.

A molecular orbital wave function must be contrived to accommodate antisymmetry and electron spin. The quantum mechanical antisymmetry principle is satisfied by taking a determinantal product of one-electron functions rather than a simple product as in Eq. (33). Since $\mathcal{H}^0$ is spin independent, electron spin is a separable quantity and may be included by adding a spin function α or β to each of the spatial orbitals. It is notationally conventional to define molecular spin orbitals as

$$\psi_i(1) = \psi_i(1)\alpha(1) \tag{34}$$

and

$$\overline{\psi_i(1)} = \psi_i(1)\beta(1) \tag{35}$$

The spatial orbital ψ_i can be used once with α spin and with β spin in a single many-electron product wave function with no violation of the Pauli exclusion principle. This is done for each spatial orbital in a spin-restricted molecular orbital wave function for a closed-shell ground state, which for a $2n$-electron system can be written

$$\Phi^r(1, 2, \ldots, 2n-1, 2n) = |\psi_1(1)\overline{\psi_1(2)} \cdots \psi_n(2n-1)\overline{\psi_n(2n)}| \tag{36}$$

A spin-restricted molecular orbital wave function for an odd-electron, open-shell system with $2n+1$ electrons is

$$\Phi^r(1, 2, \ldots, 2n-1, 2n, 2n+1)$$

$$= |\psi_1(1)\overline{\psi_1(2)} \cdots \psi_n(2n-1)\overline{\psi_n(2n)}\psi_{n+1}(2n+1)| \tag{37}$$

where there are n doubly occupied spatial orbitals and one singly occupied spatial orbital ψ_{n+1}. A spin-restricted wave function for odd-electron systems carries a special disadvantage in the calculation of spin properties such as magnetic resonance parameters; this will be taken up in more detail later.

The determination of the optimum molecular orbitals for a closed-shell system described by Eq. (36) is based on the quantum mechanical variational principle, whereby the energy expectation value of any approximate wave function can be shown to be greater than the true energy of the system,

$$\langle \Phi | \mathcal{H}^0 | \Phi \rangle \geq \langle \Psi^0 | \mathcal{H}^0 | \Psi^0 \rangle \tag{38}$$

The optimum approximate wave function of a particular form can then be found by energy minimization,

$$\delta \langle \Phi | \mathcal{H}^0 | \Phi \rangle = 0 \tag{39}$$

The determination of the optimum molecular orbitals follows a method due to Roothaan.[5] The molecular orbitals are first expanded as linear combinations of a basis set of atomic orbitals ϕ_μ centered on constituent atoms,

$$\psi_i = \sum_\mu^m c_{\mu i}\phi_\mu \tag{40}$$

where the $c_{\mu i}$ are linear expansion coefficients. The basis set expansion can be carried out to any accuracy desired compatible with the numerical tractability. Here we distinguish as alternatives a valence basis set, consisting of valence orbitals associated with each atom, and a minimal basis set, consisting of the least number of orbitals on each atom required to describe the atomic ground state (usually the valence orbitals and inner shell orbitals), and refer to any basis with a greater number of functions as an extended basis.

With the molecular orbitals expanded as linear combinations of atomic orbitals, the calculation of the optimum molecular orbitals for the system reduces to determination of the matrix of linear expansion coefficients. Using Eq. (40) and substituting Eq. (36) into Eq. (37) subject to orthonormality constraints among the molecular orbitals, we arrive at the Roothaan equations for the $c_{\mu i}$,

$$\sum_{\nu=1}^{m} (F_{\mu\nu} - \varepsilon_i S_{\mu\nu}) c_{\mu i} = 0, \qquad \mu = 1, m \tag{41}$$

where $S_{\mu\nu}$ is the overlap integral given by

$$S_{\mu\nu} = \langle \phi_\mu | \phi_\nu \rangle \tag{42}$$

and $F_{\mu\nu}$ is the Fock element

$$F_{\mu\nu} = H_{\mu\nu}^{\text{core}} + \sum_{\lambda\sigma} P_{\lambda\sigma} [(\mu\nu|\lambda\sigma) - \tfrac{1}{2}(\mu\sigma|\lambda\nu)] \tag{43}$$

Here $H_{\mu\nu}$ are one-electron core integrals

$$H_{\mu\nu}^{\text{core}} = \left\langle \phi_\mu(1) \left| \frac{\hbar}{2m} \nabla^2 - \sum_N Z_N e^2 r_N \right| \phi_\nu(1) \right\rangle \tag{44}$$

and the $(\mu\nu|\lambda\sigma)$ are two-electron repulsion integrals of the form

$$(\mu\nu|\lambda\sigma) = \langle \phi_\mu(1)\phi_\nu(1) | e^2 r_{12}^{-1} | \phi_\lambda(2)\phi_\sigma(2) \rangle \tag{45}$$

The density matrix $P_{\lambda\sigma}$ is defined as

$$P_{\lambda\sigma} = 2\sum_i c_{\lambda i} c_{\sigma i} \tag{46}$$

and is seen to depend on the $c_{\mu i}$, so Eqs. (41) must be solved iteratively. This procedure constitutes the self-consistent-field molecular orbital method for closed-shell ground states.

The spin properties of molecules often depend upon quantum mechanical exchange effects, where the potential experienced by an α electron is different from the potential experienced by a β electron at the same position. This may be due to nuclear–electron spin polarization for a closed-shell system in a magnetic field or electron spin polarization due to a net imbalance of α and β electrons in an odd-electron system. In order to describe this, one must introduce electron correlation. This is, strictly speaking, beyond the orbital approximation. However, this phenomenon may be treated within an orbital approximation by allowing different orbitals for different spins. This leads to a spin-unrestricted molecular orbital wave function, which for a $(2n+1)$-electron system can be written

$$\Phi^u(1, 2, \ldots, 2n-1, 2n, 2n+1)$$

$$= |\psi_1^\alpha(1)\overline{\psi_1^\beta(2)} \cdots \psi_n^\alpha(2n-1)\overline{\psi_n^\beta(2n)}\psi_{n+1}^\alpha(2n+1)| \tag{47}$$

where ψ_i^α and ψ_i^β have different spatial domains. The fact that an orbital wave function is strictly inappropriate for describing this phenomenon in an unperturbed system manifests itself in the fact that Φ^u is not an eigenfunction of the S^2 operator. For a perturbed system, such as a molecule in a magnetic field, S^2 no longer commutes with the Hamiltonian and an unrestricted or split shell function is permissible. In the areas relevant to this chapter, this problem requires attention especially for unperturbed ground states of odd-electron systems and the calculation of ESR hyperfine constants.

The determination of optimum orbitals in a spin-unrestricted framework was described first by Pople and Nesbet.[6] The procedure is closely parallel to the Roothaan method for closed shells, and involves simultaneous solution of the equations

$$\left.\begin{array}{l} \sum_\nu (F_{\mu\nu}^\alpha - \varepsilon_i S_{\mu\nu})c_{\mu i}^\alpha = 0, \\[2mm] \sum_\nu (F_{\mu\nu}^\beta - \varepsilon_i S_{\mu\nu})c_{\mu i}^\beta = 0, \end{array}\right\} \qquad \mu = 1, m \tag{48}$$

where

$$F_{\mu\nu}^\alpha = H_{\mu\nu} + \sum_{\lambda\sigma} [P_{\lambda\sigma}(\mu\nu|\lambda\sigma) - P_{\lambda\sigma}^\alpha(\mu\sigma|\lambda\nu)] \tag{49}$$

and likewise for $F_{\mu\nu}^\beta$. Equations (48) are coupled via the density matrix

$$P_{\lambda\sigma} = P_{\lambda\sigma}^\alpha + P_{\lambda\sigma}^\beta \tag{50}$$

where

$$P_{\lambda\sigma}^\alpha = \sum_i c_{\mu i}^\alpha c_{\nu i}^\alpha \tag{51}$$

and likewise for $P_{\lambda\sigma}^\beta$. As in the spin-restricted case, these equations must be solved iteratively by self-consistent field procedures.

As mentioned above, a spin-unrestricted wave function of the form given by Eq. (48) is an eigenfunction of S_z but not of the spin-squared operator S^2. Instead, the spin-unrestricted function contains components of several pure spin states[7]

$$\Phi^u = \sum_{m=0}^{n\beta} c_{s'+m}\Phi_{s'+m} \tag{52}$$

where $s' = \frac{1}{2}(n^\alpha - n^\beta)$ and $\Psi_{\sigma'+m}$ is a eigenfunction of $\mathbf{S}^2$ with multiplicity $2(s'+m)+1$. It is possible in principle to select from Φ^u the wave functions corresponding to a pure spin state S of the system by application of an appropriate projection operator $\mathcal{O}_s$,

$$\mathcal{O}_s \Phi^u = \prod_{k \neq s} \{A_k / [s(s+1) - k(k+1)]\} \Phi^u = \Phi_s \tag{53}$$

where A_k is an annihilation operator

$$A_k = \mathbf{S}^2 - k(k+1) \tag{54}$$

Each A_k removes the spin component $2k+1$ from Φ^u. The extended Hartree–Fock method and alternant molecular orbital theory involve determination of optimum orbitals on the basis of the application of the variational principle to $\mathcal{O}_s \Phi^u$,

$$\delta \langle \mathcal{O}_s \Phi^u | \mathcal{H}^0 | \mathcal{O}_s \Phi^u \rangle = 0 \tag{55}$$

Since $\mathcal{O}_s \Phi^u$ is a multideterminant wave function, the optimization procedure is rather arduous and the extent of numerical error introduced by using Φ^u directly as an approximation to the correct wave function becomes a matter of interest. Another approximate alternative is to optimize orbitals using the single determinant and then project out the contaminating spin states or at least annihilate the largest contaminant. These alternatives have been well studied, and an assessment of the state of current practice will be considered in the context of the appropriate spin properties in succeeding sections.

3.2. Approximate Molecular Orbital Theory

This volume deals with topics in molecular quantum mechanics treated by approximate methods, and the purview of this chapter is restricted essentially to the approximate molecular orbital calculation of NMR and ESR parameters. The details of the various alternative methods have been dealt with in considerable length in another chapter, and here we merely identify the methods relevant to the calculation of spin properties for cross reference and discuss basic features of the methods, particularly bearing upon their ability to describe spin properties.

The hierarchy of approximate molecular orbital theory begins with Hückel theory for pi electrons[8] and extended Hückel theory for all electrons.[9] The Hückel theories operate in the framework of molecular orbital theory with $F_{\mu\nu}$ in Eq. (49) not explicitly calculated but estimated empirically or semiempirically, i.e., replaced by an effective Hamiltonian element $H_{\mu\nu}^{\text{eff}}$,

$$F_{\mu\nu} \leftarrow H_{\mu\nu}^{\text{eff}} \tag{56}$$

Since no electron repulsion terms are introduced explicitly in $H_{\mu\nu}^{\text{eff}}$, determination of Hückel molecular orbitals involves a single diagonalization of $\mathbf{H}^{\text{eff}}$ and produces an independent electron wave function. The many-electron product wave functions developed from products of Hückel molecular orbitals are *de facto* spin-restricted, since, from the Pople–Nesbet equations (48), if electron repulsion is neglected, $F_{\mu\nu}^{\alpha} = F_{\mu\nu}^{\beta}$, $\psi_i^{\alpha} = \psi_i^{\beta}$, and Eq. (47) reduces to the spin-restricted equation (36). Thus spin polarization is neglected and spin densities calculated from Hückel theories may be seriously in error. For example, all spin densities in an open-shell system treated by Hückel theory are positive, whereas both positive and negative spin densities are observed experimentally.

Approximate self-consistent field molecular orbital methods involve explicit calculation of the Fock matrix elements $F_{\mu\nu}$ with approximations introduced in the evaluation of the core integrals $H_{\mu\nu}$ and the electron repulsion integrals $(\mu\nu|\lambda\sigma)$. One important class of approximate self-consistent field methods has developed with the additional neglect of differential overlap,

$$\phi_\mu(1)\phi_\nu(1) = \phi_\mu^2(1)\,\delta_{\mu\nu} \tag{57}$$

including PPP theory for π-electron systems[10] and CNDO,[11] INDO,[12] MINDO,[13] and NDDO[14] for valence electron systems. The various acronyms stand for methods with various levels of the neglect of differential overlap imposed.

The PPP and CNDO methods involve complete neglect of differential overlap. These methods have limited capabilities in describing spin polarization phenomena. The coupling of electron and nuclear spins in an open-shell system or perturbed closed-shell system involves in part a contact mechanism involving the spin density directly at the various nuclear positions. An important class of molecules in chemical systems comprises those that are planar and unsaturated. The PPP method treats only the unsaturated electrons in π-type orbitals explicitly, whereas the nuclei lie in the σ system, orthogonal to the π system. Thus the electron–nuclear hyperfine interaction requires a further approximate treatment, known as the McConnell relation, discussed further in Section 7. In CNDO theory the valence orbitals of the σ system are included, but with the strict neglect of differential overlap, coupling between the σ and π systems is not accommodated. Thus the CNDO method can also produce serious errors in describing spin polarization and negative spin densities.

The lowest level of approximate self-consistent field molecular orbitals wherein spin properties can be generally accommodated is the INDO method, with intermediate neglect of differential overlap. The INDO method neglects differential overlap except in one-center products $\phi_\mu\phi_\nu$ involving different atomic orbitals on the same atom. These integrals provide a means for spin polarization and thus a framework with at least the general important consequences of spin phenomena included. The INDO method is currently the most extensively used approximate molecular orbital method for the calculation of magnetic resonance parameters. With the INDO approximations the $F_{\mu\nu}$ elements analogous to Eq. (49) become

$$F_{\mu\mu} = U_{\mu\mu} \sum_{\lambda}^{A} [P_{\lambda\lambda}(\mu\mu|\lambda\lambda) - P_{\lambda\lambda}^{\alpha}(\mu\lambda|\mu\lambda)] + \sum_{B \neq A} (P_{BB} - Z_B)\gamma_{AB}, \quad \mu \in A$$

(58)

$$F_{\mu\mu}^{\alpha} = (2P_{\mu\nu} - P_{\mu\nu}^{\alpha})(\mu\nu|\mu\nu) - P_{\mu\nu}^{\alpha}(\mu\mu|\nu\nu), \quad \mu \neq \nu \in A$$

(59)

$$F_{\mu\nu}^{\beta} = \tfrac{1}{2}(\beta_A^0 + \beta_B^0)S_{\mu\nu} - P_{\mu\nu}^{\alpha}\gamma_{AB}, \quad \mu \in A, \quad \nu \in B$$

(60)

and likewise for the elements of $F_{\mu\nu}^{\beta}$. The parameters $U_{\mu\mu}$, one-center integrals $(\mu\nu|\lambda\sigma)$, two-center Coulomb repulsion integrals γ_{AB}, and β_A^0 are specified in Ref. 12. Overall the results of INDO calculations have been shown to be reasonably good for electron density, as evidenced by electric dipole moments and spin properties as discussed in the following section. The results for molecular geometry are close to observed values for a number of cases but there are notable exceptions. Bond energies and ionization energies are poorly accommodated. These shortcomings have recently been elucidated quantitatively by Goddard and co-workers[80] by comparison with nonempirical studies.

The original INDO parametrization has been modified extensively by Dewar and co-workers[13] to form the MINDO method, which gives improved results for ionization energies and thermochemical properties. The NDDO method (neglect of diatomic differential overlap) has been implemented by several groups and can of course adequately describe spin properties. Results are expected to be forthcoming from this vantage point.

3.3. Perturbation Theory

Having considered above the production of electronic wave functions for unperturbed states of many-electron polyatomic molecular systems, we turn to general aspects of the calculation of molecular properties of a perturbed system as encountered in magnetic resonance phenomena. This involves quantum mechanical perturbation theory, of which there are a number of alternative methodologies. We consider here only two: conventional Raleigh–Schrödinger sum-over-states perturbation theory and self-consistent-field finite perturbation theory. In the following paragraphs we review the basic prescription involved in each approach and consider briefly their relative merits. For detailed derivations the reader is referred to the recent review article on perturbation theory by Hirschfelder *et al.*[15] and the original work on finite perturbation theory by Karplus and Kolker[16] and Pople *et al.*[17]

Sum-over-states perturbation theory (SOS-PT) is applicable to a problem for which the Hamiltonian operator can be written

$$\mathcal{H} = \mathcal{H}^0 + \mathcal{H}' \tag{61}$$

where $\mathcal{H}'$ is the perturbation operator and $\mathcal{H}^0$ is the zeroth-order Hamiltonian defined as in equation (8) and for which a set of suitable solutions Φ_k is available for the zeroth-order problem. The wave-functions for the perturbed system and particularly the coefficients such as $\Psi^{(100)}$ in Eq. (20) can be expanded in the set of known functions Φ_k:

$$\Psi^{(100)} = \sum_{k \neq 0}^{\infty} c_k^{(100)} \Phi_k \tag{62}$$

where it can be shown that

$$c_k^{(100)} = -(E_k - E_0)^{-1} \langle \Phi_k | \mathcal{H}^{(100)} | \Phi_0 \rangle \tag{63}$$

where

$$E_k = \langle \Phi_k | \mathcal{H}^0 | \Phi_k \rangle \tag{64}$$

Expressions for the various magnetic resonance parameters follow from substituting expressions of the form of Eq. (62) into Eq. (22), (23), (26), and (27).

One of the main difficulties in this approach arises from the fact that the sum-over-states in Eq. (62) may be slowly converging and good descriptions of the high-energy discrete states and also continuum states are generally not available. A number of studies of magnetic resonance parameters based on this approach avoid this problem by means of the closure approximation. The matrix sum rule of quantum mechanics for generalized operators P and Q amounts to

$$\sum_k \langle \Psi_0 | P | \Psi_k \rangle \langle \Psi_k | Q | \Psi_0 \rangle = \langle \Psi_0 | PQ | \Psi_0 \rangle \tag{65}$$

as long as $P|\Psi_k\rangle$ and $Q|\Psi_k\rangle$ can be expanded in terms of the complete set of functions $\{\Psi_k\}$. The closure approximation amounts to choosing an average excitation energy ΔE such that

$$\sum_k (E_k - E_0)^{-1} \langle \Psi_0 | P | \Psi_k \rangle \langle \Psi_k | Q | \Psi_o \rangle$$
$$\cong \Delta E^{-1} \sum_k \langle \Psi_0 | P | \Psi_k \rangle \langle \Psi_k | Q | \Psi_0 \rangle \tag{66}$$

which, using the matrix sum rule, reduces further as

$$\Delta E^{-1} \sum_k \langle \Psi_0 | P | \Psi_k \rangle \langle \Psi_k | Q | \Psi_0 \rangle = \Delta E^{-1} \langle \Psi_0 | PQ | \Psi_0 \rangle \tag{67}$$

There is of course no *a priori* way to calculate the effective excitation energy and in practice this quantity is usually treated as a disposable parameter and is chosen empirically for optimum agreement between theory and experiment.

Finite perturbation theory (FPT) is applicable to systems which can be described by a Hamiltonian of the form

$$\mathcal{H}(\boldsymbol{\lambda}) = \mathcal{H}^0 + \lambda_1 \mathcal{H}_1 + \lambda_2 \mathcal{H}_2 \tag{68}$$

where the set $\lambda_1, \lambda_2, \ldots$ are external perturbations and $\mathcal{H}_1, \mathcal{H}_2, \ldots$ are the corresponding perturbation Hamiltonian operators and for which a second-order property given by

$$[\partial^2 E / \partial \lambda_1 \partial \lambda_2]_{\lambda_1 = \lambda_2 = 0} \tag{69}$$

is of interest, as in Eqs. (24)–(25) and (28)–(29). From the Hellman–Feynman theorem it can be shown that

$$\partial E(\lambda) / \partial \lambda_1 = \langle \Psi(\lambda) | \partial \mathcal{H}(\lambda) / \partial \lambda_1 | \Psi(\lambda) \rangle \tag{70a}$$

$$= \langle \Psi(\lambda) | \mathcal{H}_1 | \Psi(\lambda) \rangle \tag{70b}$$

Then

$$[\partial^2 E / \partial \lambda_1 \partial \lambda_2]_{\lambda_1 = \lambda_2 = 0} = [(\partial / \partial \lambda_2) \langle \Psi(\boldsymbol{\lambda}) | \mathcal{H}_1 | \Psi(\boldsymbol{\lambda}) \rangle]_{\boldsymbol{\lambda} = 0} \tag{71a}$$

$$= [(\partial / \partial \lambda_2) \langle \Psi_1(\lambda_2) | \mathcal{H}_1 | \Psi(\lambda_2) \rangle]_{\lambda_2 = 0} \tag{72b}$$

Thus for the calculation of a second-order property using finite perturbation theory, one first calculates $\Psi(\lambda_2)$ beginning with a Hamiltonian of the form

$$\mathcal{H}(\lambda_2) = \mathcal{H}^0 + \lambda_2 \mathcal{H}_2 \tag{72}$$

This may be carried out in the context of molecular orbital theory, recognizing that for spin-dependent perturbations a spin-unrestricted functional form is necessary and for field- and angular momentum-dependent perturbations a complex dimension to the orbitals must be permitted. The second-order property follows as the first derivative of the expectation value of $\mathcal{H}_1$ over the perturbed wave function $\Psi(\lambda_2)$ evaluated by numerical methods at $\lambda_2 = 0$.

Finite perturbation theory is at present the main procedure for numerical calculation, especially of nuclear magnetic resonance parameters. Ditchfield *et al.*[18] have worked out a formal comparison of infinite and finite perturbation theory and have shown that the results of the two methods may differ significantly if configuration interaction is important for the unperturbed function employed in the sum-over-states. Finite perturbation theory is especially advantageous for approximate molecular orbital calculations of magnetic resonance parameters since the procedure can be reduced to a well-defined canonical form and disposable parameters may be chosen for optimum agreement between calculated and observed spin properties on prototype systems.

4. NMR Shielding Constants and Chemical Shifts

The energy of a free magnetic nucleus in an externally applied magnetic field H is

$$E = -\boldsymbol{\mu}_N \cdot \mathbf{H} \tag{73}$$

where μ_N is the magnetic moment representative of the nuclear spin state $|I_N\rangle$,

$$\mu_N = \gamma_N \hbar I_{Nz} \tag{74}$$

where γ_N is the gyromagnetic ratio of nucleus N. The transition energy for a nuclear spin excitation is

$$h\nu = \gamma_N \hbar H \tag{75}$$

When a molecule is placed in a uniform external magnetic field, electronic currents are induced. The field at a particular magnetic nucleus of the molecule is thus a sum of the pure homogeneous external field $\mathbf{H}$ and a local field H_N^{ind} due to induced currents. The latter is proportional to the applied field,

$$\mathbf{H}_N^{\text{ind}} = -\boldsymbol{\sigma}_N \cdot \mathbf{H} \tag{76}$$

where the proportionality constant $\boldsymbol{\sigma}_N$, the nuclear shielding constant, is a tensor quantity since the induced field vector $\mathbf{H}_N^{\text{ind}}$ is not necessarily parallel to $\mathbf{H}$. In liquids and gases, the molecules are rapidly tumbling with respect to the external field and only the average value of the shielding tensor,

$$\sigma_N = \tfrac{1}{3}(\sigma_{Nxx} + \sigma_{Nyy} + \sigma_{Nzz}) \tag{77}$$

can be determined. The field at the nuclear position is

$$\mathbf{H}_N = \mathbf{H} - \mathbf{H}_N^{\text{ind}} = (1 - \sigma_N)\mathbf{H} \tag{78}$$

and the resonant frequency for nuclear spin excitation is

$$h\nu = \gamma_N \hbar (1 - \sigma_N) H \tag{79}$$

Nuclei in different electronic environments are characterized by different σ_N and displacements in their relative resonant frequencies. This displacement is referred to as the chemical shift. Experimentally determined chemical shifts are customarily reported as δ_N, the displacement relative to a suitable reference compound, defined as

$$\delta_N = \sigma_N - \sigma_N^{\text{ref}} \tag{80}$$

Thus the theoretical calculation of NMR chemical shifts reduces to a determination of the magnetic resonance parameter σ_N, the shielding constant for nucleus N.

4.1. Quantum Mechanical Development of σ_N

Fundamental considerations on σ_N as a magnetic resonance parameter were given in Section 2. The numerical calculation of shielding constants follows from Eq. (22) and sum-over-states perturbation theory or Eq. (24) and finite perturbation theory, and important developments in the theory of nuclear shielding constants have emerged from both points of view. Introducing a sum-over-states expansion of Eq. (62) for $\Psi^{(100)}$ and the analogous expression

$$\Psi_N^{(010)} = -\sum_{k \neq 0}^{\infty} (E_k - E_0)^{-1} \langle \Phi_k | \mathcal{H}_N^{(010)} | \Phi_0 \rangle \Phi_k \tag{81}$$

into the general expression for σ_N in Eq. (22), we have

$$\sigma_{Ntu} = \langle \Phi_0 | \mathcal{H}_{Ntu}^{(110)} | \Phi_0 \rangle$$

$$- \sum_{k \neq 0}^{\infty} (E_k - E_0)^{-1} \{ \langle \Phi_0 | \mathcal{H}_t^{(100)} | \Phi_k \rangle \langle \Phi_k | \mathcal{H}_{Nu}^{(010)} | \Phi_0 \rangle$$

$$+ \langle \Phi_0 | \mathcal{H}_{Nt}^{(010)} | \Phi_k \rangle \langle \Phi_k | \mathcal{H}_u^{(100)} | \Phi_0 \rangle \} \tag{82}$$

where the first term is the contribution to the induced magnetic field due to diamagnetic Langevin currents and the second term, where the external perturbation has coupled ground and excited states of the expansion set, is the paramagnetic contribution.

In terms of finite perturbation theory, Eq. (24) with $\lambda_1 = \mu_{Nt}$ and $\lambda_2 = H_u$ results in

$$\sigma_{Ntu} = \left[\frac{\partial}{\partial H_u} \langle \Phi(H_u) | \mathcal{H}_{Nt}^{(010)} + \mathcal{H}_{Ntu}^{(110)} H_u | \Phi(H_u) \rangle \right]_{\mathbf{H}=0} \tag{83}$$

which can be resolved into first- and second-order contributions as

$$\sigma_{Ntu} = \langle \Phi_0 | \mathcal{H}_{Ntu}^{(110)} | \Phi_0 \rangle + \left[\frac{\partial}{\partial H_u} \langle \Phi(H_u) | \mathcal{H}_{Nt}^{(010)} | \Phi(H_u) \rangle \right]_{\mathbf{H}=0} \tag{84}$$

The two terms in the equation represent the diamagnetic and paramagnetic contributions to the magnetic shielding tensor as given by finite perturbation theory. It should be noted that here the diamagnetic contribution depends only on the unperturbed wave function, whereas the paramagnetic contribution is determined solely by the perturbed wave function; the decomposition of shielding into diamagnetic and paramagnetic contributions is a consequence of the application of perturbation theory to the problem.

The earliest theoretical work on magnetic shielding constants was carried out by Ramsey[19] using SOS perturbation theory. Introducing explicit expressions for the operators $\mathcal{H}^{(110)}$, $\mathcal{H}^{(100)}$, and $\mathcal{H}^{(010)}$ and treating the diamagnetic and paramagnetic contributions separately, Eq. (82) becomes

$$\boldsymbol{\sigma}_N = \boldsymbol{\sigma}_N^d + \boldsymbol{\sigma}_N^p \tag{85}$$

where

$$\sigma_{Ntu}^{d} = \frac{e^2}{2mc^2} \left\langle \Phi_0 \left| \sum_i (\mathbf{r}_i \cdot \mathbf{r}_{iN}\delta_{tu} - r_{it}r_{iNu})r_{iN}^{-3} \right| \Phi_0 \right\rangle \tag{86}$$

and

$$\sigma_{Ntu}^{p} = -\left(\frac{e\hbar}{2mc}\right)^2 \sum_{k \neq 0} (E_k - E_0)^{-1}$$

$$\times \left\{ \left\langle \Phi_0 \left| \sum_i L_{it} \right| \Phi_k \right\rangle \left\langle \Phi_k \left| \sum_i 2L_{iNu}r_i^{-3} \right| \Phi_0 \right\rangle \right.$$

$$\left. + \left\langle \Phi_0 \left| \sum_i 2L_{iNt}r_i^{-3} \right| \Phi_k \right\rangle \left\langle \Phi_k \left| \sum_i L_{iu} \right| \Phi_0 \right\rangle \right\} \tag{87}$$

The diamagnetic contribution σ_N^d and the paramagnetic contribution σ_N^p of nuclear screening act in opposite directions and come into σ_N with opposite signs. The diamagnetic term is a molecular counterpart to the Lamb formula for screening of an atom, which is proportional to the electron density around the nucleus. The paramagnetic term may be interpreted as the impedance of the electron circulation giving rise to σ_N by the remaining nuclei, and its dependence on $\mathbf{L}_i$ and $\mathbf{L}_{iN}$ requires that the nucleus have electrons with orbital angular momentum, i.e., non-*s* electrons. Thus it is generally believed that the diamagnetic term is generally responsible for proton chemical shifts. For heavy atoms both the diamagnetic and paramagnetic terms may be large, but the paramagnetic term often dominates. The paramagnetic term is generally more sensitive to changes in electronic structure and one usually looks to this term to explain trends in chemical shifts within a series of structurally related compounds.

The sum-over-states expression given in Eqs. (85)–(87) can be reduced to integrals over molecular orbitals by forming an expansion set from determinantal products of the occupied and virtual molecular orbitals resulting from a calculation on the ground state of the system under consideration. The result is

$$\sigma_{Ntu} = \frac{e^2}{mc^2} \sum_l^{occ} \langle \psi_l | (\mathbf{r} \cdot \mathbf{r}_N \delta_{tu} - r_t r_{Nu}) r_N^{-3} | \psi_l \rangle$$

$$- \left(\frac{e\hbar}{mc}\right)^2 \sum_l^{occ} \sum_r^{vac} \Delta E_{l \to r}^{-1} \{ \langle \psi_l | L_t | \psi_r \rangle \langle \psi_r | 2L_{Nu}r_N^{-3} | \psi_l \rangle$$

$$+ \langle \psi_l | 2L_{Nt}r_N^{-3} | \psi_r \rangle \langle \psi_r | L_u | \psi_l \rangle \} \tag{88}$$

Numerical calculations based on Eq. (88) encounter several serious problems. First the description of excited states by virtual orbitals is usually a poor approximation and thus contributions to σ_N^p from the higher excited states are considerably in error. A more general problem, which applies also to Eq. (87), arises in that all distances are referred to an arbitrary origin, and unless a truly complete expansion set is used, the calculation of σ_N^p will be dependent upon the origin or gauge of the vector potential of the magnetic field.[20]

An alternative approach which avoids the difficulty mentioned above is to formulate the problem in terms of field–dependent gauge invariant atomic orbitals χ_μ, where

$$\chi_\mu = \{\exp[-(i/c)\mathbf{A}_\mu \cdot \mathbf{r}]\}\phi_\mu \qquad (89)$$

where $\mathbf{A}_\mu$ is the vector potential at nucleus μ

$$\mathbf{A}_\mu = \tfrac{1}{2}\mathbf{H} \times \mathbf{R}_\mu \qquad (90)$$

with $\mathbf{R}_\mu$ as the position vector of the atom upon which atomic orbital ϕ_μ is located. An expression for the total energy of the system can be developed as a perturbation expansion in $\mathbf{A}$, considering the total field as a combination of the external field and the field of $\boldsymbol{\mu}_N$:

$$\mathbf{A} = \tfrac{1}{2}\mathbf{H} \times \mathbf{r} + (\boldsymbol{\mu}_N \times \mathbf{r})r^{-3} \qquad (91)$$

and the shielding constant is taken as the coefficient of the term bilinear in H and μ_N.[21]

Self-consistent field perturbation theory has been used in the calculation of shielding constants by Karplus and Kolker, and extensively by Lipscomb and co-workers[22] and by Pople's group. The development described here is due to Ditchfield *et al.*[23] The diamagnetic contribution is simply an expectation value of the zeroth-order wave function, and finite perturbation theory can be used to calculate the paramagnetic term as

$$\sigma_{Ntu}^p = \left[\frac{\partial}{\partial H_u}\langle\Phi(H_u)|\mathscr{H}_{Nt}^{(010)}|\Phi(H_u)\rangle\right]_{\mathbf{H}=0} \qquad (92)$$

Thus we require $\Phi(H_\mu)$ and the derivative of its expectation value on $\mathscr{H}_{Nt}^{(010)}$.

It follows from Eq. (72) that the Hamiltonian to be used in the perturbed wave function calculation is

$$\mathscr{H}(H_u) = \mathscr{H}^0 + H_u\mathscr{H}_u^{(100)} \qquad (93)$$

and thus magnetic shielding tensors for all the nuclei in a molecule can be obtained from three perturbed wave function calculations, one for each Cartesian component of the external field $\mathbf{H}$. Using a variational approach, a self-consistent field procedure for determining $\Phi(\mathbf{H})$ in a molecular orbital approximation can be developed. The linear expansion coefficients of Eq. (40) must be allowed to be complex due to the pure imaginary nature of the perturbation. A set of Roothaan-type equations analogous to Eq. (41) results, with

$$F_{\mu\nu}(H_u) = H_{\mu\nu}^{\mathrm{core}} - iH_u[H_{\mu\nu}^{(100)}]_u$$
$$+ \sum_\lambda \sum_\sigma P_{\lambda\sigma}(H_u)[(\mu\nu|\lambda\sigma) - \tfrac{1}{2}(\mu\sigma|\lambda\nu)] \qquad (94)$$

where $H_{\mu\nu}^{core}$ and $(\mu\nu|\lambda\sigma)$ are defined as in Eqs. (44) and (45), and $P_{\lambda\sigma}$ is complex,

$$P_{\lambda\sigma} = P_{\lambda\sigma}^{Re} + iP_{\lambda\sigma}^{Im} \tag{95}$$

where $i = \sqrt{-1}$. Also

$$[H_{\mu\nu}^{(100)}]_u = \frac{e\hbar}{2mc} \langle \Phi_\mu | (\mathbf{r} \times \nabla)_u | \phi_\nu \rangle \tag{96}$$

In the presence of the perturbation, the Roothaan equations are modified only by a change in the one-electron core part of the Fock matrix. The paramagnetic contribution to the nuclear shielding constant in the FPT-MO approach is then

$$\sigma_{Ntu}^P = -\sum_\mu \sum_\nu [H_{N\mu\nu}^{(010)}]_t \{(\partial/\partial H_u)[P_{\mu\nu}^{Im}(H_u)]\}_{\mathbf{H}=0} \tag{97}$$

where

$$[H_{N\mu\nu}^{(010)}]_t = \frac{e\hbar}{2mc} \langle \Phi_\mu | (\mathbf{r}_N \times \nabla)_t \, r_N^{-3} | \phi_\nu \rangle \tag{98}$$

The corresponding equation for the diamagnetic contribution is

$$\sigma_{Ntu}^d = -\sum_\mu \sum_\nu P_{\mu\nu}^{(0)} [H_{N\mu\nu}^{(110)}]_{tu} , \tag{99}$$

where

$$[H_{N\mu\nu}^{(110)}]_{tu} = \frac{e^2}{4mc^2} \langle \phi_\mu | (\mathbf{r} \cdot \mathbf{r}_N \, \delta_{tu} - r_t \, r_{Nu}) r_N^{-3} | \phi_\nu \rangle \tag{100}$$

The derivatives of the density matrix are computed by the method of finite differences. Considerable *ab initio* calculations have been carried out in this format.[23] Finite perturbation molecular orbital theory of shielding constants can also be formulated in terms of gauge-invariant atomic orbitals, as will be discussed below.

4.2. Calculation of Shielding Constants

Our qualtitative and semiquantitative understanding of chemical shielding in polyatomic molecules was advanced considerably by early work based on approximate molecular orbital theory. In the Ramsey equation, both diamagnetic and paramagnetic terms are large for large molecules. The net shielding then comes out as the difference between two large terms, with the paramagnetic term in particular very difficult to calculate accurately. Early work, following Saika and Schlichter,[24] was based on separating local circulation around individual atoms from the overall motion of electrons in the molecule. Pople[25] formalized this approach using independent-electron molecular orbital theory, gauge-invariant atomic orbitals, and approximations generally identified with neglect of overlap. The resulting expressions for the diamagnetic and paramagnetic contribution to the shielding at nucleus N are

$$\sigma^d_{Ntu} = \frac{e^2}{mc^2}\sum_\mu P_{\mu\mu}\langle\phi_\mu|(\mathbf{r}\cdot\mathbf{r}_N\delta_{tu} - r_t r_{Nu})r_N^{-5}|\phi_\mu\rangle \tag{101}$$

and

$$\sigma^p_{Ntu} = \frac{2e^2\hbar^2}{m^2c^2}\sum_j^{occ}\sum_l^{vac}(\varepsilon_l - \varepsilon_j)^{-1}\sum_{\nu<\lambda}\sum_{\rho<\sigma}[(c_{\nu j}c_{\lambda l} - c_{\lambda j}c_{\nu l})$$

$$\times(c_{\rho j}c_{\sigma l} - c_{\sigma j}c_{\rho l})\langle\phi_\nu|L_{Nt}\,r_N^{-3}|\phi_\lambda\rangle\langle\phi_\rho|L_u|\phi_\sigma\rangle] \tag{102}$$

These equations may be simplified further by reduction of the integrals and rearrangement based on the fact that the LCAO coefficients can be identified with the various atoms of the molecule. The result is Pople's expression for the net orientationally averaged shielding constant of nucleus N

$$\sigma_N = \sigma^d_{NN} + \sigma^p_{NN} + \sum_{M\neq N}\sigma_{MN} + \sigma_N^{deloc} \tag{103}$$

where σ^d_{NN} and σ^p_{NN} are local atomic diamagnetic and paramagnetic contributions, σ_{MN} is the contribution from atom M to screening on atom N, and σ_N^{deloc} is identified with ring current effects.

The physical interpretation of nuclear screening and chemical shifts is often based on the various terms in Eq. (103). The local diamagnetic screening σ^d_{NN} comes from the circulation of electrons in the orbitals of atom N,

$$\sigma^d_{NN} = \frac{e^2}{3mc^2}\sum_\mu P_{\mu\mu}\langle\phi_\mu|r^{-1}|\phi_\mu\rangle \tag{104}$$

essentially a Lamb term for the atom in the molecule. Here increased electron population of atom N is reflected in increased $P_{\mu\mu}$, and thus increased nuclear shielding. This is exemplified by ring protons in substituted benzenes, where electron-withdrawing substituents cause deshielding with respect to electron-donating substituents. Acidic protons, as in acetic acid, are quite deshielded. This term also has been invoked to explain the decrease in ^{13}C shielding in chloromethanes as the number of chlorine substituents increases.

Local paramagnetic currents are represented by σ^p_{NN}. The paramagnetic terms in general arise from a mixing of ground and excited states by the external field, and nonzero values of the atomic integrals in Eq. (102) must involve atomic $p, d, \ldots$ orbitals. Using closure and the average energy approximation on the monatomic terms for atom N in Eq. (103), we obtain

$$\sigma^p_{NN} = -\frac{2}{3}\left(\frac{e\hbar}{mc}\right)^2\Delta E^{-1}\langle 2p_N|r^{-3}|2p_N\rangle\sum_M q_{MN} \tag{105}$$

for *p* orbitals, where

$$q_{MN} = \delta_{MN}(P_{x_M x_N} + P_{y_M y_N} + P_{z_M z_N}) - \tfrac{1}{2}(P_{y_M y_N}P_{z_M z_N} + P_{z_M z_N}P_{x_M x_N} + P_{x_M x_N}P_{y_M y_N})$$

$$+ \tfrac{1}{2}(P_{y_M z_N}P_{z_M y_N} + P_{z_M x_N}P_{x_M z_N} + P_{x_M y_N}P_{y_M x_N}) \tag{106}$$

The contribution of this term is small for proton shifts where ΔE is relatively large and becomes increasingly important for carbon to fluorine and heavy atoms. The relatively low field resonance for ^{13}C in unsaturated as compared with saturated hydrocarbons is attributed to σ^P_{NN} via a ΔE^{-1} effect. The monatomic contribution to q_{MN} indicates a charge dependence in the term, but not a simple proportionality since this can affect the size of the *p* orbitals and thus the atomic integral as well. Certain ^{13}C shifts are empirically found to vary linearly with charge. The two-atom terms in q_{MN} require multiple bonding between *N* and *M* to make a viable contribution.

The paramagnetic contribution of neighboring atoms was estimated using a classical dipolar field to give

$$\sum_{M \neq N} \sigma^P_{MN} = -\frac{1}{3N_o} \sum_{M \neq N} \sum_{tu} \chi^M_{tu}(R^2_M \delta_{tu} - 3R_{Mt}R_{Mu})R^{-5}_M \tag{107}$$

where χ is the molar magnetic susceptibility tensor, R_M is the position vector of atom *M* with respect to atom *N*, and N_0 is Avogadro's number. This effect is largest if neighboring contributions have a large and anisotropic magnetic susceptibility. For the particular case when the shielding arises from bonds with an axial symmetry,

$$\sum_{M \neq N} \sigma^P_{MN} = -\frac{1}{3N_0} \sum_{M \neq N} \Delta\chi^M(1 - 3\cos^2\theta)R^{-3}_M \tag{108}$$

where $\Delta\chi^M = \chi^M_{\parallel} - \chi^M_{\perp}$ and θ is the angle that the bond makes with the symmetry axis. This contribution is particularly important in proton shifts in linear closed-shell molecules like acetylene and hydrogen halides.

The final term in Eq. (103) is σ^{deloc}_N, arising from electron currents flowing around closed rings of atoms as in benzene and other aromatic molecules. This idea was originally involved to explain anomalously low shielding effects in benzene and related compounds and has since been developed furthur and criticized.[25]

The interpretation of experimental data on shielding constants at both a qualitative and quantitative level usually goes back to Pople's equation. Refinement and modifications have been introduced for each of the expressions. The net diamagnetic shielding constant has recently been reconsidered by Flygare[27] and a new procedure for estimating this quantity introduced. The paramagnetic contribution is currently being treated by variational perturbation techniques. Magnetic anisotropy has been refined by Buckingham and

Stiles[28] by extending the point dipole approximation to include higher magnetic multipoles. Ring current effects remain an area of active research.

Recent numerical calculations of shielding constants based on SOS perturbation theory and approximate molecular orbital theory are represented by the work of Velenik and Lynden-Bell[29] using an extended Hückel theory and by Strong *et al.*[30] using approximate SCF theory at the INDO and MINDO level; the results using MINDO are markedly better due to overestimates of ΔE by the INDO approach. As seen from Table 1, the numerical results fall short of a quantitative account of the experimental data, but are useful for semiquantitative and qualitative purposes. Strong *et al.* give contributions to shielding from individual molecular orbitals for C_2H_4 and C_2H_2.

Developments in the finite perturbation theory of shielding constants using approximate MO theory have been reported by Ellis *et al.*[31] Gauge-invariant atomic orbitals and INDO level atomic integral approximations were used. The matrix Hartree–Fock problem becomes complex, and the INDO Fock matrix elements are modified to include the perturbation. With gauge-invariant orbitals the off-diagonal core integrals become, for example,

$$H_{\mu\nu}^{Re} = \cos\left(\frac{e}{2\hbar c}\mathbf{H}\cdot\mathbf{R}_\nu\times\mathbf{R}_\mu\right)\cdot\tfrac{1}{2}(\beta_A^0+\beta_B^0)S_{\mu\nu} \tag{109}$$

and

$$H_{\mu\nu}^{Im} = \sin\left(\frac{e}{2\hbar c}\mathbf{H}\cdot\mathbf{R}_\nu\times\mathbf{R}_\mu\right)\cdot\tfrac{1}{2}(\beta_A^0+\beta_B^0)S_{\mu\nu} \tag{110}$$

for μ and ν on different atoms. $\mathbf{R}_\mu$ and $\mathbf{R}_\nu$ are position vectors for the nuclei upon which ϕ_μ and ϕ_ν are centered. Gauge invariance requires the introduction of somewhat similar premultiplication factors for the one- and two-electron atomic integrals in the calculation. The original INDO parametrization led to results which were in relatively poor agreement with experiment, but a subsequent reparametrization of the carbon integrals led to results which were in reasonable agreement with experiment. Proceeding upon this general line of development, Maciel *et al.*[32] have determined a suitable set of parameters for fluorine. Overall, it appears that with a suitable set of parameters, calculations at the INDO level of approximation are capable of accommodating the main aspects of shielding constants. Some representative results are collected in Table 2.

The literature on applications of approximate molecular orbital theory to the calculation of shielding constants in specific chemical systems is so extensive that we refer here to recent comprehensive review articles on the subject rather than the original literature: A continuing account of annual developments on shielding constants is published by Raynes[33] in the Specialist Periodical Reports of the Chemical Society. A definitive review of the theory of chemical shifts with a focus on ^{13}C has been prepared by Ditchfield and Ellis.[34]

Table 1. Calculated ^{13}C Shielding Constants and Chemical Shifts Using SOS-PT and MINDO-CI Wave Functions[a]

Compound	σ_N^d	σ_{Nxx}^p	σ_{Nyy}^p	σ_{Nzz}^p	σ_N^p	σ_N^c	σ_N	δ_N(calc.)	δ_N(exp.)
CH_4	71.4	−176.2	−176.4	−174.4	−175.7	24.0	−80.3	55.1	133.9
H_3C−CH_3	73.8	−152.3	−188.7	−189.4	−176.8	18.4	−84.6	50.8	122.8
H_2C=CH_2	80.5	−198.7	−438.1	−181.1	−272.6	58.2	−133.9	1.5	5.9
HC≡CH	87.0	−0.0	−371.5	−372.3	−248.0	54.6	−106.4	29.0	56.6
H_2C^1=C^2=CH_2									
C^1	80.5	−201.5	−376.6	−174.8	−251.0	40.9	−129.6	4.2	58.1
C^2	87.1	−183.3	−356.4	−357.2	−299.0	56.8	−155.1	19.7	−82.4
H_3C^1−C^2≡C^3H									
C^1	73.1	−160.0	−188.4	−188.9	−179.1	19.6	−86.4	49.0	126.6
C^2	87.9	−2.6	−347.7	−348.4	−232.9	45.5	−99.5	35.9	49.3
C^3	86.6	−8.9	−351.5	−352.1	−237.5	46.3	−104.6	30.8	61.5
H_1C^2−C^2≡C−CH_3									
C^1	78.3	−135.3	−173.6	−174.0	−161.0	12.7	−70.0	65.4	127.7
C^2	88.1	−4.9	−352.9	−353.3	−237.0	46.1	−102.8	32.6	55.9
C_6H_6	79.5	−300.2	−231.1	−186.6	−239.3	24.4	−135.4	0	0
Range in values	16.7	—	—	—	123.3	39.8	—	—	—

[a]From Ref. 30. σ^c is a contribution due to electron correlation effects.

Table 2. Calculated ^{13}C Shielding Constants and Chemical Shifts Using FPT and INDO Wave Functions Based on Modified Parameters[31]a

Compound	σ_N^d	σ_N^p	σ_N	δ_N (calc)	δ_N(exp)
1. CH_4	61.26	-55.20	6.06	120.62	130.8
2. $c\text{-}C_3H_6$	58.90	-65.21	-6.31	108.25	130.7
3. C_2H_6	59.78	-56.72	3.06	117.62	122.8
4. $C^*H_3CH_2CH_3$	60.05	-60.47	-0.42	114.14	113.1
5. $CH_2C^*H_2CH_3$	58.35	-55.04	3.31	117.87	112.6
6. $C^*H_3C_6H_5$	60.33	-60.74	-0.41	114.15	107.2
7. $C^*H_3CH=CH_2$	60.12	-60.07	0.06	114.50	108.2
8. $c\text{-}C_6H_{12}$	58.78	-62.41	-3.62	110.94	100.9
9. $C^*H_3C{\equiv}CH$	59.95	-58.53	1.42	115.98	—
10. $CH_3C^+(C^*H_3)_2$	60.80	-82.47	-21.67	92.89	81.3
11. $HC^+(C^*H_3)_2$	60.63	-88.71	-28.09	86.47	67.8
12. $CH_3C{\equiv}C^*H$	59.88	-132.44	-72.56	42.00	58.7
13. C_2H_2	98.87	-133.36	-74.49	40.07	54.8
14. $C^*H_2=C=CH_2\cdot$	60.44	-143.08	-82.64	31.92	54.0
15. $CH_3C^*{\equiv}CH$	57.11	-143.43	-86.32	28.24	43.0
16. $CH_3CH=C^*H_2$	59.96	-166.01	-106.05	8.51	12.2
17. C_2H_4	59.30	-168.29	-108.98	5.58	5.4
18. $CH_3C_6H_5\text{-}4\text{-}C^*$	58.49	-172.71	-114.21	0.25	4.0
19. $CH_3C_6H_5\text{-}2\text{-}C^*$	58.79	-173.47	-114.68	-0.12	0.3
20. C_5H_6	58.31	-172.87	-114.56	0.0	0.0
21. $CH_3C_6H_5\text{-}3\text{-}C^*$	58.14	-174.47	-116.32	-1.76	1.1
22. $CH_3C^*H=CH_2$	57.67	-177.83	-120.16	-5.60	-7.1
23. $CH_3C_6H_5\text{-}1\text{-}C^*$	56.74	-180.39	-123.66	-9.10	-8.2
24. $CH_3C^{*+}(CH_3)_2$	53.68	-261.26	-207.58	-93.02	-70.4
25. $HC^{*+}(CH_3)_2$	54.72	-283.34	-228.61	-114.05	-60.0
26. $CH_2=C^*=CH_2$	56.26	-232.82	-176.56	-62.07	-84.0

^{a} The nucleus N is denoted by an asterisk if necessary.

A recent review by O'Reilly[35] is especially strong on the prototype theoretical calculations of shielding constants.

5. NMR Nuclear Spin Coupling Constants

In addition to the splitting of nuclear Zeeman levels by the external magnetic field in a nuclear magnetic resonance experiment, there exists additional splitting due to the interactions of nuclear spins with one another. The nuclear spin couplings $\mathbf{K}_{MN}$ arise to first order from a direct magnetic dipole–dipole interaction, which averages to zero for tumbling molecules in liquids or gases. At second-order terms arise due to nuclear spin polarization of the electron cloud and dipolar and orbital effects, which dominate nuclear spin interactions in solution. For nuclear spin coupling, experimental measurements result in J_{MN}, related to the reduced coupling constant K_{MN} by the formula

$$\mathbf{J}_{MN} = (\hbar/2\pi)\gamma_M\gamma_N\mathbf{K}_{MN} \qquad (111)$$

Since magnetogyric ratios can be positive or negative, the use of $\mathbf{K}_{MN}$ rather than $\mathbf{J}_{MN}$ prevents confusion over the sign of the purely electronic effect. As with shielding constants, we focus on the orientationally averaged form of the reduced nuclear spin coupling tensor,

$$K_{MN} = \tfrac{1}{3}(K_{MNxx} + K_{MNyy} + K_{MNzz}) \tag{112}$$

We begin this section with a development of general quantum mechanical expressions for spin–spin coupling constants. Then each type of contribution is discussed in turn and calculations of nuclear spin coupling constant based on approximate molecular orbital theory are discussed.

5.1. Quantum Mechanical Development of $\mathbf{K}_{MN}$

The nuclear spin coupling tensors collect all terms bilinear in nuclear spins $\boldsymbol{\mu}_M$ and $\boldsymbol{\mu}_N$. The quantum mechanical expression for $\mathbf{K}_{MN}$ can be developed from Eq. (23), using Eq. (62) for $\Psi^{(100)}$ and Eq. (81) for $\Psi^{(010)}$. The result is

$$K_{MNtu} = \langle \Phi_0 | \mathcal{H}_{MNtu}^{(020)} | \Phi_0 \rangle$$

$$- \sum_{k \neq 0}^{\infty} (E_k - E_0)\{ \langle \Phi_0 | \mathcal{H}_{Mt}^{(010)} | \Phi_k \rangle \langle \Phi_k | \mathcal{H}_{Nu}^{(010)} | \Phi_0 \rangle$$

$$+ \langle \Phi_0 | \sum_i \mathbf{s}_{it} \mathcal{H}_{Mitu}^{(011)} | \Phi_k \rangle \langle \Phi_k | \sum_i \mathbf{s}_{it} \mathcal{H}_{Nitu}^{(011)} | \Phi_0 \rangle \} \tag{113}$$

The term involving $\mathcal{H}_{MN}^{(020)}$ arises from a direct nuclear magnetic dipole–dipole interaction and a nuclear dipole–electron orbital term, and the second-order terms arise from other electron-mediated nuclear spin interactions. The term in $\mathcal{H}^{(010)}$ is the orbital contribution to nuclear spin coupling and the term in $\mathcal{H}^{(011)}$ is the hyperfine contribution. Note from Eq. (16) that the electron–nuclear hyperfine operator can be resolved into a contact term and an electron–nuclear dipolar interaction term.

An alternative expression for $\mathbf{K}_{MN}$ can be obtained using finite perturbation theory. Here Eq. (25) with $\lambda_1 = \mu_N$ and $\lambda_2 = \mu_M$ results in

$$K_{MNtu} = \left[\frac{\partial}{\partial \mu_{Mu}} \langle \Psi(\mu_{Mu}) | \mathcal{H}_{MNtu}^{(020)} \cdot \mu_{Nt} + \mathcal{H}_{Nt}^{(010)} + \sum_i \mathbf{s}_{it} \mathcal{H}_{Nitu}^{(011)} | \Psi(\mu_{Mu}) \rangle \right]_{\mu_N = \mu_M = 0} \tag{114}$$

which can be resolved into first- and second-order contributions as

$$K_{MNtu} = \langle \Psi_0 | \mathcal{H}_{MNtu}^{(020)} | \Psi_0 \rangle$$

$$+ \left[\frac{\partial}{\partial \mu_{Mu}} \langle \Psi(\mu_{Mu}) \Big| \mathcal{H}_{Nt}^{(010)} + \sum_i \mathbf{s}_{it} \mathcal{H}_{Nitu}^{(011)} \Big| \Psi(\mu_{Mu}) \rangle \right]_{\mu_M = 0} \tag{115}$$

Here it should be noted that the electron-mediated orbital and hyperfine terms depend on the perturbed wave function.

The nuclear part of the first-order term involving $\mathcal{H}^{(020)}$ averages to zero for a tumbling molecule and is important only in the interpretation of solid-state, broadband NMR spectra. The electronic part of the first-order term is customarily neglected. We restrict our attention here to indirect coupling and the electron-mediated terms. Let us resolve the hyperfine operator explicitly into the Fermi contact, orbital, and dipolar contributions, each orientationally averaged [Eq. (112)],

$$K_{MN} = K_{MN}^{\text{fc}} + K_{MN}^{\text{orb}} + K_{MN}^{\text{dip}} \tag{116}$$

and consider each of the contributions to electron-mediated nuclear spin–spin coupling separately.

5.2. The Fermi Contact Term

The contact term is usually the most important of the three contributions and has received the most intensive theoretical study. As with magnetic shielding constants, the earliest theoretical work on nuclear spin coupling was carried out by Ramsey[36] using sum-over-states perturbation theory. Introducing the expression for the contact part of $\mathcal{H}^{(011)}$ into Eq. (113), we have for the orientationally averaged reduced contact coupling constant

$$K_{MN}^{\text{fc}} = (512\pi^2\beta^2/27) \sum_{k \neq 0} (E_k - E_0)^{-1} \left\langle \Phi_0 \middle| \sum_i \delta(\mathbf{r}_{iM})\mathbf{s}_i \middle| \Phi_k \right\rangle$$

$$\times \left\langle \Phi_k \middle| \sum_j \delta(\mathbf{r}_{jN})\mathbf{s}_j \middle| \Phi_0 \right\rangle \tag{117}$$

with units of cm^{-1}. Upon closure and the average energy approximation, this equation can be approximated as

$$K_{MN}^{\text{fc}} = (512\pi^2\beta^2/27) \Delta E^{-1} \left\langle \Phi_0 \middle| \sum_i \sum_j \delta(\mathbf{r}_{iM})\delta(\mathbf{r}_{jN})\mathbf{s}_i \cdot \mathbf{s}_j \middle| \Phi_0 \right\rangle \tag{118}$$

The sum-over-states expressions for K_{MN}^{fc} can be reduced to expressions over molecular orbitals in the usual manner by forming an expansion set from the occupied and virtual orbitals of the unperturbed system and working out the matrix elements. The result, following McConnell,[37] is

$$K_{MN}^{\text{fc}} = -(256\pi^2\beta^2/9) \sum_l^{\text{occ}} \sum_r^{\text{vac}} {}^3\Delta E_{l\to r}^{-1} \psi_l(\mathbf{r}_M)\psi_r(\mathbf{r}_M)\psi_l(\mathbf{r}_N)\psi_r(\mathbf{r}_N) \tag{119}$$

where ${}^3\Delta E_{l\to r}$ is the triplet excitation energy to the configuration where occupied orbital ψ_l is replaced by virtual orbital ψ_r. Note that paramagnetic excited states with unpaired spin density are required for nonvanishing values of the integrals in Eq. (119). Note also that a better approximation could be introduced at this step by allowing for configuration interaction among members of the expansion set.

Introducing an LCAO expansion for the molecular orbitals in the above equation results in

$$K_{MN}^{fc} = -\frac{256\pi^2\beta^2}{9} \sum_l^{occ} \sum_r^{vac} {}^3\Delta E_{l\to r} \sum_{\mu\nu\lambda\sigma} c_{\mu l}c_{\nu r}c_{\lambda l}c_{\sigma r}\phi_\mu(\mathbf{r}_M)\phi_\nu(\mathbf{r}_M)\phi_\lambda(\mathbf{r}_N)\phi_\sigma(\mathbf{r}_N)$$

(120)

where it should be noted that $\phi_\mu(r_M)$ and similar terms will have appreciable magnitude only for atomic s orbitals centered on the corresponding nucleus.

The calculation of the contact part of the nuclear spin coupling constants using finite perturbation theory[38] follows from the expression

$$K_{MN} = \left[\frac{\partial}{\partial\mu_M}\left\langle \Psi(\mu_M)\left|\frac{8\pi}{3}\sum_i \delta(\mathbf{r}_{iN})s_{iz}\right|\Psi(\mu_M)\right\rangle\right]_{\mu_M=0}$$

(121)

where $\Psi(\mu_M)$ is the perturbed wave function when only μ_M is present, taken from a calculation based on the Hamiltonian $\mathcal{H}^0 + (8\pi/3)\beta \sum_i \delta(\mathbf{r}_{iM})s_{iz}$. The wave function $\Psi(\mu_M)$ is calculated in the unrestricted molecular orbital form, necessary to accommodate the uneven distribution of α and β electrons induced by the perturbation. The actual self-consistent field molecular orbital calculations as described in Section 3 are modified only by a change in the one-electron core integrals, which become

$$H_{\mu\nu}^{core}(\mu_M)^\alpha = H_{\mu\nu}^{core} + (8\pi/3)\beta\mu_M\langle\phi_\mu|\delta(\mathbf{r}_M)|\phi_\nu\rangle$$

(122)

and

$$H_{\mu\nu}^{core}(\mu_M)^\beta = H_{\mu\nu}^{core} - (8\pi/3)\beta\mu_M\langle\phi_\mu|\delta(\mathbf{r}_M)|\phi_\nu\rangle$$

(123)

where $H_{\mu\nu}^{core}$ represents the one-electron terms of Eq. (60). The final expression for the reduced nuclear spin coupling constant is

$$K_{MN} = \frac{8\pi}{3}\beta \sum_{\mu\nu} \langle\phi_\mu|\delta(\mathbf{r}_N)|\phi_\nu\rangle\left[\frac{\partial}{\partial\mu_M}\rho_{\mu\nu}(\mu_M)\right]_{\mu_N=0}$$

(124)

where $\rho_{\mu\nu}$ is a spin density matrix element. The derivatives of the spin density matrix are evaluated numerically based on the method of finite differences.

5.3. The Orbital and Dipolar Terms

Perturbation theory can be used to develop expressions for the orbital and dipolar contributions to nuclear spin coupling constants in a manner completely analogous to that presented for the contact term. The orbital contribution from sum-over-states perturbation theory is

$$K_{MN}^{\text{orb}} = -(8\beta^2/3) \sum_{k \neq 0} (E_k - E_0)^{-1} \langle \phi_0 | \sum_i L_{iM} r_{iM}^{-3} | \phi_k \rangle \langle \phi_k | \sum_j L_{jN} r_{jN}^{-3} | \Phi_0 \rangle$$

(125)

Forming an expansion set of ground and excited singlet states from a molecular orbital calculation on the ground state of the unperturbed system, we obtain

$$K_{MN}^{\text{orb}} = -(16\beta^2/3) \sum_l^{\text{occ}} \sum_r^{\text{vac}} {}^1\Delta E_{l \to r} \langle \psi_l | L_M r_M^{-3} | \psi_r \rangle \langle \psi_r | L_N r_N^{-3} | \psi_l \rangle$$

(126)

Introducing an LCAO expansion for the molecular orbitals gives

$$K_{MN}^{\text{orb}} = -(16\beta^2/3) \sum_l \sum_r {}^1\Delta E_{l \to r} \sum_{\mu\nu\lambda\sigma} c_{\mu l} c_{\nu r} c_{\lambda r} c_{\sigma l} \langle \phi_\mu | L_M r_M^{-3} | \phi_\nu \rangle \langle \phi_\lambda | L_N r_N^{-3} | \phi_\sigma \rangle$$

(127)

The analogous expressions for the dipolar term is

$$K_{MN}^{\text{dip}} = (8\beta^2/3) \sum_{k \neq 0} (E_k - E_0)^{-1} \left\langle \Phi_0 \left| \sum_i (r_{iN}^2 \delta_{tu} - 3 r_{iNt} r_{iNu}) r_{iN}^{-5} \right| \Phi_k \right\rangle$$

$$\times \left\langle \Phi_k \left| \sum_j (r_{jM}^2 \delta_{tu} - 3 r_{jNt} r_{jNu}) r_{jN}^{-5} \right| \Phi_0 \right\rangle$$

(128)

which reduced to molecular orbitals is

$$K_{MN}^{\text{dip}} = \frac{4\beta^2}{3} \sum_l^{\text{occ}} \sum_r^{\text{vac}} {}^3\Delta E^{-1} \langle \psi_l | (r_M^2 \delta_{tu} - 3 r_{Mt} r_{Nu}) r_M^{-5} | \psi_r \rangle \langle \psi_r | (r_N^2 \delta_{tu} - 3 r_{Nt} r_{Nu}) r_N^{-5} | \psi_l \rangle$$

(129)

where a summation convention on the tensor suffixes is implied. This expression reduced to atomic orbitals becomes

$$K_{MN}^{\text{dip}} = (4\beta^2/3) \sum_l \sum_r {}^3\Delta E_{l \to r}^{-1} \sum_{\mu\nu\lambda\sigma} c_{\mu l} c_{\nu r} c_{\lambda l} c_{\sigma r}$$

$$\times \langle \phi_\mu | (r_M^2 \delta_{tu} - 3 r_{Mt} r_{Mu}) r_M^{-5} | \phi_\nu \rangle \langle \phi_\lambda | (r_N^2 \delta_{tu} - 3 r_{Nt} r_{Nu}) r_N^{-5} | \phi_\sigma \rangle$$

(130)

Finite perturbation theory can be used to form analogous expressions to those discussed in the previous section for the contact term. A complete and concise development of the formalism has been presented by Ditchfield and Snyder.[39]

5.4. Calculations of J_{MN}

An early approximate molecular orbital treatment of nuclear spin coupling constants was given by McConnell[37] on the Fermi contact contribu-

tion to proton coupling. Equation (119) with the average energy approximation and closure, and recognizing that the dominant contributions in the
summation come from terms involving atoms M and N, becomes

$$K_{MN}^{fc} = -(256\beta^2/9)\,^3\Delta E^{-1} P_{MN} s_M^2(\mathbf{r}_M) s_N^2(\mathbf{r}_N) \tag{131}$$

where P_{MN} is the s-orbital bond order between atoms N and M and $s_N(\mathbf{r}_N)$ is
the value of the valence s orbital of atom N at the nucleus. Equation (131)
predicts all proton couplings to be positive, which is at variance with experiment. This limitation was attributed by McLachlan to the average energy
approximation.[40]

A more general treatment of spin coupling without the average energy
approximation was given by Pople and Santry[41] using sum-over-states perturbation theory and independent-electron molecular orbital theory. Their
expression for the Fermi contact term is

$$K_{MN}^{fc} = -(256\pi^2\beta^2/9)s_M^2(\mathbf{r}_M)s_N^2(\mathbf{r}_N) \sum_l^{occ} \sum_r^{vac} {}^3\Delta E_{l\to r} c_{lM} c_{rM} c_{lN} c_{rN} \tag{132}$$

which can be simplified to

$$K_{MN}^{fc} = (64\pi^2/9)s_M^2(\mathbf{r}_M)s_N^2(\mathbf{r}_N)\pi_{MN} \tag{133}$$

where π_{MN} is the atom–atom polarizability of Hückel molecular orbital theory,

$$\pi_{MN} = -4 \sum_l^{occ} \sum_r^{vac} (\epsilon_r - \epsilon_l)^{-1} c_{lM} c_{rM} c_{lN} c_{rN} \tag{134}$$

Pople and Santry also treated the orbital and dipolar contributions to nuclear
spin coupling at a commensurate level of approximation; the resulting expressions are

$$K_{MN}^{orb} = (8\beta^2/3)\langle 2p_M|r_M^{-3}|2p_M\rangle\langle 2p_N|r_N^{-3}|2p_N\rangle\,^1\Delta E^{-1}$$
$$\times \{P_{x_M x_N}P_{y_M y_N} + P_{y_M y_N}P_{z_M z_N} + P_{z_M z_N}P_{x_M x_N}$$
$$- P_{x_M y_N}P_{y_M x_N} - P_{y_M z_N}P_{z_M y_N} - P_{z_M x_N}P_{x_M z_N}\} \tag{135}$$

and

$$K_{MN}^{dip} = (4\beta^2/25)\langle 2p_M|r^{-3}|2p_M\rangle\langle 2p_N|r^{-3}|2p_N\rangle\,^3\Delta E^{-1}$$
$$\times \{2(P_{x_M x_N}^2 + P_{y_M y_N}^2 + P_{z_M z_N}^2)$$
$$+ 3(P_{x_M x_N}P_{y_M y_N} + P_{y_M y_N}P_{z_M z_N} + P_{z_M z_N}P_{x_M x_N})$$
$$- (P_{x_M y_N}^2 + P_{y_m x_N}^2 + P_{y_M z_N}^2 + P_{z_M y_N}^2 + P_{z_M x_N}^2 + P_{x_M z_N}^2)$$
$$+ 3(P_{x_M y_N}P_{y_M x_N} + P_{y_M z_N}P_{z_M y_N} + P_{z_M x_N}P_{x_M z_N})\} \tag{136}$$

where the $P_{\mu\nu}$ are all p-orbital bond order matrix elements. The Pople–Santry
theory was applied to several small organic and inorganic systems for which

independent-electron molecular orbitals could be estimated, with results in qualitative agreement with experiment, considering the level of approximation. These calculations were very influential in the development of a descriptive molecule orbital theory of spin coupling, especially for geminal coupling constants.

The numerical results of greatest relevance here are those on CF coupling in fluoromethanes giving the calculated relative magnitudes of the Fermi contact, orbital, and dipolar contributions. The calculations are summarized in Table 3. As was expected, for proton coupling the Fermi contact term is dominant, followed in importance by the orbital term. Recent work on FF coupling by Hirao *et al.*[42] indicates that here the orbital and dipolar terms are very important and sometimes exceed the Fermi contact contribution. Thus theoretical treatment of spin coupling of heavier nuclei must consider all three contributions.

Approximate self-consistent field molecular orbital theory has recently been used extensively in the calculation of nuclear spin coupling constants. A number of studies using the CNDO and INDO methods and SOS perturbation theory have been reported. Note especially the early work of Ditchfield and Murrell.[43] A extensive characterization study has been carried out by Towl and Schaumberg.[44] They have considered contact, orbital, and dipolar terms and a number of methodological alternatives including CNDO and INDO with and without CI and adjustment of the atomic integrals $\langle s_N|\delta(\mathbf{r}_N)|s_N\rangle$ and $\langle 2p_N|r_N^{-3}|2p_N\rangle$. Overall the INDO method with CI gives the best results. For proton couplings, calculated vicinal couplings are satisfactory but geminal couplings in many cases compare badly with experiment. For proton first-row-atom couplings, the directly bonded cases are reasonably well accommodated. Long-range couplings are not in quantitative agreement, but the proper signs trends are evident. Similarly, trends in heavy atom couplings are also given. A selected summary of INDO-CI results from the Towl and Schaumberg study is given in Table 3, including a decomposition into contact, orbital, and dipolar terms. While the contact contribution is often dominant, the magnitude of the orbital and dipolar terms is appreciable in certain cases and cannot be safely neglected for heavy atom coupling.

Extensive research activity on theoretical calculation of coupling constants has been centered on the application of the INDO finite perturbation method developed by Pople *et al.*[38] Within the framework of the INDO method, the integral in Eq. (124) reduces to

$$\langle \phi_\mu|\delta(\mathbf{r}_M)|\phi_\nu\rangle = s_M^2(\mathbf{r}_M) \tag{137}$$

for all $\mu = \nu =$ valence s orbitals of atom M and zero otherwise. Thus the perturbation matrix elements in Eqs. (122) and (123) are zero except for valence s orbitals of atom M. Implementation of the perturbation simply involves addition of the quantity $h_N = (8\pi/3)\beta\mu_N s_N^2(\mathbf{r}_N)$ to the diagonal matrix

Table 3. Calculated Heavy Atom Coupling Constants Using SOS-PT and INDO-CI Wave Functions[44]

J	Molecule	Contact	Orbital	Dipolar	O+D	Total	Observed
^{1}CC	C_2H_2	69.43	4.13	2.84	6.97	76.40	171.5
	C_2H_4	29.75	−3.76	0.93	−2.83	26.92	67.6
	C_2H_6	11.99	−0.55	0.30	−0.25	11.74	34.6
^{1}CF	CH_3F	−126.30	−4.15	6.92	2.77	−123.5	−157.5
	CH_2F_2	−132.64	−12.54	4.50	−8.04	−140.7	−234.8
	CHF_3	−125.53	−17.51	3.30	−14.22	−139.7	−274.3
	CF_4	−85.33	−20.17	2.40	−17.77	−103.1	−259.2
	HFCO	−320.13	−15.83	0.15	−15.68	−335.8	−369.0
	F_2CO	−96.76	−20.74	0.89	−19.85	−116.6	−308.4
^{1}CN	H_2CN_3	8.21	−1.36	0.24	−1.12	7.09	6.7
^{1}NN	H_2CN_2	−6.12	−0.42	0.30	−0.12	−6.24	−1.60
^{1}BF	BF_3	48.88	−5.12	0.33	−4.79	44.09	—
^{1}NF	NF_3	−80.68	−3.77	5.96	2.19	−78.49	—
^{2}FF	CH_2F_2	2.90	43.72	23.72	67.44	70.35	—
	CHF_3	−1.27	22.65	19.76	42.40	41.14	—
	CF_4	−1.09	6.20	16.73	22.93	21.84	—
	F_2CO	19.18	−26.88	12.54	−14.34	4.84	—
	BF_3	−1.42	−64.18	3.72	−60.46	−61.88	—
	NF_3	−0.02	58.72	34.25	92.97	92.95	—
^{2}CN	H_2CN_2	−0.51	−1.30	0.38	−0.92	−1.43	—
^{1}FF	F_2	−470.9	1631.8	903.5	2535.3	2064.4	—

elements representing the s orbital of atom B of the core Hamiltonian for α orbitals and its subtraction for β elements. The expression for the reduced nuclear spin coupling constant at the INDO level becomes

$$K^{fc}_{MN} = [(8\pi/3)\beta]^2 s^2_M(\mathbf{r}_M) s^2_N(\mathbf{r}_N)[\partial\rho_{s_M s_M}(h_N)/\partial h_N]_{h_N=0} \tag{138}$$

Upon application of the method of finite differences, this equation can be estimated as

$$K^{fc}_{MN} = [(8\pi/3)\beta]^2 s^2_M(\mathbf{r}_M) s^2_N(\mathbf{r}_N)[\rho_{s_M s_M}(h_N)/h_N]_{h_N=0} \tag{139}$$

requiring but a single matrix Hartree–Fock calculation. The values for $s^2_N(r_N)$ are taken as disposable parameters and are determined to give best agreement between theory and experiment in a least squares sense.

A set of experimental results and FPT calculations of contact nuclear spin coupling constants at the INDO level are compared in Table 4. At this point the application of INDO-FPT to a theoretical understanding of spin coupling has been very extensive, particularly by Maciel and co-workers.[45] Solvent effects have been treated by Johnston and Barfield,[46] and the individual applications have been reviewed on an annual basis by Grinter.[47] The extension of INDO-FPT to the orbital and dipolar terms has been reported by Blizzard and Santry,[48] together with extensive applications. They find that the orbital and dipolar terms play a very important role in explaining CF and FF coupling

Table 4. *Calculated Contact Coupling Constants Using FPT and INDO Wave Functions*[38]

Molecule and coupling	Calc.	Exp.	Molecule and coupling	Calc.	Exp.
Hydrogen H—H	408.60	+280	Ethylene C—C	41.45	+34.6
Water H—O—H	−8.07	(−)7.2	Ethylene C—C	82.14	+67.6
Methane H—C—H	−6.13	−12.4	Acetylene C—C	163.75	+171.5
Ethane geminal H—C—H	−5.22	—	Ethane C—C—H	−7.20	−4.5
Methyl fluoride H—C—H	−1.86	−9.6	Ethylene C—C—H	−11.57	−2.4
Ethylene H—C—H	3.24	+2.5	Acetylene C—C—H	2.52	+49.3
Formaldehyde H—C—H	31.86	+40.2	Benzene C—C—H	−4.94	+1.0
Ethane H—C—C—H (gauche)	3.25	—	Benzene C—C—C—H	9.40	+7.4
Ethane H—C—C—H (trans)	18.63	—	Benzene C—C—C—C—H	−2.27	−1.1
Ethane H—C—C—H (average)	8.37	+8	Ammonia N—H	30.40	(+)43.6
Ethylene H—C—C—H (cis)	9.31	+11.7	Water O—H	−12.84	(−)73.5
Ethylene H—C—C—H (trans)	25.15	+19.1	Hydrogen fluoride F—H	−150.22	(−)521
Acetylene H—C—C—H	10.99	+9.5	Nitrogen trifluoride N—F	−239.16	(−)155
Allene H—C—C—C—H	−9.69	−7.0	Methyl fluoride C—F	−237.15	−158
Benzene H—C—C—H	8.15	+7.54	Methyl fluoride H—C—F	4.68	+46.4
Benzene H—C—C—C—H	2.13	+1.37	Vinyl fluoride H—C—F	16.61	+84.7
Benzene H—C—C—C—C—H	1.15	+0.69	Vinyl fluoride H—C—C—F (cis)	26.7	+20.1
Methane C—H	122.92	+125	Vinyl fluoride H—C—C—F (trans)	66.20	+52.4
Ethane C—H	122.12	+124.9	1,1-Difluoroethylene F—C—F	−13.42	+36.4
Ethylene C—H	156.71	+156.4	1,2-Difluoroethylene (cis) F—C—C—F	10.28	+18.7
Acetylene C—H	232.65	+248.7	1,2-Difluoroethylene F—C—C—F	−32.42	−124.8
Benzene C—H	140.29	+157.5			

constants, and can predominate in FF cases. The orbital term also plays a significant role in certain CC couplings. The authors report generally reasonable agreement between theory and experiment in the context of approximate molecular orbital theory. Schulman and Newton[49] have recently used this approach to study coupling constants of directly bonded carbon atoms.

In concluding this section it is important to note a significant contribution to the theory of nuclear spin coupling constants which originated in valence bond theory. Karplus[50] used valence bond theory to show that the vicinal proton spin coupling constant for an ethane-like fragment should vary with the H—C—C—H dihedral angle τ as

$$J_{MN}(\tau) = A - B \cos \tau + C \cos 2\tau \qquad (140)$$

where originally $A = 4.2$, $B = -0.5$, and $C = 4.5$. This formula has been used extensively for deducing molecular geometry from observed coupling constants. It has subsequently been shown that a similar dependence can be deduced from Pople–Santry theory. This problem has received considerable attention at the approximate SCF molecular orbital level using both SOS and FPT by Govil[51] and Barfield and Sternhell[52]; additional work has been recently reviewed in the aforementioned articles by Grinter.[47]

6. ESR g-Tensors

The energy of a free electron in an externally applied magnetic field is

$$E = -\mathbf{\mu}_e \cdot \mathbf{H} \qquad (141)$$

where $\mathbf{\mu}_e$ is the magnetic moment representative of the electron spin state $|S\rangle$,

$$\mathbf{\mu}_e = -g_e \beta \mathbf{S} \qquad (142)$$

where g is the electronic g-value, 2.0023, β is the Bohr magneton, and S is the spin angular momentum. The transition energy for a spin excitation is

$$h\nu = g_e \beta H \qquad (143)$$

where ν is the resonant frequency and h is Planck's constant.

In a molecule, the presence of an external magnetic field can induce circulation of charge as discussed in terms of nuclear shielding and chemical shifts in Section 4. For a paramagnetic molecule in an external field, the induced electronic currents give rise to a net electronic orbital angular momentum with an associated magnetic moment. Also, the spin itself creates a vestigial orbital angular momentum due to spin–orbit coupling. The spin excitation energy deviates slightly from that given by Eq. (143) to

$$h\nu = g' \beta H \qquad (144)$$

where g' is derived from a tensorial quantity determining the Zeeman splitting as function of radical orientation, conveniently written as

$$g' = g_e(1 + \Delta g) \tag{145}$$

where Δg represents the shift in the free-electron g-value. Comparison of Eq. (144) with Eq. (75) with $g_e\beta$ analogous to $\gamma_N \hbar$ shows the correspondence between σ_N in nuclear spin excitation with Δg in electron spin excitation.

Quantum Mechanical Development of the Δg Tensor

Fundamental considerations on Δg as a magnetic resonance parameter follow from the analysis presented in Section 2. The numerical calculation of Δg follows from Eq. (22) and sum-over-states perturbation theory or Eq. (24) and finite perturbation theory. Most theoretical studies of Δg to date have been based only on sum-over-states perturbation theory. Introducing a sum-over-states expansion of the form of Eq. (62) for $\Psi^{(100)}$, we obtain, upon substitution into Eq. (22), the following expression for the Cartesian components of Δg:

$$S_t \Delta g_{tu} = \left\langle \Phi_0 \left| \sum_i s_{it} \mathcal{H}_{itu}^{(101)} \right| \Phi_0 \right\rangle$$

$$- \sum_{k \neq 0}^{\infty} (E_k - E_0)^{-1} \langle \Phi_0 | \mathcal{H}_t^{(100)} | \Phi_k \rangle \left\langle \Phi_k \left| \sum_i s_{iu} \mathcal{H}_{iu}^{(001)} \right| \Phi_0 \right\rangle \tag{146}$$

In terms of finite perturbation theory, Eq. (24), with $\lambda_1 = \mu_e$ and $\lambda_2 = H_u$, we have

$$S_t \Delta g_{tu} = \left[\frac{\partial}{\partial H_u} \left\langle \Psi(H_u) \left| \sum_i s_{it} \mathcal{H}_{tu}^{(101)} \cdot H_u + \sum_i s_{it} \mathcal{H}_{it}^{(001)} \right| \Psi(H_u) \right\rangle \right]_{H=0} \tag{147}$$

which can be resolved into first- and second-order contributions as

$$S_t \Delta g_{tu} = \left\langle \Psi_0 \left| \sum_i s_{it} \mathcal{H}_{tu}^{(101)} \right| \Psi_0 \right\rangle + \left[\frac{\partial}{\partial H_u} \left\langle \Psi(H_u) \left| \sum_i s_{it} \mathcal{H}_{it}^{(001)} \right| \Psi(H_u) \right\rangle \right]_{H=0} \tag{148}$$

Most theoretical work to date on g-tensors has been based on sum-over-states perturbation theory. Substituting detailed expressions for the operators $\mathcal{H}^{(101)}$ and $\mathcal{H}^{(001)}$ into Eq. (146), we obtain the following general expression for Δg:

$$S_t \Delta g_{tu} = \frac{e^2}{4\hbar mc} g\beta \left\langle \Phi_0 \left| \sum_N \sum_i Z_N (\mathbf{r}_{iN} \cdot \mathbf{r}_i \delta_{tu} - r_{it} r_{iNu}) r_{iN}^{-3} s_{it} \right| \Phi_0 \right\rangle$$

$$- g\beta^2 \sum_{k \neq 0} (E_k - E_0)^{-1} \left\langle \Phi_0 \left| \sum_N \sum_i s_{it} z_N L_{iNt} r_{iN}^{-3} \right| \Phi_k \right\rangle \left\langle \Phi_k \left| \sum_i L_{iu} \right| \Phi_0 \right\rangle \tag{149}$$

The first term is found to be negligibly small and is neglected in qualitative and simple quantitative treatments.[53] It must be retained, however, if gauge invariance is to be strictly maintained. The essential features of g-tensor shifts are contained in the second-order term, with contributions identified with the induced orbital motion of the unpaired electron and with spin–orbit coupling.[54] Thus the Δg values are greater for systems where the induced electron current about a particular axis is larger and for systems where the spin–orbit coupling constants are greater. Positive and negative shifts from the free-electron g-value can be interpreted in terms of electrons and holes, respectively.[55]

It is common practice in numerical calculations to make the further assumption[54]

$$\mathcal{H}_{ti}^{(001)} = \sum_N s_{it} \xi_{iN}(r_{iN}) L_{iNt} \tag{150}$$

where $\xi_{iN}(r_{iN})$ is an empirical or semiempirical spin–orbit coupling constant. Using this formula and a molecular orbital wave function, the second-order part of Eq. (149) reduces to [55]

$$\Delta g_{tu} = \sum_i^{m-1} \Delta E_{i \to m} \left\langle \psi_m \left| \sum_N L_{Nt} \right| \psi_i \right\rangle \left\langle \psi_i \left| \sum_N \xi_N L_{Nu} \right| \psi_m \right\rangle$$
$$- \sum_{j=m+1}^n \Delta E_{m \to j}^{-1} \left\langle \psi_m \left| \sum_N L_{Nt} \right| \psi_j \right\rangle \left\langle \psi_j \left| \sum_N \xi_N L_{Nu} \right| \psi_n \right\rangle \tag{151}$$

where m, i, j, and n refer to singly occupied, doubly occupied, vacant, and the highest vacant molecular orbitals in the ground-state configuration. In reducing this expression to atomic orbitals, we use the further simplification

$$\left\langle \psi_i \left| \sum_N \xi_N(r_N) L_N \right| \psi_i \right\rangle \cong \zeta_N \langle \phi_\mu | L_N | \phi_\mu \rangle, \qquad \mu \in N \tag{152}$$

where ζ_N is the usual spin–orbit coupling constant for atom N, since $\xi(r_N)$ is large only in the vicinity of atom N.

Numerical calculations based on this approach have been reported for several radicals by Morikawa *et al.*[56] using INDO molecular orbital theory. The results are summarized in Table 5. The spin–orbit coupling approximation has been rigorously studied for H_2^+ by de Montgolfier and Harriman[53] and found to lack essential features. Overall Δg values have not been quantitatively studied as extensively using approximate molecular orbital theory as any of the other magnetic resonance parameters, and more attention is warranted at this point. INDO–finite perturbation theory has been formulated for g-tensors, but has not yet been implemented for numerical calculation.[57]

7. ESR Electron–Nuclear Hyperfine Tensors

In addition to the splitting of electron Zeeman levels by the external magnetic field in an electron spin resonance experiment, there exists additional

Table 5. Calculated Principal Values of the Electronic g-Tensor Using SOS-PT and INDO Wave Functions[a]

Radical	g_{xx}	g_{yy}	g_{zz}	g_{av}
CH_3CO	1.9983	2.0022	2.0035	2.0013
	(1.9964)	(2.0019)	(2.0040)	(2.0008)
HCN^-	1.9996	2.0026	2.0030	2.0017
				(2.0022)
NH_2CO	2.0001	2.0023	2.0038	2.0021
				(2.0017)
CO_2^-	1.9966	2.0024	2.0048	2.0013
	(1.9973)	(2.0016)	(2.0032)	(2.0007)
NO_2	1.9908	2.0024	2.0067	1.9999
	(1.9910)	(2.0020)	(2.0062)	(1.9997)
$CH_2=CH$	2.0021	2.0023	2.0023	2.0022
				(2.0022)
Phenyl	2.0020	2.0023	2.0024	2.0023
	(2.0014)	(2.0023)	(2.0034)	(2.0024)

[a] Experimental data in parentheses. From Ref. 56.

splitting due to electron–nuclear hyperfine interactions. The hyperfine interactions arise to first order from a contact mechanism due to the finite probability of finding electrons directly at the nuclear positions and longer range electron spin–nuclear spin magnetic dipole–dipole interactions. At second order, additional terms arise due to spin–orbit coupling and electron–orbital nuclear spin interactions, and are known as pseudo-hyperfine constants. We begin this section with a development of first general quantum mechanical expressions for hyperfine coupling constants. Then isotropic and anisotropic coupling are discussed in turn, beginning with considerations on the coupling mechanism and proceeding to a discussion of numerical calculations and selected prototypical case studies and examples of applications.

7.1. Quantum Mechanical Development of T_N

The hyperfine tensors T_N collect all terms bilinear in electron and nuclear spins. Using an expansion of the form of Eq. (81) for $\Psi^{(010)}$, an expression for T_N can be developed from Eq. (27) as

$$S_t T_{Ntu} = \left\langle \Psi_0 \left| \sum_i s_{it} \mathcal{H}_{Ntu}^{(011)} \right| \Psi_0 \right\rangle$$

$$- \sum_{k \neq 0}^{\infty} (E_k - E_0)^{-1} \langle \Psi_0 | \mathcal{H}_{Nt}^{(010)} | \Phi_k \rangle \langle \Phi_k | \sum_i s_{iu} \mathcal{H}_{iu}^{(001)} | \Phi_0 \rangle \qquad (153)$$

where the first term is the hyperfine tensor and the second term is the pseudo-hyperfine term. Application of finite perturbation theory to the problem gives an analogous expression. Equation (29) with $\lambda_1 = \mu_e$ and $\lambda_2 = \mu_N$ gives

$$S_t T_{Ntu} = \left[\frac{\partial}{\partial \mu_{Nu}} \left\langle \Psi(\mu_{Nu}) \left| \sum_i s_{it} \mathcal{H}_{Ntu}^{(011)} \cdot \mu_{Nu} + \sum_i s_{it} \mathcal{H}_t^{(001)} \right| \Psi(\mu_{Nu}) \right\rangle \right]_{\mu_N = 0} \quad (154)$$

which can be readily resolved into first- and second-order contributions,

$$S_t T_{Ntu} = \left\langle \Psi_0 \left| \sum_i s_{it} \mathcal{H}_{Ntu}^{(011)} \right| \Psi_0 \right\rangle$$

$$+ \left[\frac{\partial}{\partial \mu_{Nu}} \left\langle \Psi(\mu_{Nu}) \left| \sum_i s_{it} \mathcal{H}_t^{(001)} \right| \Psi(\mu_{Nu}) \right\rangle \right]_{\mu_N = 0} \quad (155)$$

Introducing explicitly the operator $\mathcal{H}_N^{(011)}$, we can decompose the first-order hyperfine contribution further into an isotropic contribution a_N and an anisotropic contribution $\mathbf{t}_N$. These quantities are measured and calculated in an unreduced form, and we maintain this convention here. Thus we write

$$g \beta \gamma_N \hbar S_t^{-1} \left\langle \Psi_0 \left| \sum_i s_{it} \mathcal{H}_{Ntu}^{(011)} \right| \Psi_0 \right\rangle = a_N + t_{Ntu} \quad (156)$$

where

$$a_N = g \beta \gamma_N \hbar \frac{4\pi}{3} S_z^{-1} \left\langle \Psi_0 \left| \sum_i 2 \mathbf{s}_{iz} \delta(\mathbf{r}_{iN}) \right| \Psi_0 \right\rangle \quad (157)$$

and

$$t_{Ntu} = \tfrac{1}{2} g \beta \gamma_N \hbar S_z^{-1} \left\langle \Psi_0 \left| \sum_i 2 \mathbf{s}_{iz} (r_{iN}^2 \delta_{tu} - 3 r_{iNt} r_{iNu}) r_{iN}^{-5} \right| \Psi_0 \right\rangle \quad (158)$$

with z as the axis of electron spin quantization. The isotropic hyperfine term a_N arises from the contact interaction and the anisotropic hyperfine tensor $\mathbf{t}_N$ is a consequence of electron–nuclear magnetic dipolar interactions. The pseudo-hyperfine term is only of limited interest, and has been treated by Lefebvre.[58] We restrict our attention here to first-order terms.

The calculation of a_N and $\mathbf{t}_N$ using molecular orbital theory proceeds along the following lines. First the wave function Ψ_0 for the paramagnetic ground state of the system under consideration is obtained, using some form of open-shell methodology. On the basis of the spin-unrestricted wave function as specified in Eqs. (47), the expectation values of Eqs. (157) and (158) reduce to

$$\left\langle \Psi \left| \sum_i 2 \mathbf{s}_{iz} \delta(\mathbf{r}_{iN}) \right| \Psi \right\rangle = \sum_{\mu\nu} \rho_{\mu\nu} \phi_\mu^*(\mathbf{r}_N) \phi_\nu(\mathbf{r}_N) \quad (159)$$

and

$$\left\langle \Psi_0 \left| \sum_i 2s_{iz} (r_i^2 \delta_{tu} - 3r_{iNt}r_{iNu}) r_{iN}^{-5} \right| \Psi_0 \right\rangle$$

$$= \sum_{\mu\nu} \rho_{\mu\nu} \left\langle \phi_\mu \left| r_N^2 \delta_{tu} - 3r_{Nt}r_{Nu} \right| \phi_\nu \right\rangle \qquad (160)$$

where $\rho_{\mu\nu}$ is an element of the spin density matrix,

$$\rho_{\mu\nu} = P_{\mu\nu}^\alpha - P_{\mu\nu}^\beta \qquad (161)$$

and $\phi_\mu(r_N)$ is the density at the nucleus N. Note that only s-type atomic orbitals have a finite density directly at the nuclear positions. The final expressions for the isotropic and anisotropic elements of the electron–nuclear hyperfine tensor are thus

$$a_N = (4\pi/3)g\beta\gamma_N\hbar S_z^{-1} \sum_{\mu\nu} \rho_{\mu\nu}\phi_\mu^*(\mathbf{r}_N)\phi_\nu(\mathbf{r}_N) \qquad (162)$$

$$t_N = -\tfrac{1}{2}g\beta\hbar\gamma_N S_z^{-1} \sum_{\mu\nu} \rho_{\mu\nu} \left\langle \phi_\mu \left| r_N^{-5} (r_N^2 \delta_{tu} - 3r_{Nt}r_{Nu}) \right| \phi_\nu \right\rangle \qquad (163)$$

The anisotropic hyperfine coupling constants are obtained as eigenvalues of the hyperfine tensors calculated with respect to the molecular coordinate frame. The corresponding eigenvectors define the orientation of the principal axes of the hyperfine tensor with respect to the molecular frame. The anisotropic term is pertinent only to spectra of an oriented system and averages to zero for a molecule in solution.

7.2. Isotropic Hyperfine Coupling

The electron–nuclear contact interaction gives rise to isotropic hyperfine coupling constants. Contact interaction requires an unpaired spin density in s-type orbitals associated with magnetic nuclei. The electronic structure of free radicals is often discussed in terminology derived from descriptive structural chemistry, where the unpaired electron is associated with a radical site, i.e., a particular valence orbital or hybrid orbital on one or another atom in the principal valence structures. If the nucleus of the atom is magnetic and the radical site has s character, contact interaction and isotropic hyperfine splitting of spectral lines arise directly. A typical case in point is $\cdot CF_3$, where the radical site is essentially a carbon sp^3 hybrid orbital and $a_C = +92.7\ G$.[59]

Descriptive molecular orbital theory of free radicals places the unpaired electron in the highest occupied molecular orbital, and can formally be regarded as a spin-restricted account of the electronic structure. In many cases the highest occupied molecular orbital can be roughly identified with the radical site, such as in methyl radical and ethyl radical, where the radical site is the π orbital of an sp^2-hybridized carbon atom. Significant hyperfine coupling is observed to all protons of both systems[60] and the manner in which this comes about serves to illustrate the two main mechanisms for the transmission of unpaired spin from the radical site into s-type orbitals: spin polarization and spin delocalization.

Spin polarization is responsible for hyperfine coupling to magnetic nuclei that lie in the nodal plane of a radical site, such as the protons in methyl radical and the methylene protons in ethyl radical. Weissman[61] and McConnell[62] both noted that the observed a_N values for protons were roughly proportional to the π-electron spin densities for the system calculated from simple Hückel molecular orbital theory in a π-electron basis. The quantitative statement of the relation was given by McConnell as

$$a_N = Q_N \rho^{\pi}_{\mu\mu}, \qquad \mu \in N \tag{164}$$

Empirically Q_N was found to be constant within useful limits near the symmetry-determined value for benzene anion, 22.5 G.

The mechanism by which unpaired spin is transmitted from the π system into the σ system of a planar unsaturated free radical was first described by McConnell[62] and Bershon[63] using valence bond arguments and by Weissman[61] using molecular orbital theory and configuration interaction. Consider a C—H fragment described by the spin-restricted single-determinant function

$$\Phi_0 = |\sigma\bar{\sigma}\pi| \tag{165}$$

where σ is a C—H bonding orbital with contributions from the σ atomic orbitals of carbon and the hydrogen 1s function and π is an unhybridized $2p_\pi$ atomic orbital. The spin density at the hydrogen nucleus is identically zero for this function, since α-spin density from σ is exactly canceled by β-spin density from $\bar{\sigma}$. An improved description of the system is provided by perturbation theory, where the electron repulsion operator $\sum_{i<j} e^2 r_{ij}$ is taken as the perturbation operator; this is equivalent to a perturbation theory approximation to configuration interaction. The relevant mixing as far as spin density is concerned is

$$\Psi = \Phi_0 + C_1\Phi_1 \tag{166}$$

where Φ_1 is a spin-correct excited state resulting from a $\sigma \to \sigma^*$ excitation from the C—H bonding σ orbital to the C—H antibonding σ^* orbital,

$$\Phi_1 = 6^{-1/2}\{2|\sigma\sigma^*\pi| - |\sigma\bar{\sigma}\pi| - |\bar{\sigma}\sigma^*\pi|\} \tag{167}$$

and

$$c_1 = -(E_1 - E_0)^{-1}\left\langle \Phi_1 \left| \sum_{i<j} e^2 r_{ij}^{-1} \right| \Phi_0 \right\rangle = \Delta E_{\sigma \to \sigma^*}^{-1} \tfrac{3}{2}(\sigma\pi/\pi\sigma^*) \tag{168}$$

where the rightmost expression in Eq. (168) follows from algebraic reduction.

The spin density at nucleus N is thus

$$\rho(r_N) = 2\Delta E^{-1}_{\sigma \to \sigma^*} (\sigma\pi|\pi\sigma^*)\sigma(\mathbf{r}_N)\sigma^*(\mathbf{r}_N) \qquad (169)$$

which can be written in a form which displays the π-electron spin density explicitly as

$$\rho(r_N) = 2\Delta E^{-1}\sigma(\mathbf{r}_N)\sigma^*(\mathbf{r}_N)\sum_\mu\sum_\nu (\sigma\phi_\mu|\phi_\nu\sigma^*)\rho^\pi_{\mu\nu} \qquad (170)$$

Then, using Eq. (162), we can write

$$a_N = \sum_\mu\sum_\nu Q^N_{\mu\nu}\rho_{\mu\nu} \qquad (171)$$

where

$$Q_{\mu\nu} = (16\pi/3)g\beta\gamma_N\hbar\,\Delta E^{-1}\sigma(\mathbf{r}_N)\sigma^*(\mathbf{r}_N)(\sigma\phi_\mu|\phi_\nu\sigma^*) \qquad (172)$$

which is small unless $\phi_\mu = \phi_\nu$ in the exchange coupling integral; hence

$$a_N \cong Q^N_{\mu\mu}\rho^\pi_{\mu\mu}. \qquad (173)$$

The introduction of correlation of α and β electrons by perturbative configuration interaction has produced a spin polarization of the electrons. Noting that

$$\sigma_N(\mathbf{r}_N)\sigma^*(\mathbf{r}_N) = [s_C(\mathbf{r}_N) + \lambda s_H(\mathbf{r}_N)][s_C(\mathbf{r}_N) - \mu s_H(\mathbf{r}_N)] \qquad (174)$$

where λ and μ are related to LCAO expansion coefficients and are introduced to display phasal relationships between carbon s orbital S_C and hydrogen s orbital S_H explicitly, we obtain the proportionalities

$$\sigma_N(\mathbf{r}_C)\sigma^*(\mathbf{r}_C) \propto s^2_C(\mathbf{r}_C) \qquad (175)$$

and

$$\sigma_N(\mathbf{r}_H)\sigma^*(\mathbf{r}_H) \propto -s^2_H(\mathbf{r}_H) \qquad (176)$$

and thus Q^C and a_C are positive and Q^H and a_H are negative. The unpaired spin in the carbon $2p_\pi$ orbital essentially induces a net parallel unpaired spin in the carbon σ orbitals in accordance with Hund's rule; the concentration of parallel spin around the carbon leads to a concentration of net antiparallel spin about the proton and net negative spin density at the hydrogen nucleus. This theoretical account of the mechanism of hyperfine coupling in π radicals by no means guarantees the constancy of Q^N for a particular nucleus, but this has emerged empirically, with some significant exceptions. More elaborate expressions of the proportionality have been developed to handle these cases.[64]

The highest occupied molecular orbital differs from the radical site of a valence structure in that the molecular orbital can be delocalized over the entire molecule. Spin delocalization occurs when the highest occupied molecular orbitals has s character from a number of magnetic nuclei. The highest occupied molecular orbital (HOMO) of ethyl radical is essentially a $2p_\pi$ orbital from the methylene carbon, but also contains an admixture of the methyl group orbital of π symmetry,

$$\psi_{\text{HOMO}} = 2p_{\pi_{\text{C}}} + \lambda[2 \cdot 1s_{\text{H}_1} - 1s_{\text{H}_2} - 1s_{\text{H}_3}] \tag{177}$$

where H_1, H_2, and H_3 are methyl protons. Thus the highest occupied molecular orbital is delocalized into the β-methyl group, and the methyl protons gain unpaired spin density by a direct spin delocalization, and the spin density of the β protons will be positive.

For magnetic nuclei in a three-dimensional polyatomic molecule a combination of spin delocalization and spin polarization may be responsible for contact hyperfine interactions. Note that a spin-restricted function can accommodate spin delocalization, but spin polarization requires a wave function including some form of electron correlation. This can be in the form of configuration interaction on the basis of an expansion set of restricted functions, relaxation of spin restrictions on the single-determinant function as in the spin-unrestricted form, or going to a spin-extended multideterminant wave function. Approximate molecular orbital theory at the INDO level of approximation of principal concern here uses spin-unrestricted single-determinant molecular orbital wave functions.

7.3. Calculations of Isotropic Hyperfine Constants

The calculation of isotropic hyperfine coupling constants has been attempted by each of the quantum mechanical methods developed for the theoretical treatment of molecular structure. Early studies were carried out using Hückel theory[8] and approximate SCF molecular orbital theory in the π-electron approximation,[65] linking π-electron spin densities to a_N using the McConnell relation. Discrepancies between theory and experiment were treated by including additional terms in an expression based on Eq. (162) or variations on the general theme using perturbation theory. Considerable interesting research was produced in this area, and the subject of π-electron calculation of hyperfine coupling constants has been treated at length in recent review articles and texts.[8,65] We thus turn our attention to calculations beyond the π-electron approximation.

As mentioned, the lowest level of approximate molecular orbital theory wherein one can hope to obtain a general account of spin properties is the INDO method,[12] due particularly to the fact that one-center atomic exchange integrals, the main vehicle for spin polarization, are included. Isotropic hyperfine coupling constants were the first spin property to be studied by this method[12]; the most extensive studies of isotropic constants to date are based on this approach,[66] using the unrestricted Hartree–Fock method for obtaining wave functions for paramagnetic molecules.

The approximations involved in the INDO calculation of molecular orbital wave functions were reviewed in Section 3, and for the calculation of a_N based on spin densities calculated by the INDO method we need to reduce the expression for a_N in Eq. (162) in a manner consistent with the wave function calculation. Thus it is assumed that for the INDO calculation of a_N, all contributions to the double summation over μ and ν in Eq. (162) are neglected unless both ϕ_μ and ϕ_ν are centered on atom N. Of the atomic functions centered on atom N, only s functions have nonvanishing densities at the nucleus and contribute to a_N. Thus

$$\sum_{\mu\nu} \rho_{\mu\nu}\phi_\mu(\mathbf{r}_N)\phi_\nu(\mathbf{r}_N) \cong \rho_{s_N s_N}|\phi_{s_N}(\mathbf{r}_N)|^2 \tag{178}$$

and for INDO calculations

$$a_N = \{(4\pi/3)g\beta\gamma_N\hbar\langle S_z\rangle^{-1}|\phi_{s_N}(\mathbf{r}_N)|^2\}\rho_{s_N s_N} \tag{179}$$

where $\rho_{s_N s_N}$ is the unpaired electron population of the valence s orbital on atom N and $|\phi_{s_N}(\mathbf{r}_N)|^2$ is the density of the valence s orbital evaluated at the nucleus. The question of inner shells is neglected in valence electron calculations and the anomalous behavior of Slater orbitals of the nucleus complicates the analytical evaluation of $|\phi_{s_N}(\mathbf{r}_N)|^2$, so for practical purposes this quantity is treated as a semiempirical parameter and is evaluated for optimum agreement between calculated and observed a_N using least squares methods on as large a data base as possible.

The level of agreement obtained between theory and experiment for INDO calculations of a_N in organic free radicals is summarized in Tables 6 and 7. The values of ^{1}H, ^{14}N, and ^{19}F isotropic constants calculated using the INDO method compare reasonably well with experiment, while the results for ^{17}O, based on a small sample, are inferior. INDO parameters and calculations of a_N for molecules containing second row atoms have been reported by Benson and Hudson,[67] with quite good results for ^{35}Cl couplings, $\sigma = 1.9$ G and $r = 0.96$ for ten points. Annihilation of the principal contaminating spin component in the unrestricted wave function leads to no significant improvement in the correlation between calculated and observed values.[68] The ^{15}N results have been refined by Hirst.[79] Overall the quality of the theoretical results on the INDO molecular orbital calculation of isotropic constants is suitable for use of the method in the analysis and interpretation of ESR spectra. Selected results are given in Table 5.

The calculated spin densities and isotropic hyperfine coupling constants using approximate molecular orbital theory at the INDO level of approximation reflect quantitatively the characteristic features of the electronic structure of free radicals. The calculated spin density of methyl radical involves spin polarization. The unit spin density in the carbon $2p_\pi$ orbital polarizes parallel unpaired spin in the carbon σ orbitals. The excess of α spin around carbon

Table 6. Calculated Isotropic Hyperfine Coupling Constants for ^1H Using INDO Wave Functions[66]

Radical	Atom or group	a_N, G Calc.	a_N, G Exp.	Radical	Atom or group	a_N, G Calc.	a_N, G Exp.
Methyl	—	−22.4	(−)23.04	Stilbene$^-$	1	−3.7	(−)1.90
Fluoromethyl	—	−7.8	(−)21.10		2	2.0	(+)0.8
Difluoromethyl	—	21.9	(+)22.20		3	−3.9	(−)3.80
Ethyl	CH_2	−20.4	(−)22.38		4	1.9	(+)0.32
	CH_2	27.6	(+)26.87		5	−3.4	(−)2.96
Vinyl	α	17.1	(+)13.40		7	−5.2	(−)4.36
	β_1	55.1	(+)65.00	Biphenylene$^-$	1	0.2	(+)0.21
	β_2	21.2	(+)37.00		2	−2.1	(−)2.86
Formyl	—	32.7	(+)16.10	Azulene$^-$	1	0	(+)0.27
Ethynyl	1	−14.6	(−)13.93		2	−3.0	(−)3.95
Allyl	1′	−14.9	(−)14.83		4	−7.0	(−)6.22
	2	6.9	(+)4.06		5	3.9	(+)1.34
Phenyl	2	18.7	(+)19.50		6	−9.4	(−)8.82
	3	6.1	(+)6.50	Fluoranthene$^-$	1	−4.4	(−)3.90
	4	3.9	—		2	2.2	(+)1.30
Cyclopentadienyl	—	−4.8	(−)5.60		3	−6.4	(−)5.20
Tropyl	—	−3.2	(−)3.95		7	0.2	—
Benzyl	−CH_2	−17.0	(−)16.35		8	−0.9	—
	2	−6.4	(−)5.14	Benzonitrile$^-$	2	−3.3	(−)3.63
	3	3.6	(+)1.75		3	1.1	(+)0.30
	4	−5.6	(−)6.14		4	−8.0	(−)8.42
Phenoxy	2	−4.1	(−)6.60	Phthalonitrile$^-$	3	1.5	(+)0.33
	3	2.2	(+)1.96		4	−4.0	(−)4.24
	4	−3.4	(−)10.40	Isophthalonitrile$^-$	2	1.5	(+)0.08
Cyclohexadienyl	CH_2	97.6	(+)47.71		4	−7.6	(−)8.29
	2	−11.1	(+)8.99	Terephthalonitrile$^-$	5	2.6	(+)1.44
					—	−1.0	(−)1.59

Compound	Position		
Perinaphthenyl$^-$	3	5.51	(+)2.65
	4	-9.8	(-)13.04
Benzene$^-$	1	-7.5	(-)7.30
	2	4.3	(+)2.80
Cyclooctatetraene$^-$	—	-3.6	(-)3.75
trans-Butadiene$^-$	—	-2.6	(-)3.21
	1	-9.8	(-)7.62
	1'	-10.3	(-)7.62
	2	-0.8	(-)2.79
Naphthalene$^-$	1	-5.3	(-)4.90
	2	-0.9	(-)1.83
Anthracene$^-$	1	-2.7	(-)2.74
	2	-0.6	(-)1.51
	9	-6.8	(-)5.34
Anthracene$^+$	1	-2.9	(-)3.08
	2	-0.6	(-)1.38
	9	-6.6	(-)6.49
Phenanthrene$^-$	1	-4.6	(-)3.60
	2	1.2	(+)0.72
	3	-3.8	(-)2.88
	4	0.6	(+)0.32
	9	-5.0	(-)4.32
Pyrene$^-$	1	-5.5	(-)4.75
	2	2.5	(+)1.09
	4	-1.9	(-)2.08

Compound	Position		
1,2,4,5-Tetracyanobenzene$^-$	—	2.2	(+)1.11
p-Nitrobenzonitrile$^-$	2	1.8	(+)0.76
Nitrobenzene$^-$	3	-3.5	(-)3.12
	2	-3.6	(-)3.39
	3	1.9	(+)1.09
m-Dinitrobenzene$^-$	4	-3.8	(-)3.97
	2	0.4	(+)3.11
	4	-7.8	(-)4.19
	5	3.2	(+)1.08
p-Dinitrobenzene$^-$	—	-1.0	(-)1.12
m-Fluoronitrobenzene$^-$	2	-3.7	(-)3.30
	4	-3.7	(-)3.30
	5	1.8	(+)1.10
	6	-3.4	(-)3.00
p-Fluoronitrobenzene$^-$	2	-3.8	(-)3.56
	3	2.2	(+)1.16
3,5-Difluoronitrobenzene$^-$	2	-3.5	(-)3.26
	4	-3.6	(-)3.98
	3	-1.9	(-)3.65
o-Benzosemiquinone$^-$	4	0.2	(+)0.95
	—	-0.9	(-)2.37
p-Benzosemiquinone$^-$	—	2.4	(+)0.79
2,5-Dioxo-1,4-semiquinone^{3-}	—	-1.0	(-)3.23
1,4-Naphthosemiquinone$^-$	2	0.6	(+)0.65
	5	-0.1	(-)0.51
	6		

Table 7. *Calculated Heavy Atom Hyperfine Coupling Constants Using INDO Wave Functions*[66]

Radical	Atom or group	a_N, G Calc.	a_N, G Exp.
(a) ^{14}N Couplings			
Benzonitrile⁻	—	2.4	(+)2.15
Phthalonitrile⁻	—	1.9	(+)1.80
Isophthalonitrile⁻	—	1.3	(+)1.02
Terephthalonitrile⁻	—	2.0	(+)1.81
1,2,4,5-Tetracyanobenzene⁻	—	1.4	(+)1.15
p-Nitrobenzonitrile⁻	CN	1.1	(+)0.76
	NO₂	4.7	(+)7.15
Nitrobenzene⁻	—	7.1	(+)10.32
m-Dinitrobenzene⁻	—	0.5	(+)4.68
p-Dinitrobenzene⁻	—	−0.0	(−)1.74
m-Fluoronitrobenzene⁻	—	6.6	(+)12.60
p-Fluoronitrobenzene⁻	—	7.1	(+)9.95
3,5-Difluoronitrobenzene⁻	—	6.1	(+)8.09
Pyrazine⁻	—	8.3	(+)7.21
N,N-Dihydropyrazine⁺	—	7.8	(+)7.60
Pyridazine⁻	—	7.7	(+)5.90
s-Tetrazine⁻	—	5.8	(+)5.28
1,5-Diazanaphthalene⁻	—	5.9	(+)3.37
Phthalazine⁻	—	0.3	(+)0.88
Quinoxaline⁻	—	7.3	(+)5.64
Dihydroquinoxaline⁺	—	7.7	(+)6.65
Phenazine⁻	—	7.2	(+)5.14
1,4,5,8-Tetraazaanthracene⁻	—	3.3	(+)2.41
p-Dicyanotetrazine⁻	Ring	5.9	(+)5.88
	CN	−0.9	(−)0.16
p-Nitrobenzaldehyde⁻	—	−0.5	(+)5.83
p-Cyanobenzaldehyde⁻	—	1.0	(+)1.40
4-Cyanopyridine⁻	Ring	8.3	(+)5.67
	CN	2.7	(+)2.33
(b) ^{19}F Couplings[a]			
Fluoromethyl	—	71.3	(+)64.30
Difluoromethyl	—	87.1	(+)84.20
Trifluoromethyl	—	159.5	(+)142.40
Monofluoroacetamide	—	34.4	54.60
Difluoroacetamide	1′	31.5	75.00
	1	39.0	75.00
m-Fluoronitrobenzene⁻	—	−4.0	(−)3.70
p-Fluoronitrobenzene⁻	—	6.3	(+)8.41
3,5-Difluoronitrobenzene⁻	—	−3.8	(−)2.73
(c) ^{17}O Couplings			
p-Benzosemiquinone⁻	—	−8.7	(−)9.53
1,4-Naphthosemiquinone⁻	—	−9.3	(−)8.58
9,10-Anthrasemiquinone⁻	—	−9.9	(−)7.53
2,5-Dioxo-1,4-semiquinone³⁻	—	−3.6	(−)4.57
Nitrobenzene⁻	—	−4.3	(−)8.84

[a] See also Ref. 79.

leaves a net spin of opposite sign in the vicinity of the peripheral hydrogen atoms. The calculated isotropic hyperfine constants are $a_C = 45$ G and $a_H = -22.4$ G using the proportionality constants given in Table I of Ref. 66, as compared to observed values of $(+)38$ G and $(-)23$ G.[60]

The INDO-calculated spin density for ethyl radical involves both spin polarization and spin delocalization. Spin polarization leads to positive spin density in the methylene carbon $1s$ orbital and negative spin density in methylene hydrogens just as in the methyl radical. The net spin density in the methyl carbon is also negative due to spin polarization, but the methyl hydrogens are positive in an orientationally averaged methyl group. Examination of the results from a single calculation with one of the methyl hydrogens coincident with the plane defined by the methylene group shows that spin polarization is not significant at this distance from the radical site, and the methyl spin density is thus due to spin delocalization. For the methyl group, a_C and a_H are calculated to be -12 and $+28$ G, with observed values of 14 and 29 G, respectively.[60]

The overall capabilities and limitations of INDO molecular orbital theory for calculating isotropic hyperfine coupling constants may be ascertained in part from a consideration of the statistics of Ref. 66, but further consideration of prototype systems is necessary to point out certain shortcomings. The limitation on the calculation of proton constants is best illustrated by the benzyl radical. The benzyl radical has been treated by every quantum mechanical method capable of dealing with open-shell systems, and collected works on the system have been reviewed in several recent papers.[69,70] The INDO results are representative enough and are given in Table 6. Note that overall there is good quantitative agreement between calculated and observed values but that the ratio of *ortho* a_H to *para* a_H is out of order. This result is typical of quite a number of molecular orbital studies at both the pi-electron and valence electron level and its origins have been discussed in terms of geometrical considerations and the quality of the wave function; current thinking points to a combination of the two factors. The most recent extensive studies on isotropic constants in the benzyl radical[71] in general have been based on energy-optimized geometries rather than standard geometries as in earlier work; the difference is sometimes significant. Note, however, in this regard that while approximate molecular orbital calculations at the INDO level have overall been quite successful, it is not unusual to find systems where predictions of molecular geometry are in error and studies along this line must be carried out in the proper perspective.

A related problem arose in early INDO studies of the cyclohexadienyl radical,[66] where the standard geometry gave the wrong electronic ground state. For systems with low-lying excited states, the approximate nature of calculations at this level can lead to erroneous predictions of the state of lowest energy. The case of an orbitally degenerate ground state as found in Jahn–

Table 8. Calculated Hyperfine Coupling Tensors (MHz) for Methyl Radical and Malonic Acid Radical Using INDO Molecular Wave Functions[17]

Coupling constant		INDO-MO	Ab initio MO+CI	Experiment
^1H	a_H	−65.0	−122.7	−64.6
	t_{xx}	−45.0	−43.6	—
	t_{yy}	2.1	−7.8	1.2
	t_{zz}	42.9	51.4	—
	$(t_{xx}+t_{zz})/2$	−1.0	—	−0.6±0.8
CH_3				
^{13}C	a_C	120.8	393.8	107.4
	t_{xx}	−75.0	−53.1	—
	t_{yy}	150.0	103.2	126.1
	t_{zz}	−75.0	−50.1	—
	$(t_{xx}+t_{zz})/2$	−75.0	−51.6	−63.5
^1H	a_H	−51	—	−59
	t_{xx}	−38	—	−32
	t_{yy}	1.6	—	1
	t_{zz}	36	—	31
$HC(OOOH)_2$				
^{13}C	a_c	106	—	93
	t_{xx}	−60	—	−70
	t_{yy}	122	—	120
	t_{zz}	−62	—	−50

Teller molecules has been treated by optimizing the Jahn–Teller half-states separately and averaging the results.

The limitations and typical errors mentioned above notwithstanding, INDO molecular orbital theory has now been used successfully for the calculation of isotropic hyperfine constants in a wide variety of cases. As evidenced by Refs. 72 and 73, this methodology has become a convenient tool for spectral analysis and interpretation in the hands of ESR spectroscopists.

7.4. Anisotropic Hyperfine Coupling

Electron–nuclear magnetic dipole–dipole interaction gives rise to anisotropic hyperfine coupling constants. Whereas the isotropic constants discussed above are essentially a measure of the unpaired spin density in s orbitals centered on the magnetic nucleus, the anisotropic constants reflect the spin population of valence $p, d, f, \ldots$ orbitals centered on the magnetic nucleus and of all orbitals centered on neighboring atoms. Determination of the anisotropic hyperfine coupling constants requires measuring the ESR spectrum of the system in the crystalline state as a function of the orientation of the crystallographic axes with respect to the external field. Although the analysis is thus more difficult and tedious than for solution spectra, considerable additional information is obtained on both the electronic and molecular structure of the sample, since a number of orbitals can contribute to each constant and the orientation of the tensorial axes can be used to estimate intramolecular bond angles and the relative orientation of molecules in the crystal.

Some of the earliest experimental measurements of anisotropic hyperfine coupling constants were obtained in studies of radiation-damaged malonic acid crystals.[74] The radicaloid species was identified as $HC(COOH)_2$ with the observed anisotropic hyperfine splittings associated with the interaction of an unpaired electron essentially localized in the $2p_\pi$ orbital of the central carbon atom with the magnetic nucleus of the α-hydrogen atom. The isotropic hyperfine coupling constant was found to be $a_H = -59$ MHz and the anisotropic constants t_{xx}, t_{yy}, and t_{zz} were -32, $+1$, and $+31$ Mz, respectively.

Early theoretical studies on anisotropic hyperfine coupling in prototype systems were aimed at explaining these results: McConnell and Strathdee[75] studied the C—H fragment with unit unpaired spin density in a carbon $2p_\pi$ orbital. Here the calculation of each anisotropic constant according to Eq. (163) reduces to a single term involving integrals of the form $\langle 2p_\pi | r_N^{-3}(3\cos^2\theta - 1) | 2p_\pi \rangle$. Contributions to the integrand will be zero when $(3\cos^2\theta - 1)$ is zero, and this defines as a nodal surface a cone of half-angle $\cos^{-1}(1/3)$. When $\mathbf{H}_0$ is parallel to the internuclear line (z direction), the carbon $2p_\pi$ orbital is almost entirely within the positive region of the cone. This gives rise to a large and positive interaction, identified with the experimentally observed t_{zz}. When $\mathbf{H}_0$ is parallel to the nodal plane and perpendicular to the C—H bond, the orbital lies in the negative region of the cone and the interaction is large and negative, and identified with the experimentally observed t_{xx}. With $\mathbf{H}_0$ parallel to the $2p_\pi$ orbital axis the nodal cone intersects the region of high electron density and thus the interaction would be numerically small, t_{yy}. This pattern of anisotropic hyperfine coupling constants, t_{xx} large and negative, t_{yy} near zero, and t_{zz} large and positive, has now been observed in a number of systems and is considered diagnostic of hyperfine interactions of unpaired electrons with α-hydrogen nuclei in π-electron organic radicals.

7.5. Calculations of Anisotropic Hyperfine Constants

Recent quantitative approximate molecular orbital calculations on anisotropic hyperfine coupling constants were reported by Loew and co-workers[76] using extended Hückel-theory valence-electron self-consistent field studies at the INDO level of approximation. Beveridge and McIver[77] described the INDO methodology appropriate for anisotropic constants and reported results for CH, CH_3 and the radicaloid derivatives of malonic acid, and results on a series of seven organic and inorganic radicals were reported by Morikawa *et al.*[78]

The calculation of anisotropic constants using INDO theory involves calculation of matrix elements of the form $\langle \phi_\mu | (r_N^2 \delta_{tu} - 3r_{Nt} r_{Nu}) r_N^{-5} | \phi_\nu \rangle$ at a level of approximation commensurate with the integral evaluation in the INDO wave function determination. There are one-, two-, and three-center integrals of this form involved. Since all polycenter electron repulsion integrals are neglected in the INDO wave function calculation, the analogous dipole–dipole interaction integrals are assumed negligibly small. This leaves only two-center integrals with an orbital at one atom interacting with a nucleus of another atom and one-center integrals with electrons associated with an atom interacting with the nucleus of that atom. These integrals can be evaluated analytically or else estimated semiempirically; best agreement with experiment was obtained with a more semiempirical result.

INDO approximate molecular orbital results are presented for CH_3 in Table 8 along with a comparison with results from an *ab initio* SCF-MO-CI calculation. Note that the coupling constants are given in MHz rather than as in Tables 6 and 7. The approximate molecular orbital results give a reasonably good account of the experimental data. The results are better than the nonempirical *ab initio* calculations, but the limited agreement for the latter is in part due to difficulties in describing the region near the nuclei with no disposable parameters. The calculations of Morikawa *et al.*[78] include σ and π radicals and the results are encouraging. An extensive assessment of the capabilities and limitations of INDO in treating anisotropic constants is required before conclusive statements can be made regarding its worth.

The calculations of Loew and co-workers[76] were concerned with an application of the approximate molecular orbital theory of electron magnetic resonance parameters to a study of conformational aspects of oxidized and reduced rubredoxins. The results of calculations on possible conformers were compared with experiment in an effort to provide a microscopic interpretation of the observed results.

References

1. W. Kauzmann, *Quantum Chemistry*, Academic Press, New York (1957), Chapter 8.
2. C. P. Schlicter, *Principles of Magnetic Resonance*, Harper and Row, New York (1963).
3. R. McWeeney, *Spins in Chemistry*, Academic Press, New York (1970).
4. F. L. Pilar, *Elementary Quantum Chemistry*, McGraw-Hill, New York (1968).
5. C. C. J. Roothaan, *Rev. Mod. Phys.* **23**, 69 (1951).
6. J. A. Pople and R. K. Nesbet, *J. Chem. Phys.* **22**, 571 (1964).
7. F. Sasaki and K. Ohno, *J. Math. Phys.* **4**, 1140 (1943).
8. A. Streitwieser, *Molecular Orbital Theory for Organic Chemists*, Wiley, New York (1959).
9. R. Hoffman, *J. Chem. Phys.* **39**, 1397 (1963).

10. R. Pariser and R. G. Parr, *J. Chem. Phys.* **2**, 466 (1953); J. A. Pople, *Proc. Phys. Soc. (Lond.)* **81**, 305 (1955).
11. J. A. Pople and G. A. Segal, *J. Chem. Phys.* **44**, 3289 (1966).
12. J. A. Pople, D. L. Beveridge, and P. A. Dobosh, *J. Chem. Phys.* **47**, 2026 (1967); R. N. Dixon, *Mol. Phys.* **12**, 83 (1967).
13. N. C. Baird and M. J. S. Dewar, *J. Chem. Phys.* **50**, 1262 (1969).
14. J. A. Pople, D. P. Santry, and G. A. Segal, *J. Chem. Phys.* **43**, 5129 (1965).
15. J. O. Hirschfelder, W. Byers-Brown, and S. T. Epstein, *Adv. Quant. Chem.* **1**, 255 (1964).
16. M. Karplus and H. J. Kolker, *J. Chem. Phys.* **35**, 2235 (1961); **38**, 1263 (1963).
17. J. A. Pople, J. W. McIver, Jr., and N. S. Ostlund, *J. Chem. Phys.* **49**, 2960 (1968).
18. R. Ditchfield, N. S. Ostlund, J. N. Murrell, and M. A. Turpin, *Mol. Phys.* **18**, 433 (1970).
19. N. F. Ramsey, *Phys. Rev.* **77**, 567 (1950); **78**, 699 (1950); **86**, 243 (1952).
20. S. T. Epstein, *J. Chem. Phys.* **42**, 2897.
21. J. D. Memory, *Quantum Theory of Magnetic Resonance Parameters*, McGraw-Hill, New York (1968).
22. W. N. Lipscomb, *Advan. Magn. Resonance* **2**, 137 (1966).
23. R. Ditchfield, D. P. Miller, and J. A. Pople, *J. Chem. Phys.* **53**, 613 (1970).
24. A. Saika and C. P. Schlicter, *J. Chem. Phys.* **22**, 26 (1954).
25. J. A. Pople, *J. Chem. Phys.* **37**, 53 (1962); **37**, 60 (1962); *Mol. Phys.* **7**, 301 (1964); *Disc. Faraday Soc.* **34**, 7 (1962).
26. J. I. Musher, *Adv. Magn. Resonance* **2**, 177 (1966).
27. W. H. Flygare, *Chem. Rev.* **74**, 653 (1974).
28. A. D. Buckingham and P. J. Stiles, *Mol. Phys.* **24**, 99 (1972).
29. A. Velenik and R. M. Lynden-Bell, *Mol. Phys.* **19**, 371 (1970).
30. A. B. Strong, D. Ikenberry, and D. M. Grant, *J. Magn. Resonance* **9**, 145 (1973).
31. P. D. Ellis, G. E. Maciel, and J. W. McIver, Jr., *J. Am. Chem. Soc.* **94**, 4069 (1972).
32. G. E. Maciel, J. C. Dallas, R. L. Elliot, and H. C. Dorn, *J. Am. Chem. Soc.* **95**, 5857 (1973).
33. W. T. Raynes, in *Nuclear Magnetic Resonance* (Riki Harris, ed.), Specialist Periodical Reports, The Chemical Society, London, Vol. 1 (1972), p. 1; Vol. 2 (1973), p. 1; Vol. 3 (1974), p. 1.
34. R. Ditchfield and P. D. Ellis, Theory of ^{13}C chemical shifts, in *Topics in ^{13}C NMR Spectroscopy* (G. C. Levy, ed.), Wiley, New York (1974), Vol. 1.
35. D. E. O'Reilly, *Prog. NMR Spectry.* **2**, 1 (1967).
36. N. F. Ramsey, *Phys. Rev.* **91**, 303 (1953); N. F. Ramsey and E. M. Purcell, *Phys. Rev.* **85**, 143 (1952).
37. H. M. McConnell, *J. Chem. Phys.* **24**, 460 (1956).
38. J. A. Pople, J. W. McIver, Jr., and N. S. Ostlund, *J. Chem. Phys.* **49**, 2965 (1968).
39. R. Ditchfield and L. C. Synder, *J. Chem. Phys.* **56**, 5823 (1972).
40. A. D. McLachlan, *J. Chem. Phys.* **32**, 1263 (1960).
41. J. A. Pople and D. P. Santry, *Mol. Phys.* **8**, 1, 269 (1964).
42. K. Hirao, H. Nakatsuji, H. Kato, and T. Yonezawa, *J. Am. Chem. Soc.* **94**, 4078 (1972); **95**, 31 (1973).
43. R. Ditchfield and J. N. Murrell, *Mol. Phys.* **14**, 481 (1968); J. N. Murrell, *Prog. NMR Spectry.* **2**, 1 (1967).
44. A. D. C. Towl and K. Schaumberg, *Mol. Phys.* **22**, 49 (1971).
45. G. E. Maciel, J. W. McIver, Jr., N. S. Ostlund, and J. A. Pople, *J. Am. Chem. Soc.* **92**, 1, 11, 4151, 4497, 4056 (1970); G. Maciel, *J. Am. Chem. Soc.* **93**, 4375 (1971); P. D. Ellis and G. E. Maciel, **92**, 5829 (1970); K. D. Summerhays and G. E. Maciel, *Mol. Phys.* **24**, 913 (1972).
46. M. Johnston and M. Barfield, *J. Chem. Phys.* **54**, 3083 (1971); M. Barfield, *Chem. Rev.* **73**, 53 (1973).
47. R. Grinter, in *Nuclear Magnetic Resonance* (R. K. Harris, ed.), Specialist Periodical Reports, The Chemical Society, London, Vol. 1 (1972), Vol. 2 (1973), p. 50, Vol. 3 (1974), p. 50.
48. A. C. Blizzard and D. P. Santry, *J. Chem. Phys.* **55**, 950 (1971).
49. J. M. Schulman and M. D. Newton, *J. Am. Chem. Soc.* **96**, 6295 (1974).
50. M. Karplus, *J. Chem. Phys.* **30**, 11 (1959); *J. Am. Chem. Soc.* **85**, 2870 (1963).

51. G. Govil, *Indian J. Chem.* **9**, 824 (1971); *Mol. Phys.* **21**, 953 (1971).
52. M. Barfield and S. Sternhell, *J. Am. Chem. Soc.* **94**, 1905 (1972).
53. Ph. de Montgolfier and J. E. Harriman, *J. Chem. Phys.* **55**, 5262 (1971).
54. A. J. Stone, *Proc. Roy. Soc. (Lond.) A* **271**, 424 (1963).
55. P. W. Atkins and M. C. R. Symons, *The Structure of Inorganic Radicals*, Elsevier Publishing Co., Amsterdam (1967).
56. T. Morikawa, O. Kikuchi, and K. Someno, *Theor. Chim. Acta (Berl.)* **22**, 224 (1971).
57. J. W. McIver, Jr., private communication.
58. R. Lefebvre, *Mol. Phys.* **12**, 417 (1967).
59. R. W. Fessenden and R. H. Schuler, *J. Chem. Phys.* **43**, 2704 (1965).
60. R. W. Fessenden, *J. Phys. Chem.* **71**, 74 (1967).
61. S. I. Weissman, *J. Chem. Phys.* **22**, 1135 (1954).
62. H. M. McConnell, *J. Chem. Phys.* **24**, 632, 764 (1956); H. M. McConnell and D. B. Chestnut, *J. Chem. Phys.* **28**, 107 (1958).
63. R. Bershon, *J. Chem. Phys.* **24**, 1066 (1956).
64. J. P. Colpa and J. R. Bolton, *Mol. Phys.* **6**, 273 (1963); A. D. McLachlan, *Mol. Phys.* **3**, 233 (1960).
65. A. T. Amos, *Adv. Atomic Mol. Phys.* **1**, 196 (1963); A. Carrington, *Quart. Rev.* **17**, 67 (1963).
66. J. A. Pople, D. L. Beveridge, and P. A. Dobosh, *J. Am. Chem. Soc.* **90**, 4201 (1968).
67. H. G. Benson and A. Hudson, *Theor. Chim. Acta (Berl.)* **23**, 259 (1971).
68. D. L. Beveridge and P. A. Dobosh, *J. Chem. Phys.* **48**, 5532 (1968).
69. A. Carrington and I. C. P. Smith, *Mol. Phys.* **9**, 137 (1965); D. L. Beveridge and E. Guth, *J. Chem. Phys.* **55**, 458 (1971).
70. M. Raimondi, M. Simonetta, and G. F. Tantardini, *J. Chem. Phys.* **56**, 5091 (1972).
71. R. V. Lloyd and D. E. Wood, *J. Am. Chem. Soc.* **96**, 659 (1974).
72. R. V. Lloyd and D. E. Wood, *J. Am. Chem. Soc.* **96**, 659 (1974).
73. C. M. Bogan and L. D. Kispert, *J. Phys. Chem.* **77**, 1491 (1973).
74. H. M. McConnell, C. Heller, T. Cole, and R. W. Fessenden, *J. Am. Chem. Soc.* **82**, 776 (1960); A. Hersfeld, J. R. Morton, and D. H. Whiffen, *Mol. Phys.* **4**, 327 (1961).
75. H. M. McConnell and J. Strathdee, *Mol. Phys.* **2**, 129 (1959).
76. G. H. Loew, M. Chadwick, and D. A. Steinberg, *Theor. Chim. Acta (Berl.)* **33**, 125 (1974); G. H. Harris and D. Lo, *Theor. Chim. Acta (Berl.)* **33**, 137, 147 (1974); **32**, 217 (1974).
77. D. L. Beveridge and J. W. McIver, Jr., *J. Chem. Phys.* **54**, 4681 (1971).
78. T. Morikawa, O. Kikuchi, and K. Someno, *Theor. Chim. Acta (Berl.)* **24**, 393 (1972).
79. D. M. Hirst, *Theor. Chim. Acta* **20**, 292 (1971).
80. W. R. Wedt and W. A. Goddard III, *J. Am. Chem. Soc.* **96**, 1689 (1974).

The Molecular Cluster Approach to Some Solid-State Problems

Richard P. Messmer

1. Introduction

The quantum theory of the solid state is a vast subject which covers manifold phenomena, from optical properties to superconductivity, and from magnetism to surface properties. It is this richness of phenomena, together with its many important practical applications, which has made this subject such an active and fertile area of investigation for over forty years. Unfortunately this very diversity results in the lack of a completely coherent presentation of the theory from a set of fundamental principles—as the perusal of the standard texts of the subject will show. For example, there is at present no practical general way to formulate a theory of the electronic properties of a solid made up of a collection of atoms at arbitrary spatial positions. If these atoms, on the other hand, fall on the vertices of a lattice which can be repeated periodically through all of space, then one may take advantage of this symmetry to formulate a theory—the energy band theory—for the electronic structure of this "perfect" crystalline solid. Ideally one would like to have a practical general approach to the electronic structure of solids for which the "perfect" crystalline solid would be a special case of the "imperfect" or amorphous solid. The lack of such a general theoretical approach must be considered as one of the outstanding challenges which solid-state theory poses to the theorist.

Richard P. Messmer • General Electric Corporate Research and Development, Schenectady, New York

If one considers this problem from the point of view of molecular theory, one is led to the following analogy. When the benzene molecule is treated by the molecular orbital theory—which is analogous to the energy band theory of solids, as will be discussed later—the symmetry of the molecule greatly reduces the labor involved in its solution, as is well known. If, however, a highly distorted benzene molecule is treated, even though the labor is increased, a solution still can be obtained within the context of molecular orbital theory. The highly distorted benzene molecule is analogous to an amorphous solid, since in each case the symmetry of the system is lost. However, contrary to the molecular case, where molecular orbital theory can be used in both cases in the solid-state problem the energy band theory cannot be applied, irrespective of the amount of labor employed, to the amorphous solid case, but only to the high-symmetry "crystalline" case. There is, of course, one fundamental difference between the molecular orbital treatment of benzene and the band theory treatment of a crystalline solid; namely, in molecular systems one works in "real space," whereas in the crystalline solid because of the translational symmetry, "k-space" or "reciprocal space" proves to be extremely useful. As a result of the latter fact, many of the concepts in the theory of crystalline solids are based on "k-space."

At first sight, the above discussion may appear obvious and therefore unimportant. It certainly may be obvious that the loss of (translational) symmetry in a very large system such as a solid makes an otherwise soluble problem insoluble from a practical standpoint; whereas for a small system such as a molecule, the loss of symmetry does not make the problem insoluble. This statement is, however, more than a comparison of relative size and computational difficulty. It derives its importance from the fact that as a consequence of this situation a different *conceptual framework* must be used in order to treat imperfect or amorphous solids than is used for perfect crystalline solids. In particular, k-space concepts useful in crystalline solids are no longer valid for imperfect solids and thus one must substitute real-space concepts. The lack of a common conceptual framework between systems that have perfect periodicity and those that do not is very important and arises as a problem again and again in discussions of a number of phenomena, some of which are treated below.

In the remainder of the introduction, the theory of energy bands for perfect crystalline solids will be outlined, followed by a general discussion of some of the fundamental problems for which lattice periodicity is completely or partially absent. In the latter category are such problems as isolated defects and impurities in solids, amorphous solids, and the surface of solids, and the interaction of these surfaces with molecules or atoms. The present author has recently reviewed some of the progress of theory in treating the first two problems,[1] and thus the present chapter will be concerned primarily with the theoretical treatment of some problems associated with surfaces.

1.1. Perfect Crystalline Solids and the Bloch Theorem

Let us for the moment consider the treatment of the bulk electronic properties of a solid in order to avoid all problems associated with the surfaces. In this case we will assume we can write a wave function ψ for the system as a single Slater determinant composed of one-electron orbitals ϕ_i, analogous to the corresponding approach for molecules. However, in contrast to a molecule, for a finite solid there will be of the order of $\sim 10^{23}$ electrons and a corresponding number of orbitals. However, to eliminate surface effects, the solid must be extended to infinity in all directions. Clearly for an arbitrary solid this conceptual framework presents us with a completely intractable problem in the general case. If, however, we deal with a perfect crystalline solid, then due to its symmetry it will belong to a *space group* and the one-electron orbitals will transform according to irreducible representations of this space group. This is completely analogous to the molecular case in which the one-electron orbitals transform according to the various irreducible representations of the *point group* of the molecule.

In mathematical terms, let us consider the orbitals of the solid to be represented by a linear combination of atomic orbitals, as is common in many applications to molecular problems; thus

$$\phi_j(\mathbf{r}) = \sum_{l=0}^{\infty} \sum_{\beta\gamma} a_{l\beta\gamma}(j)\chi_{\beta\gamma}(\mathbf{r} - \mathbf{R}_l - \mathbf{R}_\beta), \qquad j = 1,2,\ldots,\infty \tag{1}$$

where j labels the individual eigenfunctions and where the sum on l is over all unit cells of the crystal, and the finite sums on β and γ are over all the atoms in the unit cell and over all the atomic-like orbitals on these atoms, respectively. (See Fig. 1.)

If the Schrödinger equation, for which $\phi_j(\mathbf{r})$ is a solution, is given by

$$[-\nabla^2 + V(\mathbf{r}) - \varepsilon_j]\phi_j(\mathbf{r}) = 0 \qquad \text{(in rydbergs)} \tag{2}$$

then due to the fact that we have a regular periodic lattice, the potential is periodic, i.e., $V(\mathbf{r} + \mathbf{R}_l) = V(\mathbf{r})$. From this it can be shown, as demonstrated by Bloch in 1928,[2] that

$$\phi_j(\mathbf{r} + \mathbf{R}_l) = [\exp(i\mathbf{k}_j \cdot \mathbf{R}_l)]\phi_j(\mathbf{r}) \tag{3}$$

Equation (3) is a statement of Bloch's theorem. $\mathbf{k}$ is a vector in reciprocal space and $\mathbf{R}_l$ is a translation lattice vector in direct space. From these considerations Eq. (1) may be rewritten as

$$\phi_{\mathbf{k}n}(\mathbf{r}) = \sum_{\beta\gamma} c_{\beta\gamma;n}(\mathbf{k}) \sum_l [\exp(i\mathbf{k} \cdot \mathbf{R}_l)]\chi_{\beta\gamma}(\mathbf{r} - \mathbf{R}_l - \mathbf{R}_\beta) \tag{4}$$

There will be as many energy bands (labeled by n) as are given by the product of the number of atoms per unit cell and the number of atomic-like orbitals per

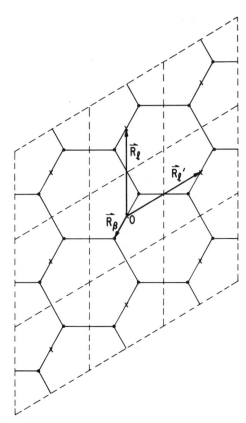

Fig. 1. A two-dimensional structure (appropriate, e.g., to a layer of graphite). The dashed lines delineate the unit cells; the ×'s mark the centers of each unit cell; 0 denotes the origin of the coordinate system. In this example there are two atoms per unit cell. The vectors $\mathbf{R}_{l'}$ and $\mathbf{R}_{l}$ are examples of translational lattice vectors relating the centers of each unit cell to the origin. $\mathbf{R}_{\beta}$ is a vector connecting the center of a unit cell to the individual atoms contained in the unit cell.

atom. Thus $\mathbf{k}$ and n serve to uniquely label the eigenfunctions of Eq. (4), just as j was a unique label in Eq. (1). Equation (4) may be written as

$$\phi_{\mathbf{k},n(\mathbf{r})} = \sum_{\beta\gamma} c_{\beta\gamma;n}(\mathbf{k}) b_{\beta\gamma\mathbf{k}} \tag{5}$$

where the $b_{\beta\gamma\mathbf{k}}$ represents the Bloch sums in Eq. (4). Equation (2) may be rewritten as

$$[-\nabla^2 + V(\mathbf{r}) - \varepsilon_{\mathbf{k},n}]\phi_{\mathbf{k},n}(\mathbf{r}) = 0 \tag{6}$$

It is now clear that through the use of translational symmetry the solution of the problem has been reduced from that of determining the solution of an infinite set of equations [see Eq. (1) and (2)] to that of determining the solution of a finite set of equations [see Eqs. (5) and (6)] for an infinite number of $\mathbf{k}$ values. In practice, however, the problem is solved for a large, finite number of $\mathbf{k}$ values to obtain the energy bands (i.e., $\varepsilon_{\mathbf{k}}$ vs. $\mathbf{k}$ for each n), since one may interpolate between calculated $\mathbf{k}$ points because $\varepsilon_{\mathbf{k}}$ is a well-behaved function of $\mathbf{k}$.

This discussion can hardly be considered to be complete; there are many gaps, which may be filled in by consulting standard texts on the subject.[3,4] The main purpose is to state the energy band problem in a context which is familiar within the usual molecular electronic structure framework.

1.2. Imperfect Solids and the Breakdown of Bloch's Theorem

There are a number of classes of problems in the solid state for which the perfect three-dimensional translational symmetry of the crystal lattice is absent. We will briefly consider three such classes.

1.2.1. Point Defects

For the purpose of illustration, consider the silicon lattice, in which each atom is tetrahedrally coordinated to four other silicon atoms. We may look upon the perfect lattice as being made up of two interpenetrating face-centered cubic (fcc) lattices, in which the first silicon atom is on one fcc lattice and its four neighbors are on the second, and so on. If, however, a point defect is created by replacing one of the silicon atoms by an impurity atom such as phosphorus or by removing an atom to form a lattice vacancy, then the lattice periodicity is interrupted and Bloch's theorem is no longer strictly valid.

If, as in the case of a P impurity in Si, the change in the lattice potential is not too great, one would expect that the Bloch functions that are solutions to the perfect lattice problem would not be greatly perturbed by the presence of the impurity. Thus these functions might serve as the basis for a perturbation treatment to investigate the effect of the impurity. The "effective mass theory"[5] makes use of this simple physical idea and is quite successful when the perturbation is small.

However, if the perturbation is not small, such as in the case of the vacancy in silicon or diamond or a nitrogen atom impurity in diamond,[6] a different approach must be adopted. In these latter cases the perturbation is sufficiently large that some quite localized electronic states associated with the defect or impurity are formed. The Bloch functions for the perfect crystal no longer provide a convenient set of basis functions from which to start. In such circumstances, it is more convenient and conceptually more appealing to work with the atomic basis functions of Eqs. (1) and (2) and from this infinite problem to factor out a small, finite cluster of atoms (a molecular cluster) that contains the defect or impurity and solve for the electronic states of these systems by using familiar molecular orbital techniques, such as the extended Hückel method,[6,7] CNDO,[8] or the SCF-$X\alpha$-scattered wave method.[9] This molecular cluster method, suggested by Messmer and Watkins in 1970,[10] has proved to be quite useful for many such problems.[1]

1.2.2. Amorphous Solids

If we continue to use silicon as the model solid, then amorphous silicon differs from crystalline silicon by the fact that although each atom continues to have four nearest neighbors in roughly tetrahedral coordination, the long-range order is absent, i.e., the atoms do not fall on a regular lattice. In this case, the complete lack of lattice periodicity and hence the breakdown of Bloch's theorem are even more serious than the case above, where only one or a small number of atoms disrupted the periodicity. Thus, here it is rather hopeless to start with Bloch functions in **k**-space and to try to use perturbation theory. Attempts[11] in this direction have been quite unsuccessful. On the other hand, methods that use real space and molecular concepts have been far more successful. A discussion of these molecular approaches has been given recently.[1] Brief mention will be made here of only one such method, which has been very useful because of its simplicity and the insight it has provided.

The basic model is based on an extension of work which was carried out by Hall[12] over twenty years ago to investigate the electronic structure of diamond from a bond-orbital point of view. The model assumes that the amorphous silicon structure can be described in terms of interacting sp^3 hybrids. Overlaps are neglected and only two matrix elements of the one-electron Hamiltonian are retained, H_1 and H_2. H_1 represents the interaction of two hybrids on a given atom and H_2 represents the interaction of two hybrids involved in a bond. Weaire[13] has used this model as applied to an infinite collection of atoms to show that the band gap in silicon should persist in the amorphous material (which is known experimentally to be the case). Such a proof is possible because of the simplicity of the model.

More recently Joannopoulos and Yndurain[14] have used finite clusters of atoms to which essentially the same model is applied but with the use of a so-called "Bethe lattice" as a boundary condition. The Bethe lattice used is an infinite connected network of atoms with fourfold coordination but with no rings of bonds. They call the method the "cluster-Bethe-lattice" method since it consists of a finite cluster of atoms in arbitrary (but definite) positions to which the Bethe lattice is attached at the periphery. Again because of the simplicity of the model a number of results can be expressed mathematically in closed form. Many possibilities exist for extending the method to incorporate a more realistic treatment of the electronic structure of the cluster and thus this method looks quite promising.

12.3. Surfaces of Solids

The study of the surfaces of solids is important from both the fundamental and practical aspects. The understanding of the electronic states at semiconductor surfaces is important, for example, to the development of certain new semiconductor devices. Heterogeneous catalysis also depends upon the electronic properties of surface atoms and is extremely important to a variety of essential processes.

In the case of semiconductors, the technologically interesting surfaces are rather well defined, especially in comparison to catalytically important surfaces. For such semiconductors and certain well-defined metal surfaces, lattice periodicity may persist in planes parallel to the surface and theories derived from modifications of band theory can be employed to investigate such ordered surfaces. The theoretical techniques which have been developed to treat such problems will be discussed in Section 2.

The surfaces important in many catalytic applications and in the oxidation of ordinary metals do not possess such ideal geometries, however. For example, many catalysts consist of small metal particles (which contain from a few atoms to many thousands of atoms, depending upon the application) supported on a refractory material such as alumina, silica, or a zeolite. The electronic structure of such surfaces is obviously best considered from the viewpoint of a molecular cluster approach using molecular orbital or other chemically oriented theories. Even for the case of surface states on perfect single-crystal faces, the molecular cluster approach can be quite useful, as shown by recent work of Goddard, McGill, and co-workers in treating silicon surface states using the generalized valence bond (GVB) method.[15] The development and application of the molecular cluster approach to surface problems will be discussed in Section 3.

There are a variety of solid-state problems for which a chemical approach using a molecular cluster can provide valuable insight and information. A number of examples have been discussed previously,[1] and in the present chapter, evidence will be presented to show that this is also the case for many surface problems.

2. Solid-State Theory Approaches to Surface Problems

2.1. The Perfect Surface

For the purpose of discussion here, a perfect crystal surface is considered to be one which has periodicity parallel to the surface plane. The periodicity may be the same as that of the underlying bulk planes or a multiple thereof. In the latter situation the surface may be "reconstructed," as occurs for certain surfaces of semiconductors.[16] One may then take advantage of this two-dimensional periodicity in order to simplify the problem (although to a lesser extent) in much the same way as three-dimensional periodicity simplified the bulk crystal problem.

2.1.1. Qualitative Models

Models which invoke a number of simplifying assumptions and which are useful for discerning qualitative information will be briefly discussed in this subsection. Those which provide for a more realistic treatment of the surface potential and are consequently computationally more difficult will be discussed in Section 2.1.2.

A thorough review of one-dimensional models appropriate to the qualitative understanding of surface states in semiconductors has been given by Davison and Levine.[17] These models usually employ potentials $V(\mathbf{r})$ in the Schrödinger equation [see Eq. (2)] which have a form shown schematically in Fig. 2a, where the lattice potential is terminated abruptly by a step at the "surface" and is replaced by a constant potential outside this surface which represents the vacuum.

A model which has been used fairly extensively to discuss the electronic properties of metal surfaces is the "jellium" model. In this model one allows a smooth potential to connect the vacuum with the bulk, but the discrete nature of the atoms in the solid is ignored. The solid is imagined to consist of an electron gas arising from valence electrons of the atoms and a uniform background of positive charge arising from the remaining ion cores of the atoms. The potential perpendicular to the surface plane in this model is shown schematically in Fig. 2b. This model can provide a crude first approximation to surface behavior for a free-electron-like metal, but obviously has rather severe limitations. The work of Lang and Kohn represents the most sophisticated use

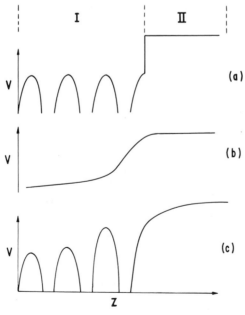

Fig. 2. Model potentials for (a) an abruptly terminated lattice, (b) the surface of "jellium," and (c) a realistic surface. Region I shows the potential of the last few atomic planes; region II is the potential outside the surface.

of this model.[18] After obtaining a self-consistent solution of the model, they calculated the effect of the discrete lattice to first order in perturbation theory in order to obtain the surface energy and work functions for a number of metals.

An alternate description of crystal termination can be given within the LCAO-tight-binding (TB) methods. The true potential at a surface should look more like Fig. 2c than Fig. 2a. One can obtain a different potential for the surface atoms as compared to the bulk in the LCAO-TB scheme by simply choosing different matrix element parameters. This has been employed frequently in the literature.[17,19] One such simple approach will be discussed here since it has also served as a vehicle to study some simple chemisorption models.

In the model of Kalkstein and Soven,[20] one assumes an infinite simple cubic lattice with one s orbital per site with a one-electron Hamiltonian $\hat{H}_0$. The matrix elements retained in the model are identical to those of simple Hückel theory. A perturbation $\hat{V}$ is introduced which cleaves the crystal into two parts and $\hat{H} = \hat{H}_0 + \hat{V}$ is the cleaved crystal Hamiltonian, which by definition can have no matrix elements between functions on different sides of the cleavage plane. The perturbation is also assumed to add a diagonal correction (in a site representation) to the surface atom matrix elements. Use is made of the translational periodicity parallel to the cleavage plane to construct Bloch-like basis functions $\phi_{\mathbf{k}_\parallel,m}$ for each of the planes parallel to the surface, i.e., $m = 0,1,2,\ldots$. The Green's function of the cleaved crystal problem is obtained and the density of states per atom for atoms lying in a given plane (e.g., the mth plane) is given by

$$\rho_m(E) = N \operatorname{Im} \sum_{\mathbf{k}_\parallel} G(m,m,\mathbf{k}_\parallel) \tag{7}$$

where N is a normalization factor, E is the energy, and $G(m,m,\mathbf{k}_\parallel)$ is a diagonal matrix element of the Green's operator in the chosen basis. If for the moment we write the one-electron solution of the problem in simple LCAO form as

$$\psi_k = \sum_n a_{kn}\phi_n \tag{8}$$

with corresponding eigenvalues ε_k, then the density of states may be written as

$$\rho_m(E) = \sum_k |a_{km}|^2 \delta(E - \varepsilon_k) \tag{9}$$

which more clearly shows the relationship between $\rho_m(E)$ and the wave functions. Equation (7) was used to evaluate the atomic density of states for atoms in various planes from the surface inward; this function was then compared to the corresponding function for the perfect crystal. Due to the fact that it took many planes for $\rho_m(E)$ to attain its bulk form, the authors concluded that the surface states can extend rather deeply into the crystal.

The above is of course a one-band model and as such is appropriate to a description of a simple metal. Recently, however, a simple two-band model has been discussed by Ho *et al.*[21] for which the bulk crystal and surface Green's functions are obtained in analytic form. This model should prove quite useful for qualitative discussions of the surface states of semiconductors.

2.1.2. Quantitative Models

Turning now to more realistic descriptions of the potential in the surface region, we will briefly describe a model used by Appelbaum and Hamann[22] which has been rather successfully used to investigate surface states on the various faces of silicon.[23,24] The method considers a semiinfinite solid and makes use of the fact that the disturbance produced by the creation of a surface must die off on moving into the bulk of the solid. Hence the potential after a certain number of atomic layers from the surface must be the same as in the perfect bulk crystal. Let us assume this number of layers is n; then a "matching plane" is set up n layers from the surface and the bulk Bloch functions are assumed to be a proper description of the solid on the bulk side of the matching plane. On the surface side of the matching plane use is made of the two-dimensional periodicity parallel to the surface plane and functions similar to those discussed above (actually, in this case they are expressed as a Fourier series) are used to describe the wave functions. The determination of the proper electronic states of the system is then accomplished by demanding that the functions and their first derivatives be continuous at the matching plane.

Thus far this method has made use of pseudo- or model potentials to describe the atomic potentials and has been applied to self-consistently determine the electronic structure of the Na (100) surface,[22] the Si (111) and Si (100) surfaces,[23,24] and the Ge (111) surface.[24] The theoretical results[23] for the Si (111) surface predict several bands of surface states, one of which corresponds to the classical picture of an array of "dangling bonds" pointing out of the surface. The surface layer is predicted to relax inward and the energy positions of the other surface state bands seem to be supported by experiment.[25]

Because atomic model potentials and Fourier expansions are part of the present computational procedure in this method it is not clear that it can be extended to provide a useful approach for transition metals or other solids which possess localized d orbitals.

The other quantitative models which have been proposed to date do not consider a semiinfinite crystal, but rather a crystal "slab" or "film" consisting of a number of atomic planes parallel to the crystal face of interest. For reasons of computational tractability, the number of atom layers which can be considered is of the order of a dozen or so in most applications. Only those methods that use a realistic potential (see Fig. 2c) will be discussed.

The first of these methods and one of the few for which calculations have been published is that of Alldredge and Kleinman.[26] They have investigated surface states on Li (100),[26] Al (110),[27] and Al (111),[28] using a pseudopotential approach. These materials are not the most interesting as far as surface state properties are concerned and no experimental information is available with which to compare their results. The more interesting transition metals cannot be treated by such a pseudopotential approach.

Methods which are applicable to transition metal surfaces and employ a realistic potential have been proposed by Beeby,[29] Kasowski,[30] Kar and Soven,[31] and Kohn.[32] They all make use of multiple scattering approaches which are closely related to the KKR method[33] of energy bands in bulk solids. The approaches of Kar and Soven[31] and Kohn[32] are essentially the same; the former authors present numerical results for a single layer of Cu atoms, however. Kasowski is the only other of these authors to present numerical results. For the single layer of Cu atoms the results of Kasowski and of Kar and Soven disagree. The latter authors carried out an independent check of their results using a parametrized LCAO method. Hence there appears to be an error in the calculations of Kasowski, although its source is at present not clear.

2.2. Surface–Adsorbate Interactions

In this section we will consider some attempts to extend the methods of the last section from the case of a perfect surface to that of a surface with chemisorbed atoms.

2.2.1. Qualitative Models

The jellium model has been used to discuss the chemisorption of hydrogen on metals using a number of simplifying assumptions.[34] More recently it has been applied rigorously to H, Li, and O chemisorption by solving the equations numerically.[35] By comparing the two calculations for H, it is apparent that the approximations used in the first calculation cannot be justified. No experimental information seems to exist for low-coverage chemisorption of H, Li, or O on a free-electron-like metal, and hence the most appropriate comparison between these calculations and experiment cannot be made. A comparison with data on transition metals shows that the sign of the calculated dipole moment and the trend in binding energies are roughly in accord with experiment.[35]

Actually one advantage of the jellium model of chemisorption in comparison to the more quantitative solid-state approaches is the fact that one considers a single atom interacting with the surface. Any of the methods which take advantage of the two-dimensional periodicity of the surface plane and underlying crystal have to treat a periodic array of chemisorbed atoms.

A model which has been used and discussed by many workers[36–41] is based on the Anderson Hamiltonian originally used as a model to discuss magnetic impurities in alloys.[42] Several excellent discussions of this work are available,[38,39,41] and hence it will not be dwelt on here. The techniques of Kalkstein and Soven have been combined with this model by Schrieffer and co-workers to discuss the changes in local density of states at various planes in the crystal which result from chemisorption.[41] Such studies have concluded that the chemisorptive bond is a fairly localized phenomenon[41]—consistent with usual chemical intuition. An interesting aspect of the techniques which are employed in this model is their similarity both formally and physically with much of the early treatment of unsaturated hydrocarbons.[43,44]

A simple model of H chemisorption on graphite proposed by Messmer *et al.*[45] will be discussed here, since the basic principles are quite similar to those employed by some of the more quantitative treatments later made for other systems. A single plane (only the basal surface plane) is considered, since the distance between such planes is relatively large in graphite (3.4 Å). A small segment of the surface layer is shown in Fig. 1. In order to take advantage of the periodicity in the surface plane, one must consider the interaction of a periodic array of H atoms (the adsorbate layer) with the surface. However, one can discuss the effect of coverage by changing the density of hydrogen atoms in the adsorbate layer, e.g., by going from one hydrogen per four carbon atoms to one hydrogen per eight carbon atoms. In the actual calculations only the latter was considered, although two different positions of the hydrogens relative to the carbon lattice were considered. Bloch functions were constructed and an energy band calculation was performed using the extended Hückel method to evaluate matrix elements between atomic functions.[46] The size of the resultant secular equation which must be solved, in the case of one hydrogen per eight carbons is 33×33 with four basis functions on carbon. If one wishes to consider H chemisorption on a slab or thin film of graphite using n planes of graphite, the size is $(1 + 4mn) \times (1 + 4mn)$ for the case of one hydrogen per m carbons. Hence in order to consider the low-coverage limit in which there is little adsorbate–adsorbate interaction a rather healthy sized problem is generated.

In the model calculation for the case of one hydrogen per eight carbons,[45] changes in the total density of states upon H chemisorption were calculated, which in principle could be compared with experimental photoemission results, although in fact no such experimental information is available.

2.2.2. Quantitative Models

The semiinfinite crystal approach of Appelbaum and Hamann[22] mentioned in Section 2.1.2 has been used to investigate the interaction of H atoms with the (111) surface of Si. The calculations were carried out self-consistently with a hydrogen atom saturating each of the dangling bonds on the surface. The original band of surface states, which arose from the dangling bonds and was situated energetically in the band gap,[47] disappears upon saturation with hydrogen atoms. However, a new band of states, which is related to the Si–H bonds, is found in the valence band. The calculated bond length and force constant of these Si–H bonds were found to be close to the corresponding quantities for the silane molecule.

As mentioned previously, this method involves the choice of a matching plane in determining the wave functions. The deeper this matching plane is into the bulk, the more work the computations are. The sensitivity of results to the placement of this plane was also considered. This variable produces nonnegligible effects, but further development of the method may improve the situation. Several other applications of this method to chemisorption are in progress.

Of the methods that employ the crystal slab approach, the only one that has been applied to a chemisorption problem thus far is that due to Kasowski.[48] He considered a slab of five layers of Ni atoms representing a (100) face of the crystal on which an adsorbate layer of O atoms or Na atoms was chemisorbed. The periodicity in the adsorbate layer was chosen such that there was one adsorbate atom for every Ni atom in the plane and the adsorbate layer is placed so that it fits into the empty fourfold sites of the (100) face. The calculations were not carried out self-consistently, however. For the case of O on Ni a band of states appeared which was associated with the nonbonding electrons of oxygen. The position of this band compared favorably to results of experiments of Hagstrum and Becker.[49] No attempt was made, however, to calculate the density of states and compare it with the detailed photoemission results available for oxygen chemisorbed on Ni.[50] Also, the effect of self-consistency might be expected to be rather significant for an adsorbate as electronegative as oxygen. Undoubtedly future work along this line will consider such effects and provide much useful information.

3. Molecular Cluster Approach to Surface Problems

A more chemical approach to the problem of studying the interaction of adsorbates with surfaces is to focus on the local aspects of the problem. This can be accomplished by studying the bonding of one or a few adsorbate atoms or molecules to a cluster of substrate atoms. In many cases, such as metal particles on a support[51] or an irregular surface,[52] this may be even more relevant to the actual situation than considering an ideal surface with periodic behavior. Even for experiments performed on ideal surfaces, the cluster approach can provide a useful model to interpret a number of experimental data.

In order to consider a cluster of sufficient size (e.g., 5–25 atoms) interacting with an adsorbate, the traditional *ab initio* methods of quantum chemistry are far too cumbersome and computationally expensive, especially if one wishes to consider transition metal atoms. Hence, one must depend on either semiempirical schemes such as the extended Hückel[53] (EH) or complete neglect of differential overlap[54] (CNDO) methods, or the self-consistent field-$X\alpha$-scattered wave (SCF-$X\alpha$-SW) method.[55,56] The latter method is a "first principles" approach, in contrast to the semiempirical schemes, but is computationally much faster than the *ab initio* methods.[57]

3.1. Nonmetals

3.1.1. EH and CNDO Calculations

Apparently the first application of the EH and CNDO methods in chemisorption to obtain binding curves between adatoms and clusters of atoms representing the substrate was the work of Bennett *et al.*[58,59] It was realized by these workers that the substrate should be chosen such that in the limit of one surface atom and one adatom, the EH and CNDO methods should provide an adequate description of the chemical bond in the resultant diatomic molecule. The obvious first choice, considering the available literature, was a substrate of carbon atoms, the basal plane of graphite. The reason is that these methods had been devised and parametrized for calculations of organic molecules and that the interplanar separation in graphite is sufficiently large that a single layer was expected to be a good approximation to the substrate.

Initially, the EH method was used to investigate the binding curves for a hydrogen atom interacting at various sites on a 16-atom or 32-atom representation of the graphite substrate.[58] However, the chemisorption of electronegative atoms, such as O, N, and F, could not be properly described within a non-self-consistent EH framework. Consequently the CNDO/2 method was used[59] to investigate the binding of H, C, O, N, and F atoms at various sites on an 18-atom representation of the basal plane of graphite. A criterion for the determination of matrix elements of atoms at the periphery of the substrate was also introduced to minimize edge effects. Some binding curves from this work are shown in Fig. 3. Only the relative binding energies are considered to be significant, rather than the absolute values, due to the approximate nature of the method.

These calculations showed that a definite specificity exists with regard to the energetically most favorable binding site, depending on the particular adatom. Graphite is not a particularly interesting substrate, however. It would thus be very desirable to obtain similar information for metal substrates, particularly transition metals, where a great deal of experimental data are available. Unfortunately the parametrization of the EH and CNDO methods for such systems is not straightforward.

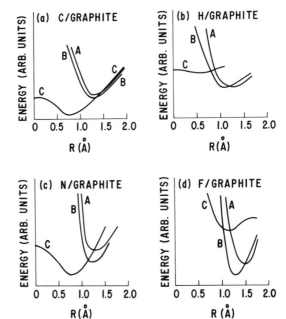

Fig. 3. Relative binding energies of chemisorption of atoms on graphite calculated by the CNDO/2 method for (a) carbon, (b) hydrogen, (c) nitrogen, and (d) fluorine atoms. Curve A: binding of adatom directly above a carbon atom of substrate; curve B: binding of adatom between two carbon atoms; curve C: binding of adatom at center of hexagonal ring of carbon atoms.

Before going on to consider the work which has investigated the interaction of adsorbates with clusters of metal atoms, it is important to assess the progress which has been made in studying metal clusters *per se*. This is considered in the next section.

3.2. Metal Clusters

3.2.1. EH and CNDO Calculations

The first calculations which attempted to determine the electronic properties of metal clusters were made by Baetzold.[60] He used extended Hückel and CNDO methods to investigate silver and palladium clusters. The electronic structures of silver clusters containing up to 20 atoms were calculated by the EH method, but due to the fact that the CNDO calculations are computationally more demanding, only clusters of up to eight atoms were considered in this case. Unfortunately, the author did not compare his distribution of energy levels with the total density of states (DOS) of the bulk solids as determined from an accurate band structure calculation, which would have provided an immediate assessment of how closely the electronic structure of a small metal cluster resembles that of the bulk solid. The strictly energetic information obtained by a comparison to the bulk DOS of course can be supplemented by comparison of the atomic orbital character of the cluster wave functions with the orbital character of the corresponding energy band states. The author did discuss the effect of cluster size on the lowest ionization potential (IP) and on the binding energy per atom, finding for the Ag clusters that the IP decreased toward the bulk value with larger clusters and that the binding energy per atom increased with cluster size—not, however, in a simple monotonic fashion. Similar studies were later carried out for clusters of other metals.[61]

Blyholder[62] later considered clusters of nickel atoms using CNDO. The largest cluster treated contained 13 atoms—which in an fcc metal such as Ni represents an atom and its 12 nearest neighbors in the lattice. The CNDO parameters were chosen by treating an octahedral Ni_6 cluster and trying to match selected quantities of the bulk such as the binding energy per atom, Fermi level, and d-band width. A distribution of the orbitals as a function of energy for the 13-atom cluster was presented and a discussion of the atomic character of the wave functions was given. Again, no direct comparison with bulk band structure or density of states was given.

Extended Hückel calculations on Ni clusters have been published by Fassaert *et al.*[63] (up to 12-atom clusters) and more recently by Anderson and Hoffmann[64] (a nine-atom cluster). Figures 4a and 4b compare the calculated density of states (DOS) from an EH calculation for 13 Ni atoms[65] using the Anderson and Hoffmann parameters with that obtained from the CNDO results of Blyholder[62] and a recent accurate bulk band structure calculation.[66] Since the extended Hückel and CNDO methods are spin-restricted methods, only the $X\alpha$–spin-restricted results and only the majority-spin DOS results of Callaway and Wang are used for comparison. The densities of states for the metal clusters are obtained by replacing each discrete eigenvalue by a Gaussian with a width of 0.01 Ry, weighted by the degeneracy of the orbital. In Fig. 4a, a comparison is presented of the bulk band DOS with the EH and CNDO results. The cluster results have been shifted so as to line up the Fermi levels in order to provide a meaningful comparison. For the CNDO calculation we see that there is not a high density of states at the Fermi energy E_F, in contrast to the bulk situation, and that the high DOS in the CNDO result occurs at an energy for which the DOS in the bulk is rapidly decreasing.

For the EH case, there is a high DOS near E_F but the d-band width is much narrower than in the bulk case; the s-d hybridization responsible for the long tail far from E_F in the bulk case is almost absent in the EH cluster calculation. The differences between the CNDO and EH calculations for Ni_{13} are quite remarkable. In Fig. 4b a comparison of the bulk band DOS and the $X\alpha$–SW DOS for Ni_{13} is given. The similarity is rather striking.

Baetzold and Mack,[67] using the EH method, have recently calculated the electronic properties of a number of metal clusters, including Cu_{19} and Pd_{19} clusters. They obtained d-band widths of ~ 0.2 and ~ 0.3 eV, respectively, whereas the d-band widths in the bulk are of the order of a few eV. They state: "In most properties, the clusters in the size range reported here are different from the bulk metals."

These semiempirical calculations hence raise the following questions: (i) How fundamentally different are the electronic structures of isolated small metal clusters (i.e., <20 atoms) and the bulk metal? (ii) How many metal atoms must a cluster contain for its electronic structure to have some resemblance to the bulk? (iii) Are semiempirical molecular orbital calculations such as EH and CNDO really capable of describing the electronic structure of metal clusters?

If small clusters have fundamentally different properties than bulk metals, this is potentially very important to catalysis which uses small metal clusters (usually on a support, however) as active centers. Furthermore, it would suggest that a great deal of research investigating reactions on surfaces of bulk metals may be irrelevant to the understanding of those catalytic reactions that use small clusters (i.e., nearly atomically dispersed).

If it takes a large number of atoms (≥ 100) to begin to see some of the bulk metal properties, this would eliminate the possibility of a theoretical treatment of chemisorption and surface reactions carried out on various crystal faces by using a molecular orbital treatment on a reasonably sized cluster.

These first two points seem almost to have been ignored by several workers, who merely proceed ahead to study chemisorption on small clusters using semiempirical techniques and compare their results with experiments on bulk single-crystal surfaces. Some of this work will be discussed later.

The last question was rather difficult to answer until recently. The difficulty arose from three factors: (1) *Ab initio* calculations are virtually impossible to carry out on transition metal clusters due to the extreme complexity and expense; (2) there existed no better or reliable computational alternative to EH or CNDO for such systems; and (3) no experimental data on isolated, well-defined metal clusters are available.

This situation has changed now quite dramatically due to the self-consistent-field–Xα–scattered wave (SCF–Xα–SW) method of Slater and Johnson,[68] which enables one to calculate reliable electronic structures for clusters of transition metal atoms. Such calculations will be discussed in the next section, after which we shall return to the three questions posed above, and try to provide the best answers consistent with present knowledge.

3.2.2. SCF–Xα–SW Calculations

The SCF–Xα–SW method is relatively new as far as computational methods in theoretical chemistry are concerned. It has, however, been fully described by Slater[69,70] and Johnson[71] and summaries of applications have been given.[70–72] Further discussion of the method is given in another chapter in this treatise by Connolly.[73] Hence we will consider here only applications to metal clusters and in the next section discuss applications to the interaction of adsorbates with metal atoms.

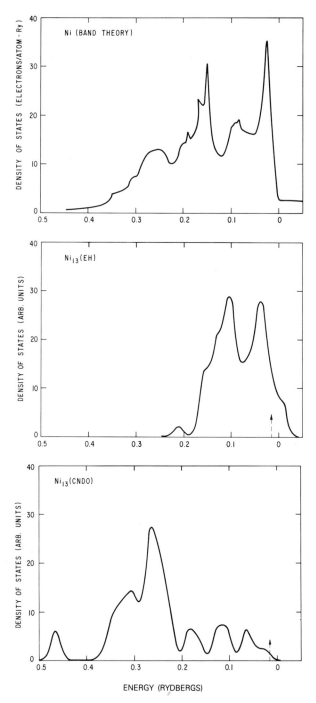

Fig. 4a. Comparison of densities of states for Ni as determined by bulk band structure[66] extended Hückel calculation for Ni_{13} (see Ref. 65), and CNDO/2 calculation for Ni_{13} cluster[62] (bottom).

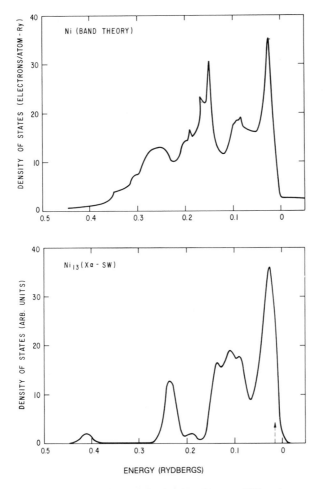

Fig. 4b. A comparison of the density of states of Ni as determined from a bulk band structure calculation[66] (top) and a
13-atom Ni cluster using the SCF–Xα–SW method (bottom).

In order to assess the differences between the electronic structure of small
metal clusters and the bulk metal as well as to ascertain the efficacy of EH and
CNDO calculations for treating metal clusters, calculations[74] for Cu, Ni, and
Pd using the SCF–Xα–SW method will be presented.

The electronic energy levels for simple cubic Cu_8 are shown in Fig. 5a.[75]
All of the levels shown arise from the ten $3d$ electrons and one $4s$ electron of
each Cu atom, plus some contribution from the Cu virtual $4p$ orbitals. The
highest occupied level is the t_{1u} level at -0.411 Ry which is fully occupied,
resulting in zero net spin polarization for the Cu_8 cluster. The first unoccupied
level is t_{2g} at -0.265 Ry. The Fermi level E_F separates the filled orbitals from
the empty ones; although it is somewhat arbitrarily positioned in this case, it is
consistent with the more accurately determined position in Ni_8.

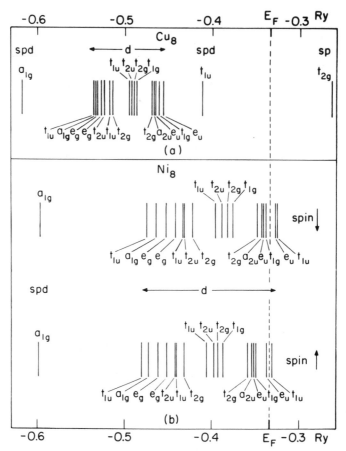

Fig. 5. SCF–Xα electronic energy levels of simple cubic metal clusters: (a) Cu$_8$; (b) Ni$_8$.

One can characterize the complete set of Cu$_8$ energies shown in Fig. 5a as a dense band of d levels bounded above and below, respectively, by the t_{1u} level at -0.411 Ry and the a_{1g} level at -0.617 Ry, both of which are predominantly s- and p-like in character, but with significant d admixture. Thus if one regards this $a_{1u}-t_{1u}$ energy interval as the precursor of the sp band in bulk crystalline copper,[76] then already in the Cu$_8$ cluster the d band is totally overlapped by the sp band as in the solid. Furthermore, the d band width for this Cu$_8$ calculation is 1.1 eV, in contrast to ~ 0.2 eV for the EH results of Baetzold and Mack for a Cu$_{19}$ cluster. SCF–Xα–SW calculations for Cu$_{13}$ give an even larger d-band width, which is nearer the bulk value. In addition, the calculations show that a Cu$_{13}$ cluster is capable of reflecting many of the features of the electronic structure of the bulk metal.

The electronic energy levels for a simple cubic Ni$_8$ cluster are shown in Fig. 5b. In comparison with the Cu$_8$ results, the Ni$_8$ d band is shifted to higher energies, significantly widened and split in energy by the net paramagnetic spin polarization, which arises because the highest occupied orbital e_u is only half-filled. E_F separates the occupied and empty spin orbitals. The exchange splitting of the a_{1g} level near -0.6 Ry is considerably less than that of the d band due to the predominantly sp character of this a_{1g} orbital. The electronic structure of the Ni_8 cluster is remarkably similar to the spin–unrestricted band structure calculated for ferromagnetic crystalline nickel by Callaway and Wang,[66] with respect to the sp–d band overlap and the presence of three distinct peaks in the majority- and minority-spin d-band densities of states. The spin magneton number (0.25) per atom in paramagnetic Ni_8 is considerably less than the value (0.54) for ferromagnetic crystalline nickel. This is consistent with the results of paramagnetic susceptibility measurements for small catalytic nickel particles.

A spin-restricted SCF–Xα–SW calculation has been made for Ni_{13} (Ref. 74) and the resultant electronic energy levels compared with those from an EH calculation[55] using the parameters of Anderson and Hoffmann. It was found in the SCF–Xα–SW calculation that the d-band width increased over that for Ni_8 (shown in Fig. 5b), becoming more like that in the bulk, and that the hybridization continued to reflect that in the bulk, in contrast to the EH results shown in Fig. 4c.

Likewise, a spin-restricted Xα calculation for Pd_{13} has been made[74] and the results for the energy levels are given in Fig. 6. Here the d-band width is

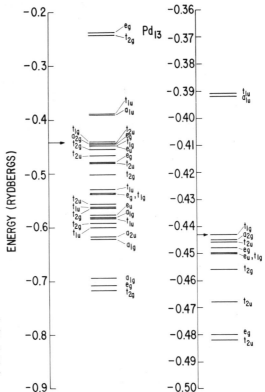

Fig. 6. SCF–Xα electronic energy levels for a cubooctahedral cluster of Pd_{13}. At the right an expanded scale shows more clearly the closely spaced levels near the Fermi level (denoted by the arrow).

larger than that for Ni_{13}, consistent with the bulk electronic structure, and there is increased *sp–d* hybridization, consistent with bulk band structure results for $Pd^{(77)}$ and $Ni^{(66)}$. From Fig. 6, it should be noted that the *d*-band width for Pd_{13} is over 3 eV, while in the EH calculation for Pd_{19} by Baetzold and Mack it was only ~ 0.3 eV.

Hence on the basis of the $X\alpha$ calculations made to date for metal clusters, one is struck by the similarities rather than the differences between the electronic structures of such clusters and the bulk metal. It is also clear that this observation cannot be made regarding the EH calculations, and in fact the contrast between the $X\alpha$ and EH molecular orbital calculations is rather profound in most cases.

Returning to the three questions posed earlier, it appears that the $X\alpha$ calculations performed thus far would suggest the following answers: (i) The electronic structures of small clusters ($\geqslant 10$ atoms) and the bulk metal are not fundamentally different; (ii) not many atoms are needed in a cluster to have a resemblance to the bulk; 8–13 atoms already show many of the bulk features; (iii) the EH and CNDO methods, at least as applied in their usual forms, seem to be incapable of describing the electronic structure of metal clusters, as evidenced by comparison of several such calculations with those performed using the much more accurate and reliable $X\alpha$ method.

Thus in future applications of the EH and CNDO methods to clusters of metal atoms and considerations of chemisorption, it would appear that the first order of business should be a careful determination of parameters that produce an electronic structure which is in substantial agreement with either existing $X\alpha$ cluster calculations or bulk band structure calculations. It may turn out that even this is quite difficult to accomplish. It is unfortunate that these semiempirical methods, which have found such great utility in organic systems and other non-transition metal systems, have such difficulties in the present applications, but it is a problem which must be squarely faced in future applications of these methods.

3.3. Metals and Adsorbates

The objective of most of the studies which investigate models for adsorbate–metal interactions is to help in the interpretation of experimental findings associated with chemisorption of gases at very low pressures ($\leqslant 1$ monolayer coverage) on rather well-defined metal surfaces—such as single-crystal faces[78] or thin metal films.[79] Before discussing some of the molecular orbital studies which have been carried out on model systems, the work of van der Avoird,[80] who has used a rather different approach, should be mentioned.

van der Avoird has investigated dissociative chemisorption of H_2 on transition and noble metals by considering a four-atom complex—the two H atoms and two metal atoms—and using a perturbation–theoretic approach devised for treating intermolecular forces by Jansen.[81] The model is quite simplistic, in that it assumes only one orbital (taken as a Gaussian) on each of the four centers. Potential surfaces for a planar complex showed that no activation energy is required for dissociation of H_2 on Pt, Pd, and Ni. More recent calculations[82] which take a more realistic form for the single functions on the atoms conclude again that dissociative chemisorption on Ni requires no activation energy, whereas on Cu an activation energy of ~ 45 kcal/mole is needed. This is qualitatively consistent with experimental findings.[83] While these results are quite interesting, there are a number of simplifying assumptions whose validity remains somewhat dubious, especially the neglect of the other electrons and orbitals on the metal atoms.

3.3.1. EH Calculations

Of the various papers which have used extended Hückel theory as applied to a cluster of metal atoms to investigate chemisorption, the work of Anders *et al.*[84,85] is particularly notable.* They have studied the chemisorption of H and N atoms on the (100) face of tungsten. The model clusters used to represent the tungsten substrate are shown in Fig. 7. The authors carried out iterative EH calculations in order to minimize the unphysically large charge buildups which inevitably occur in noniterative calculations. Their distribution of energy levels as well as the atomic character of wave functions for the W clusters reflect the electronic structure of bulk tungsten rather well, as is evidenced by comparison with the band structure calculation of Mattheiss.[86] They obtained binding curves for a hydrogen or nitrogen atom approaching perpendicularly to the surface clusters shown in Fig. 7 at the sites labeled *A*, *B*, and *C*. The hydrogen atom was found to bind more strongly at an *A* site, whereas the nitrogen atom had its energetically most stable position at a *B* site.

*It represents a possible exception to the criticism made earlier about EH calculations for metal clusters; however, this work employs a number of innovations beyond the usual scheme.

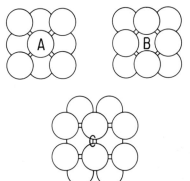

Fig. 7. Three types of binding sites used to investigate chemisorption on the (100) surface of tungsten (see Refs. 84 and 85).

For the case of hydrogen, it is known from low-energy electron diffraction (LEED) studies[87] that the H atoms adsorb on W at sites that have fourfold symmetry, which is consistent with the calculations. It is also known from experiment that H atoms have a low activation energy for diffusion across the surface[88] (~ 0.25 eV). The calculations are consistent with this if one assumes migration to be from *A* to *C* to *A* sites, since the binding energy at the *C* site for H is only ~ 0.25 eV lower than for the *A* site. It would be valuable, however, to have a more accurate study, such as an $X\alpha$ calculation, with which to compare these results.

Space does not permit a discussion of all the EH studies which have investigated chemisorption on metals. Therefore a discussion will be given of only one other recent set of calculations—those of Anderson and Hoffmann (AH).[64] Unlike the work of Anders *et al.*, AH carried out noniterative EH calculations. They investigated clusters of tungsten and nickel atoms interacting with diatomic molecules. Binding curves were not calculated. Net charge differences of well over one electron were obtained on adjacent atoms in some of the clusters. The diatomic molecule they studied for which the most reliable experimental data exist is CO. Hence we will discuss the Anderson–Hoffmann results for CO on Ni and compare their conclusions with what is known from experiment.

In the AH calculation, the CO molecule was assumed to adsorb with its internuclear axis perpendicular to the surface of the nickel cluster and with the carbon end closest to the Ni; such a geometry is supported by many experimental observations. The CO was assumed to sit directly above a single Ni atom of the cluster at a distance of 2.0 Å. The energy levels of this system were then calculated and compared with the electronic structure as deduced from photoemission experiments.

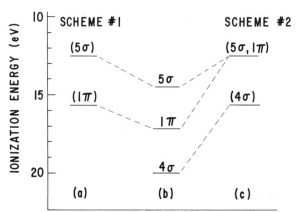

Fig. 8. Schematic representation of photoemission results for (a) CO chemisorbed on Ni, with orbital assignments due to Eastman and Cashion[50]; (b) molecular CO; (c) CO chemisorbed on Ni, with orbital assignments due to Gustafsson *et al.*[90]

Figure 8 shows a schematic representation of the energy levels as obtained by photoemission experiments of isolated CO (Fig. 8b) and the Ni + chemisorbed CO system according to Eastman and Cashion[50] (Fig. 8a). The 5σ orbital contains the lone pair of electrons on the carbon atom of CO. The levels in Fig. 8a are taken at the peak maxima of the photoemission results, and the dashed lines show the correspondence to the gas-phase CO spectrum given by Eastman and Cashion. The 4σ could not be seen in the photoemission study because the photon energy used was not sufficiently high. The shift in level positions between the gas-phase CO case and the chemisorption case is attributed to "relaxation processes." An important effect here is the better screening of the resultant hole (due to photoemission) by the CO in contact with the nickel surface.

The EH calculations find a 5σ-derived level at 14.1 eV and a 1π-derived level at 15.6 eV; this was interpreted by AH as providing support for the assignment of photoemission peaks given in Fig. 8a.

It is curious, however, that the experimental energy spacing between the 5σ and 1π levels in the case of molecular CO and in the case of chemisorbed CO and Ni should be nearly the same. Chemical intuition would suggest that the 5σ lone pair, which interacts with the metal, should be significantly lowered (i.e., give a higher ionization potential) due to its stabilizing, bond-forming interaction with the metal.[74] Calculations on Ni(CO)$_4$ suggest this also.[89]

In fact, a very recent experimental investigation by Gustafsson *et al.*,[90] which used synchroton radiation to investigate the spectrum of chemisorbed CO on Ni in a higher energy range and as a function of photon energy, suggests that scheme 1 of Fig. 8a is incorrect. No evidence was found for a third peak (which would correspond to the 4σ orbital using scheme 1) and the behavior of the peak of lower ionization energy as a function of photon energy led to the conclusion that both the 5σ and 1π-derived orbitals are contained in this first peak (see scheme 2 of Fig. 8). This new work, which is more definitive, produces an interpretation (scheme 2) which is thus in accord with chemical ideas, i.e., the 5σ orbital is appreciably stabilized with regard to the 1π and 4σ orbitals. The fact that the conclusions drawn from the EH calculations supported the incorrect assignment is somewhat disturbing. It does not give one much confidence in the ability of the method to make useful predictions for such systems. In contrast, $X\alpha$ calculations to be described in the next section have clearly found the correct qualitative behavior.

3.3.2. SCF–Xα–SW Calculations

Thus far $X\alpha$ calculations for metal cluster–adsorbate interactions have been carried out only for metal clusters of from one to five atoms. A summary will be given here of some of the results of these preliminary investigations.

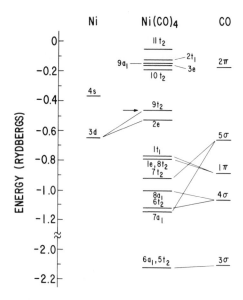

Fig. 9. SCF–Xα energy levels for the molecule Ni(CO)$_4$ calculated by Klemperer and Johnson.[89] Connecting lines show the dominant orbital contributions of CO to Ni(CO)$_4$; however, there is mixing of the 4σ and 5σ of CO in the Ni(CO)$_4$ molecule. The arrow denotes the highest occupied orbital.

a. Ni + CO. As models for the chemisorption of CO on Ni, two calculations will be discussed. The first is for Ni(CO)$_4$, which is a stable inorganic molecule. The Xα energy levels as calculated by Klemperer and Johnson[89] are shown in the center of Fig. 9. To the right in the figure are the Xα energy levels for the isolated CO molecule. The CO orbital character present in Ni(CO)$_4$ is shown by the connecting lines in the figure. We observe that the 5σ- derived levels are considerably lowered due to bond formation in Ni(CO)$_4$ and that in fact are below the 1π-derived levels, which move upward somewhat. This would appear to be indicative of what one might expect for chemisorbed CO on Ni, at least qualitatively.

The second calculation is for a cluster of five nickel atoms and a CO molecule. This calculation, made by Batra and Robaux,[91] assumes the CO molecule to be at a site analogous to that of the *B* site of Fig. 7 with four atoms in the top plane and only the one atom in the center of the lower plane. They did not investigate other types of binding sites, such as *A*- or *C*-like sites of Fig. 7, nor did they obtain binding curves as a function of CO distance from the Ni. For the only calculation which they carried out, the 5σ-derived level crossed over the 1π-derived level to give an ordering much the same as in Ni(CO)$_4$. Their calculation can hardly be considered definitive, since they failed to investigate a number of important aspects of the problem, but it is indicative of the correct qualitative behavior as compared to experiment.* Namely, the 5σ orbital of

---

*It should be pointed out that there is a discrepancy between the Xα results for Ni$_5$ as obtained by Batra and Robaux[91] and those obtained by Niemczyk.[98] Recent calculations by D. R. Salahub and the author support the results of Niemczyk, suggesting there is an error in the work of Batra and Robaux. Their error, although probably affecting the quantitative aspects of their work for Ni$_5$ + CO, should not be serious enough to change the qualitative conclusions.

CO is significantly stabilized relative to the 1π and 4σ orbitals by a bonding interaction with the Ni surface. Clearly the results from these two calculations are inconsistent with scheme 1 of Fig. 8 and provide support for scheme 2.

b. Ni, Pd, Pt + C$_2$H$_4$. Olefins are involved in a number of important catalytic reactions which take place on supported transition metal catalysts. One would therefore like to understand the interaction and bonding between the olefin (and/or resultant intermediates) with clusters of transition metal atoms in order to make progress in the understanding of such catalytic systems. As first steps to this end, a few model calculations have been carried out.

In a study by Rösch and Rhodin,[92] the bonding of ethylene to two nickel atoms was investigated for two geometries. In the first situation the axis of the Ni$_2$ was parallel to that of ethylene, with the ethylene oriented such that its π orbital would be perpendicular to the Ni$_2$ axis. This was referred to as the "di-σ" geometry. In the second situation, the ethylene interacted with one end of the Ni$_2$ with the Ni–C$_2$H$_4$ geometry similar to that in Zeise's anion (see Fig. 10). The authors found that in both cases the bonding involved principally two molecular orbitals. In one MO the π orbital of ethylene interacts with empty d orbital(s) and in the second, occupied d orbital(s) interact with the empty π^* ethylene orbital. The calculations for both model geometries predicted a decrease in the work function and a shift in the π-derived level of ethylene to higher binding energy with respect to the σ levels. Both these features are found by experiment.[93]

In a recent study,[94] the interaction of ethylene with a single atom of Ni, Pd, and Pt was investigated. The geometry was the same as that used in a previous calculation[95] for Zeise's anion (see Fig. 10), except that the three chlorine atoms are missing. The resulting orbital energies from these calculations are shown in Fig. 11. In this figure the orbitals among the species that are related have been connected by broken lines. The ethylene orbitals from $2a_g$ up to $1b_{3g}$ undergo an almost uniform shift when ethylene interacts with a Pt atom. The positions of these shifted ethylene orbitals remain relatively con-

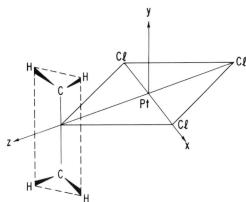

Fig. 10. Coordinate system and geometry
of Zeise's anion.

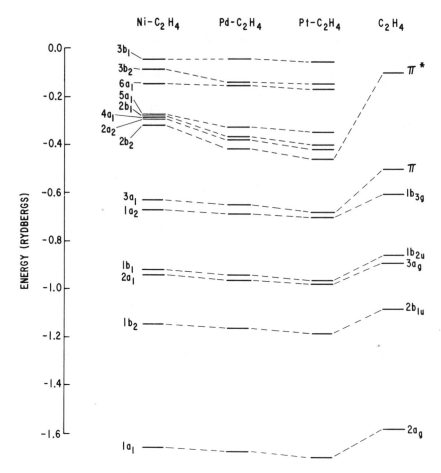

Fig. 11. Comparison of SCF–Xα electronic energy levels for Ni–C_2H_4, Pd–C_2H_4, Pt–C_2H_4 and C_2H_4. The same geometry was used for all calculations.

stant, with only a slight upward shift on going from Pt to Pd to Ni. The π and π^* levels of ethylene, on the other hand, experience rather profound changes on interacting with an atom of Pt, Pd, or Ni. The $2b_2$ and $3a_1$ orbitals are in each case primarily responsible for the bonding. The $2b_1$, $4a_1$, and $2a_2$ orbitals in each case are nonbonding orbitals, which are almost exclusively atomic d orbitals. The shift of the π level ($3a_1$) of ethylene on interacting with a metal atom is also a feature found in the photoelectron spectrum of ethylene chemisorbed on Ni.[93] The trend of increasing stability of the bonding orbitals across the series from Ni to Pt is consistent with the experimental trend in heats of adsorption for ethylene.

As discussed in a previous section, the results of Xα calculations on isolated metal clusters so far suggest that the electronic structures of these clusters are rather similar to the bulk metal. Hence, if, as suggested by experiment, very small supported metal particles have rather different catalytic behavior which cannot be explained strictly on the basis of increased surface area, the possibility suggests itself that the support may be involved.

A very crude model for the effect of an electronegative support has been investigated for the $Pt-C_2H_4$ complex.[94] This involves a comparison between the electronic structure of the $Pt-C_2H_4$ and $Pt-C_2H_4Cl_3^-$ complexes.[95] The latter is Zeise's anion shown in Fig. 10. The σ orbitals (due to ethylene) of $Pt-C_2H_4$, i.e., the orbitals $1a_1$ up to $1a_2$, remain essentially unchanged when the chlorine atoms are added to form $Pt-C_2H_4Cl_3^-$. A very striking effect occurs, however, for the orbitals that are derived from the π and π^* levels of ethylene; these orbitals are significantly lowered in energy in $Pt-C_2H_4Cl_3^-$ as compared to $Pt-C_2H_4$.

Thus the chlorine atoms serve to withdraw electrons from the Pt atom and promote the bonding between ethylene and Pt by allowing the $4a_1$ and $5a_1$ orbitals to become bonding rather than nonbonding orbitals. This provides a rather simple but nonetheless instructive example of the possible effect of the surrounding environment on metal–adsorbate bonding in the case of small metal clusters. For large metal particles on a support, however, this electron-withdrawing effect of a support may be completely screened out from the surfaces of these particles by the electrons of the intervening metal atoms. Thus for these large particles the main effect on catalytic behavior might be determined by the surface area.

A recent calculation by Yang and Johnson[96] on a model of an iron–sulfur protein, ferredoxin, shows that the sulfur atoms may be thought of as a support for the Fe atoms. The possible analogies with the effects described above are of great interest.

c. Ni + O. Two models of the Ni + chemisorbed oxygen system have been discussed. The first model,[97] nominally for a high coverage of oxygen uses a complex of one Ni atom and six oxygen atoms in an NiO_6^{-10} complex. The electronic energy levels of this model have been compared to both UPS and XPS photoemission data for this system with rather reasonable agreement. The similarity of the photoemission results to the energy levels of an NiO_6^{-10} cluster suggests that in certain cases of high oxygen coverage on the Ni surface the oxygen atoms penetrate the Ni lattice somewhat to form an incipient nickel oxide.

A second study, by Niemczyk,[98] uses a cluster of five nickel atoms and a single O atom interacting at a B-like site (Fig. 7). This is the geometry deduced by some recent LEED experiments. The author obtained a binding curve for the Ni_5-O system in order to compare with the value deduced by the LEED work; agreement was reasonable. Similar calculations were also carried out for Ni_5-S and Ni_5-Se.

4. Summary

In this chapter, we have attempted to point out the fact that a molecular cluster approach can be used to investigate a number of solid-state problems. Particular emphasis was placed on the theoretical problems of solid surfaces and the interactions of these surfaces with adsorbates. These latter problems have received relatively little attention until quite recently, a fact which is amply reflected in the chronology of the references presented here. This new interest in the understanding of surface phenomena has been spurred by the development of ultrahigh-vacuum techniques and the concurrent development of many new experimental methods for the investigation of surfaces.

It is obvious from our discussion of the current literature on the subject that the understanding of the electronic structure aspects of surfaces and chemisorption is presently rather crude. Nonetheless, new theoretical treatments, such as the Xα method, promise to provide the tools by which significant progress can be made in the future.

Indeed, the study of surface phenomena represents one of the new frontiers at the border of physics and chemistry, which should provide exciting challenges and many surprises to confront our traditional ideas of bonding and structure.

ACKNOWLEDGMENTS

The author wishes to express his gratitude to Prof. K. H. Johnson and Dr. D. R. Salahub for many stimulating conversations and the benefit of their comments on a preliminary draft of this manuscript. This work was supported in part by the U.S. Air Force Office of Scientific Research (AFSC) under Contract No. F44620–72–C–0008.

References

1. R. P. Messmer, in *Theoretical Chemistry*, MTP Rev. of Phys. Chem. Vol. 1, Ser. 2 (C. A. Coulson and A. D. Buckingham, eds.), Butterworths, London (1975), p. 219.
2. F. Bloch, *Z. Physik* **55**, 555 (1928).
3. J. M. Ziman, *Principles of the Theory of Solids*, University Press, Cambridge (1963).
4. J. C. Slater, *Quantum Theory of Molecules and Solids*, Vol. 2, *Symmetry and Energy Bands in Crystals*, McGraw-Hill, New York, (1965).
5. W. Kohn, in *Solid State Physics* (F. Seitz and D. Turnbull, eds.), Academic Press, New York (1957), Vol. 5, p. 258.
6. R. P. Messmer and G. D. Watkins, *Phys. Rev. B* **7**, 2568 (1973).
7. R. P. Messmer and G. D. Watkins, *Radiation Effects* **9**, 9 (1971).
8. M. R. Hayns, *Phys. Rev. B* **5**, 697 (1972).
9. G. D. Watkins and R. P. Messmer, *Phys. Rev. Lett.* **32**, 1244 (1974).

10. R. P. Messmer and G. D. Watkins, *Phys. Rev. Lett.* **25**, 656 (1970).
11. M. H. Cohen, *Physics Today* **24**(5), 26 (1971).
12. G. G. Hall, *Phil. Mag.* **43**, 338 (1952).
13. D. Weaire, *Phys. Rev. Lett.* **26**, 1541 (1971).
14. J. D. Joannopoulas and F. Yndurain, *Phys. Rev. B* **10**, 5164 (1974).
15. W. A. Goddard III and T. C. McGill, private communication.
16. P. W. Palmberg, *Surface Sci.* **11**, 153 (1968).
17. S. G. Davison and J. D. Levine, in *Solid State Physics* (H. Ehrenreich, F. Seitz, and D. Turnbull, eds.), Academic Press, New York (1970), Vol. 25, p. 1.
18. N. D. Lang and W. Kohn, *Phys. Rev. B* 3, 1215 (1971).
19. J. D. Levine and P. Mark, *Phys. Rev.* **182**, 926 (1969).
20. D. Kalkstein and P. Soven, *Surface Sci.* **26**, 85 (1971).
21. W. Ho, S. L. Cunningham, W. H. Weinberg, and L. Dobrzynski, *Phys. Rev. B* **12**, 3027 (1975).
22. J. A. Appelbaum and D. R. Hamann, *Phys. Rev. B* **6**, 2166 (1972).
23. J. A. Appelbaum and D. R. Hamann, *Phys. Rev. Lett.* **31**, 106 (1973).
24. J. A. Appelbaum, G. A. Baraff, and D. R. Hamann, to be published.
25. J. E. Rowe and H. Ibach, *Phys. Rev. Lett.* **31**, 102 (1973).
26. G. P. Alldredge and L. Kleinman, *Phys. Rev. Lett.* **28**, 1264 (1972).
27. E. Caruthers, L. Kleinman, and G. P. Alldredge, *Phys. Rev. B* **9**, 3325 (1974).
28. E. Caruthers, L. Kleinman, and G. P. Alldredge, *Phys. Rev. B* **9**, 3330 (1974).
29. J. L. Beeby, *J. Phys. C: Solid State Phys.* **6**, 1242 (1973).
30. R. V. Kasowski, *Phys. Rev. Lett.* **33**, 83 (1974).
31. N. Kar and P. Soven, *Phys. Rev. B* **11**, 3761 (1975).
32. W. Kohn, *Phys. Rev. B* **11**, 3756 (1975).
33. J. Korringa, *Physica* **13**, 392 (1947); W. Kohn and N. Rostoker, *Phys. Rev.* **94** 1111 (1954).
34. J. R. Smith, S. C. Ying, and W. Kohn, *Phys. Rev. Lett.* **30**, 610 (1973).
35. N. D. Lang and A. R. Williams, *Phys. Rev. Lett.* **34**, 531 (1975).
36. T. B. Grimley, *J. Vac. Sci. Technol.* **8**, 31 (1971).
37. D. M. Newns, *Phys. Rev.* **178**, 1123 (1969).
38. J. W. Gadzuk, in *Surface Physics of Crystalline Materials*, (J. M. Blakely, ed.), Academic Press, New York (1975).
39. J. R. Schrieffer, *J. Vac. Sci. Technol.* **9**, 561 (1972).
40. T. L. Einstein and J. R. Schrieffer, *Phys. Rev. B* **7**, 3629 (1973).
41. J. W. Davenport, T. L. Einstein, J. R. Schrieffer, and P. Soven, in *The Physical Basis of Heterogeneous Catalysis*, Plenum Press, New York (1975), p. 295.
42. P. W. Anderson, *Phys. Rev.* **124**, 41 (1961).
43. C. A. Coulson, *Proc. Camb. Phil. Soc.* **36**, 201 (1940).
44. C. A. Coulson, and H. C. Longuet-Higgins, *Proc. Roy. Soc. A* **191**, 39 (1947).
45. R. P. Messmer, B. McCarroll, and C. M. Singal, *J. Vac. Sci. Technol.* **9**, 891 (1971).
46. R. P. Messmer, *Chem. Phys. Lett.* **11**, 589 (1971).
47. J. A. Appelbaum and D. R. Hamann, to be published.
48. R. V. Kasowski, *Phys. Rev. Lett.* **33**, 1147 (1974).
49. H. D. Hagstrum and G. E. Becker, *Phys. Rev. Lett.* **22**, 1054 (1969).
50. D. E. Eastman and J. K. Cashion, *Phys. Rev. Lett.* **27**, 1520 (1971).
51. J. H. Sinfelt, in *Annual Review of Materials Science* (R. A. Huggins, R. H. Bube, and R. W. Roberts, eds.), Annual Reviews, Palo Alto, California (1972), Vol. 2, p. 641.
52. B. Lang, R. W. Joyner, and G. A. Somerjai, *Surface Sci.* **30**, 454 (1972).
53. M. Wolfsberg and L. Helmholz *J. Chem. Phys.* **20**, 837 (1953); R. Hoffmann, *J. Chem. Phys.* **39**, 1397 (1963).
54. J. A. Pople, D. P. Santry, and G. A. Segal, *J. Chem. Phys.* **43**, S129 (1965).
55. J. C. Slater, *J. Chem. Phys.* **43**, S228 (1965).
56. K. H. Johnson, *J. Chem. Phys.* **45**, 3085 (1966).
57. K. H. Johnson and F. C. Smith, Jr., *Phys. Rev. B* **5**, 831 (1972).
58. A. J. Bennett, B. McCarroll, and R. P. Messmer, *Surface Sci.* **24**, 191 (1971).
59. A. J. Bennett, B. McCarroll, and R. P. Messmer, *Phys. Rev. B* **3**, 1397 (1971).

60. R. C. Baetzold, *J. Chem. Phys.* **55**, 4363 (1971).
61. R. C. Baetzold, *J. Catal.* **29**, 129 (1973).
62. G. Blyholder, *Surface Sci.* **42**, 249 (1974).
63. D. J. M. Fassaert, H. Verbeek, and A. van der Avoird, *Surface Sci.* **29**, 501 (1972).
64. A. B. Anderson and R. Hoffmann, *J. Chem. Phys.* **61**, 4545 (1974).
65. R. P. Messmer, C. W. Tucker, Jr., and K. H. Johnson, *Chem. Phys. Lett.* **36**, 425 (1976).
66. J. Callaway and C. S. Wang, *Phys. Rev. B* **7**, 1096 (1973).
67. R. C. Baetzold and R. E. Mack, *J. Chem. Phys.* **62**, 1513 (1975).
68. J. C. Slater and K. H. Johnson, *Phys. Rev. B* **5**, 844 (1972).
69. J. C. Slater, in *Advances in Quantum Chemistry* (P. O. Löwdin, ed.), Academic Press, New York (1972), Vol. 6, p. 1.
70. J. C. Slater, *Quantum Theory of Molecules and Solids*, Vol. 4, *The Self-Consistent Field Theory of Molecules and Solids*, McGraw-Hill, New York (1974).
71. K. H. Johnson, in *Advances in Quantum Chemistry* (P. O. Löwdin, ed.), Academic Press, New York (1973), Vol. 7, p. 143.
72. K. H. Johnson, J. G. Norman, Jr., and J. W. D. Connolly, in *Computational Methods for Large Molecules and Localized States in Solids* (F. Herman, A. D. McLean, and R. K. Nesbet, eds.), Academic Press, New York (1973), p. 161.
73. J. W. D. Connolly, in: *Modern Theoretical Chemistry, Vol. 7: Semiempirical Methods of Electronic Structure Calculation, Part A: Techniques* (G. A. Segal, ed.), Plenum, New York (1977), Chap. 4.
74. R. P. Messmer, S. K. Knudson, K. H. Johnson, J. B. Diamond, and C. Y. Yang, *Phys. Rev. B* **13**, 1396 (1976).
75. K. H. Johnson, J. B. Diamond, R. P. Messmer, and S. K. Knudson, General Electric Technical Report No. 74CRD052, Schenectady, New York (1974).
76. G. A. Burdick, *Phys. Rev.* **129**, 138 (1963).
77. F. M. Mueller, A. J. Freeman, J. O. Dimmock, and A. M. Furdyna, *Phys. Rev. B* **1**, 4617 (1970).
78. G. A. Somorjai and L. L. Kesmodel, in *Surface Chemistry and Colloids*, MTP Rev. of Phys. Chem., Vol. 7, Ser. 2, in press.
79. J. R. Anderson (ed.), *Chemisorption and Reactions on Metallic Films*, Academic Press, New York (1971), Vol. 1.
80. A. van der Avoird, *Surface Sci.* **18**, 159 (1969).
81. L. Jansen, in *Advances in Quantum Chemistry* (P. O. Löwdin ed.), Academic Press, New York (1965), Vol. 2, p. 119.
82. H. Deuss and A. van der Avoird, *Phys. Rev. B* **8**, 2241 (1973).
83. S. J. Holden and D. R. Rossington, *J. Catal.* **4**, 403 (1965).
84. L. W. Anders, R. S. Hansen, and L. S. Bartell, *J. Chem. Phys.* **59**, 5277 (1973).
85. L. W. Anders, R. S. Hansen, and L. S. Bartell, *J. Chem. Phys.* **62**, 1641 (1975).
86. L. F. Mattheiss, *Phys. Rev.* **139**, A1893 (1965).
87. P. J. Estrup and J. Anderson, *J. Chem. Phys.* **45**, 2254 (1966).
88. R. Gomer, R. Wortman, and R. Lundy, *J. Chem. Phys.* **26**, 1147 (1957).
89. W. G. Klemperer and K. H. Johnson, to be published.
90. T. Gustafsson, E. W. Plummer, D. E. Eastman, and J. L. Freeouf, *Bull. Am. Phys. Soc.* **20**, 304 (1975).
91. I. P. Batra and O. Robaux, *J. Vac. Sci. Technol.* **12**, 242 (1975).
92. N. Rösch and T. N. Rhodin, *Phys. Rev. Lett.* **32**, 1189 (1974).
93. J. E. Demuth and D. E. Eastman, *Phys. Rev. Lett.* **32**, 1123 (1974).
94. R. P. Messmer, in *The Physical Basis of Heterogeneous Catalysis*, Plenum Press, New York (1975), p. 261.
95. N. Rösch, R. P. Messmer, and K. H. Johnson, *J. Am. Chem. Soc.* **96**, 3855 (1974).
96. C. Y. Yang and K. H. Johnson, to be published.
97. R. P. Messmer, C. W. Tucker, Jr., and K. H. Johnson, *Surface Sci.* **42**, 341 (1974).
98. S. J. Niemczyk, *J. Vac. Sci. Technol.* **12**, 246 (1975).

7

Electron Scattering

Donald G. Truhlar

1. Introduction

This chapter discusses some approximate methods for the calculation of electron scattering cross sections of molecules.

There has been much progress in the development of computational methods for electron scattering over the last five or so years. Entirely new methods have been developed, old methods have been improved, and both have been applied more systematically than previously. Some of the approaches which have been made to the calculation of electron scattering cross sections will be discussed in this chapter, and special emphasis will be placed on those aspects of the approximations which naturally link up with the approximation methods used for bound-state calculations which are discussed in the rest of this treatise. Representative but not exhaustive references will be given. We will treat the electronic part of the wave function, but the rotational and vibrational motions involved in electron–molecule scattering will not be discussed in detail. While many aspects of the approximate theories used for the electronic part of the bound-state problems are useful for electron scattering, some more specialized techniques and approximations are also useful.

In theoretical chemistry, more attention is directed to electron–molecule scattering than to electron–atom scattering. Thus we will emphasize methods which have been shown to be useful for electron–molecule scattering or which show promise of future usefulness for electron–molecule scattering calculations. It would be particularly appropriate if the methods discussed here were all illustrated by applications to electron–molecule scattering. However, some

Donald G. Truhlar • Department of Chemistry and Chemical Physics Program, University of Minnesota, Minneapolis, Minnesota

of the techniques which show the most promise for future electron–molecule scattering calculations have so far only been applied to electron–atom scattering or have been applied in a much more definitive fashion to electron–atom scattering. Thus in many cases the best illustrations must be taken from electron–atom scattering. Of course electron–molecule scattering involves treating both the electronic part and the internuclear part and the latter has no analog in electron–atom scattering.

In a few cases involving electron scattering by hydrogen and helium atoms essentially exact cross sections have been computed. But most electron scattering calculations contain approximations of uncertain validity. Nevertheless, it is often possible to judge the general reliability of various approximation schemes for various accuracy requirements, targets, types of information about scattering processes, and energy ranges. A method which is successful for one application will often not be satisfactory for another and *vice versa*. Thus an understanding of the uses of electron scattering cross sections and of the characteristics of the different energy ranges is a necessary prerequisite for understanding the usefulness of the various approximation schemes.

A few of the more important applications of electron scattering cross sections are electron-impact spectroscopy, radiation chemistry, aeronomy and other studies involving atmospheres of the earth and other planets, astrophysics, the study of the sun's corona, electron-drift experiments and gaseous electronics, the study of laboratory discharges and plasmas, fusion, and lasers. Laboratory measurements on electron impact processes have been thoroughly reviewed by Massey *et al.*[1,2] and Christophorou.[3]

Several textbooks and monographs with an appreciable emphasis on the theory of electron scattering are available[4–11] and should be consulted for an introduction to the field.

Both for classifying the physical processes occurring and for sorting the appropriate computational procedures it is useful to distinguish five ranges of impact energy (very low, low, intermediate, high, and very high).

The low-energy region is characterized by $T_i < U_i$, where T_i is the initial translational energy of the scattering electron and U_i is the ionization energy of the interacting target electrons. In this energy range the scattering electron and the target electrons are best treated on an equivalent basis since they have roughly equivalent energies. Thus the procedure for calculating the total wave function may resemble a bound-state calculation for the composite system of target plus scattering electron more than at higher energies, where the scattering electron may be usefully treated on a nonequivalent basis due to its higher energy. When $T_i = 1$ eV the de Broglie wavelength is 12.3 Å. This is very large (larger than the dimensions of the target for most targets of interest in this chapter) and indicates the need of a quantum mechanical treatment and that classical mechanical concepts should be used with utmost caution if at all. In the low-energy region, elastic and inelastic scattering occur and, provided too many channels are not open, variational minimum principles may be used to perform accurate calculations. (Since these principles require explicit inclusion of all open channels in the trial scattering wave function, they are not of practical value at higher energies.) At this low energy it is the usual practice to expand the total scattering wave function in terms of eigenfunctions of the total angular momentum. Although this is more complicated for electron–molecule scattering than electron–atom scattering, it still provides enough simplifications that it is often worth attempting even if further approximations are required. Less than ten angular momenta need to be treated by full calculations using low-energy methods. If angular momenta above the lowest few values are needed at all (e.g., to adequately include large-impact-parameter collisions to converge the differential cross section at small scattering angles), they can usually be treated by perturbation theory or other high-energy approximations. For the lower angular momenta the close coupling method and the matrix close coupling (also called algebraic close coupling) method are often used; in the language of bound-state calculations, these are continuum configuration interaction methods. Resonances (temporary negative ions for electron–neutral scattering or temporary electron attachment for electron–ion scattering) are often important at low energies, particularly shape resonances at initial or final translational energies up to a few eV and Feshbach resonances near thresholds.* Since a resonance may be visualized as an electron temporarily bound to some state of the target, bound-state calculational procedures can sometimes be used almost without modification for the description of resonances.

At very low energies, e.g., thermal energies, additional simplifications occur. For very low energies all scattering is elastic. The de Broglie wavelength at $T_i = 0.026$ eV is 76 Å. When the de Broglie wavelength is much greater than the dimensions of the target, all elastic scattering occurs only in the component at the wave function corresponding to zero orbital angular momentum of the scattering electron, i.e., in the s wave, and the s wave is sensitive only to the spherically symmetric part of the interaction potential. This yields an isotropic differential cross section. In the limit of zero translational energy, the scattering information is all contained in the scattering length† and at very low energies

---

*"Shape resonances" are resonances which can be described reasonably correctly by a model involving scattering of the electron by an effective potential due to its interaction with the target in some unperturbed or perturbed state. "Feshbach resonances" require a target-excitation mechanism for thier description. They involve temporary binding of the scattering electron to some excited state of the target when the electron is incident on the target in some lower energy state.

†The "scattering length" is the limit as the incident momentum tends to zero of the s-wave phase shift divided by the momentum in atomic units. Generalizations of the scattering length definition to higher partial waves are not needed in this chapter.

the effective range formalism[12] may be used to express the deviation from the zero-energy limit. (This formalism, which is not discussed in this chapter, is also useful at higher energies for high orbital angular momentum of the scattering electron.) An upper bound on the scattering length may be obtained by an extension of the upper bound theorem used for bound-state energies.[13]

The intermediate–energy region is roughly characterized by $U_i < T_i < 15U_i$. The precise location of the intermediate energy range thus depends on the target and the process considered, but it may often be considered as about 10–150 eV. The de Broglie wavelength of an electron varies from 3.9 Å at 10 eV to 1.0 Å at 150 eV. In this energy region elastic and inelastic scattering and ionization may occur and many channels are open. Thus the variational minimum principles which are useful in the low-energy region are not useful here. There are some resonances, but most scattering is nonresonant. The optical electric dipole selection rules that hold in the high-energy limit are not in effect here and this is the region useful for electron-impact spectroscopy.[14,15] The plane-wave perturbation theories, which become accurate at higher energies where the interaction potential is small compared to the initial kinetic energy, are not yet accurate because the incident electron is not well described by a plane wave unless its initial energy is higher and because target charge polarization effects are hard to treat by perturbation theory at intermediate energy. The intermediate-energy region is the most difficult region to treat and is the subject of much current interest. Most of this work involves modifying the low-energy or high-energy methods so that they will be more appropriate for this energy region.

In the high-energy region corresponding to $U_i \ll T_i \ll$ few keV, the de Broglie wavelength of an electron varies from about 1.0 to 0.3 Å. Thus classical pictures are more appropriate than at lower energies. The usual theory is perturbation theory with a plane wave (or a Coulomb wave) for the zeroth-order free wave and no interaction potential (or a pure Coulombic interaction) in the zeroth-order Hamiltonian. The perturbation series is the well-known Born series (explained in introductory texts) and its modifications. The first term in the series is the first Born approximation. It consists of an integral over the unperturbed initial and final target wave functions, the initial and final plane waves, and the interaction potential. Thus intermediate target states (which are necessary to represent target charge polarization) and distortion of the scattering electron's wave function by the interaction are neglected in the zeroth-order wave function. Contrary to popular assumption, there is no proof that the high-energy limit of the differential cross section is correctly given by the first Born approximation and there is some evidence that it is not.[16] In many respects the angular momentum l of the scattering electron plays a role similar to the energy. Just as the high-l components of the low-energy scattering wave function can often be treated using high-energy methods, the low-l components (e.g., the s wave) of the high-energy scattering wave function should be treated for many purposes using low-energy methods. This is necessary because these low-l components are not prevented by the centrifugal potential from penetrating into regions where the interaction potential is not negligible with respect to even high initial kinetic energies. But the high-energy scattering is often treated without making a partial wave decomposition of the wave function into its components with various l or total angular momenta so the requirement for a more accurate treatment at low l is often ignored.

The very-high-energy region includes energies above a few keV. The usual energy for electron diffraction experiments is 40 keV, where the de Broglie wavelength of an electron is 0.06 Å. Some special techniques have been developed for this energy region.[2,11] For example, the atoms of a molecule may sometimes be considered to scatter independently. Some high-resolution spectroscopy may be done in this energy region,[17] although selection rules identical to the optical ones are valid to a good approximation.

The nuclei may be treated as point charges at energies as high as 3 MeV.[18] When electrons with energies of 15 MeV and higher collide with an atom, a significant contribution to the scattering is made by electrons that have penetrated the nucleus.[18,19] Such energies are out of the range considered in this chapter.

This chapter is concerned with the $(1e, 1e)$ scattering process where an electron scatters from a gas-phase target A

$$e^- + A \rightarrow e^- + A$$

(where A need not be neutral). In this collision process, the state of A may remain the same (elastic scattering) or it may change (inelastic scattering). The $(1e, 2e)$ and $(1e, \text{many } e)$ electron-impact ionization processes are closely related but are more complicated due to the presence of three or more bodies in the final state. The $(0e, 1e)$ photoionization process

$$h\nu + A \rightarrow A^+ + e^-$$

is easier to treat than the $(1e, 1e)$ scattering process, but because it involves a free electron in the final state it provides an interesting link between bound states and scattering states. Recently there has been much progress in treating photoionization using bound-state techniques.[20,21] These treatments can profitably be compared to earlier treatments using scattering techniques[22,23] but photoionization will not be discussed much more in this chapter. Another interesting electron scattering problem not treated here is free–free radiation[24]:

$$e^- + A \rightarrow e^- + A + h\nu$$

For electron scattering the center of mass may be assumed to coincide with the center of mass of the nuclei and in the center-of-mass system (barycentric system) the wave function for electron scattering by an N-electron target depends on relative nuclear coordinates $\mathbf{R}_1, \mathbf{R}_2, \ldots, \mathbf{R}_{\nu-1}$ (for a molecule with ν nuclei), on the coordinates $\mathbf{r}_1, \mathbf{r}_2, \ldots, \mathbf{r}_{N+1}$ of the electrons with respect to the center-of-mass of the nuclei, and on spin coordinates. The electronic part of the wave function can be expanded in a set of basis functions which depend on electronic coordinates and spins (and thus are determined by the Schrödinger equation with nuclear kinetic energy terms omitted) and depend parametrically on relative nuclear coordinates and spins. This is the fixed-nuclei formalism and it is analogous to the electronically adiabatic separation of electronic and nuclear motion which is usually made for treating bound states.* In this chapter we will not explicitly indicate the parametric dependence of the electronic wave function and the electronic basis functions on the nuclear coordinates and nuclear spins.

The fixed-nuclei approach offers important simplifications in general but not at large distances r of the scattering electron from the target (greater than about 10 bohrs),[28] especially for polar molecules,[29] because of the long-range nature of their interactions with electrons, and for very low energies ($T_i \lesssim 0.1$ eV). The breakdown of the fixed-nuclei approach in these cases (large r and low T_i) is similar to its breakdown in Rydberg bound states, where the molecular framework may be pictured as rotating appreciably during an orbit of the distant and slowly moving electron. For these cases, it is more appropriate to use the laboratory-frame approach in which the electron is considered to be interacting with the molecule in a given rotational–vibrational state. Scattering calculations have been performed using both formalisms. Procedures necessary for interrelating them and for using the fixed-nuclei approach at small r and the laboratory-frame approach at large r have also been developed.[28,30] In this chapter we use the fixed-nuclei approach for consistency with the rest of the volume.

2. Explicit Inclusion of Electronic Excitations

2.1. Expansions Including Free Waves

2.1.1. Coupled Equations

First we consider methods based on expanding the electronic wave function in a suitable set of functions, including products of target eigenstates and continuum functions (free waves) for the scattering electron. These free

---

*For bound states this is usually called the Born–Oppenheimer adiabatic approximation[25] and for electron scattering it is sometimes called the adiabatic or the adiabatic-nuclei approximation.[26] The former nomenclature is sometimes used in electron scattering but it should not be confused with the Born–Oppenheimer exchange approximation for electron scattering[27] or the adiabatic polarization assumption of electron scattering (see Section 4).

waves are included in order to satisfy scattering boundary conditions, and the variational methods of scattering theory are applicable to trial functions which satisfy scattering boundary conditions [scattering boundary conditions are defined by Eqs. (1) and (8)–(11) below]. Some methods formulated entirely in terms of square-integrable basis functions are considered in Section 2.2.

Variational principles are not used for electron–impact ionization since the correct boundary conditions are not completely known, due to the presence of two free electrons in the final state.[31] Further, the correct wave function for any electron collision at energies above the target's ionization potential must include some terms corresponding to such two-free-electron final states. Nevertheless, these terms are usually neglected except when electron-impact ionization is the quantity of interest, and ionization cross sections are usually calculated using perturbation theory with no attempt at a complete expansion of the final state wave function.[32]

The most general possible trial function for electron scattering by an N-electron target which satisfies correct scattering boundary conditions (neglecting ionization channels) may be written

$$\psi^P(x_1, x_2, \ldots, x_{N+1}) = \hat{O}\left[\sum_{i=1}^{P} X_{ip}(r_{N+1}) f_i(x_1, x_2, \ldots, x_N, \hat{\mathbf{r}}_{N+1}, \sigma_{N+1}) \right.$$

$$\left. + \sum_{m=1}^{M} c_{mp} W_m(x_1, x_2, \ldots, x_{N+1}) \right] \qquad (1)$$

where $\hat{O}$ is a permutation operator (which may be the antisymmetrizer or a partial antisymmetrizer) acting on the $\{x_i\}_{i=1}^{N+1}$; x_i is the set of coordinates $(\mathbf{r}_i, \sigma_i)$ or equivalently $(r_i, \hat{\mathbf{r}}_i, \sigma_i)$; $\mathbf{r}_i$ is the spatial coordinate of electron i; r_i is the distance $|\mathbf{r}_i|$ from the center of mass to electron i; $\hat{\mathbf{r}}_i$ is the unit vector from the center of mass to electron i; σ_i is the spin coordinate of electron i; p and i are sets of channel indices, where p designates a particular choice of boundary conditions and i designates an arbitrary channel; P and M are constants representing the numbers of terms retained in each sum in the trial function; W_m is a square-integrable basis function; f_i is

$$\Phi_i(x_1, x_2, \ldots, x_N) Y_{l_i m_i}(\hat{\mathbf{r}}_{N+1}) \chi_{m_i^s}(\sigma_{N+1})$$

or a linear combination of such products involving degenerate Φ_i, where Φ_i is an eigenfunction of the channel Hamiltonian H_i defined in terms of the total Hamiltonian H by

$$H_i(x_1, x_2, \ldots, x_N) = \lim_{r_{N+1} \to \infty} [H(x_1, x_2, \ldots, x_{N+1}) - T(\mathbf{r}_{N+1})] \qquad (2)$$

where T is the kinetic energy of the scattering electron defined by

$$T(\mathbf{r}_{N+1}) = p_{N+1}^2 / 2m \qquad (3)$$

and

$$H_i(x_1, x_2, \ldots, x_N)\Phi_i(x_1, x_2, \ldots, x_N) = E_i\Phi_i(x_1, x_2, \ldots, x_N) \tag{4}$$

$$T(\mathbf{r}_{N+1})X_{ip}(r_{N+1})Y_{l_im_i'}(\hat{\mathbf{r}}_{N+1}) \xrightarrow[r_{N+1}\to\infty]{} \frac{\hbar^2 k_i^2}{2m}X_{ip}(r_{N+1})Y_{l_im_i'}(\hat{\mathbf{r}}_{N+1}) \tag{5}$$

where the total energy is

$$E = E_i + \frac{\hbar^2 k_i^2}{2m} \tag{6}$$

$$= E_i + T_i \tag{7}$$

and $Y_{l_im_i'}$ is a spherical harmonic; $\chi_{m_i^s}$ is a spin wave function; X_{ip} is the channel radial function, which is square-integrable if k_i is imaginary and may be assumed to satisfy the following scattering boundary condition for the N_{op} open channels, $i = 1, 2, \ldots, N_{op}$, for which k_i is real:

$$X_{ip}(r) \underset{r_{N+1}\to\infty}{\sim} Y_{ip}(r; \tau) \tag{8}$$

$$Y_{ip}(r; \tau) = \alpha_{0ip}(\tau)A_{i0}(r; \tau) + \alpha_{1ip}(\tau)A_{i1}(r; \tau) \tag{9}$$

$$A_{i\beta}(r; \tau) \underset{r\to\infty}{\sim} r^{-1}a_0^{-1/2}\sin(\Theta_i + \tfrac{1}{2}\beta\pi + \tau) \tag{10}$$

$$\Theta_i = k_ir - \tfrac{1}{2}l_i\pi + Zk_i^{-1}\ln 2k_ir + \arg\Gamma(l_i + 1 - iZk_i^{-1}) \tag{11}$$

Here α_{0ip} and α_{1ip} are coefficients for channel i depending on the particular choice p of boundary condition; τ is an arbitrary phase angle; a_0 is the unit of length, which is taken as the bohr; Z is the net charge of the target—the sum of the nuclear charges of the atoms minus the number of bound electrons, i.e.,

$$Z = z_1 + z_2 + \cdots + z_\nu - N \tag{12}$$

and Γ is the gamma function.

Notice that the first sum in Eq. (1) consists of terms representing a scattering electron, represented by X_{ip} and the $\hat{\mathbf{r}}_{N+1}$-dependent and σ_{N+1}-dependent parts of F_i, moving in the field of one or more target eigenstates Φ_i. The dummy index i labels the channels and, for the particular choice of boundary conditions $\alpha_{0ip} = \delta_{ip}$, p denotes the initial channel (the initial channel was designated i in Section 1). In general, a particular choice of boundary conditions singles out one channel, here called p, for special treatment. The second sum in Eq. (1) consists of terms which represent the $(N+1)$-electron system without necessarily singling out the scattering electron for nonequivalent treatment. Thus mathematically this sum is used to obtain a more complete expansion of the total wave function. Physically it may be used to represent compound states of the scattering electron plus target or to include correlation of scattering and target electrons.

Illustrative examples of the form (1) for scattering wave functions are given in detail for various atoms elsewhere.[33–35]* A general formulation for electron scattering by atoms or ions with any number of incomplete subshells is given by Smith and Morgan.[38] It should be noted that terms with square-integrable X_{ip} may be moved from the first sum to the second sum in (1) without changing the total wave function but involving a change in P, which is therefore not a well-defined quantity for a given trial wave function. Where such terms are placed is a matter of convenience. We will place them in the second sum.

The scattering cross sections may be expressed[39,40] in terms of the asymptotic form of the wave function ψ^p (by using Green's theorem it is possible to write the asymptotic form of the wave function in terms of an integral over the wave function). To calculate cross sections it is convenient to define three $N_{op} \times N_{op}$ matrices which are independent of τ and the particular choice of scattering boundary conditions on the wave function. The reactance matrix $\mathbf{R}$ is defined by

$$\mathbf{R} = \mathbf{V}^{1/2}[(\sin \tau)\boldsymbol{\alpha}_0 + (\cos \tau)\boldsymbol{\alpha}_1][(\cos \tau)\boldsymbol{\alpha}_0 - (\sin \tau)\boldsymbol{\alpha}_1]^{-1}\mathbf{V}^{-1/2} \qquad (13)$$

where $\mathbf{V}$ is the diagonal velocity matrix, i.e.,

$$V_{ij} = (\hbar k_i/m)\delta_{ij} \qquad (14)$$

and

$$(\boldsymbol{\alpha}_\beta)_{ij} = \alpha_{\beta ij}, \qquad \beta = 0, 1 \qquad (15)$$

Although $\boldsymbol{\alpha}_0$ and $\boldsymbol{\alpha}_1$ depend on the particular choice of boundary condition, $\mathbf{R}$ as defined above is independent of this choice.

Then the scattering matrix S is related to the reactance matrix and the diagonal unit matrix $\mathbf{I}$ by

$$S = (\mathbf{I} - i\mathbf{R})^{-1}(\mathbf{I} + i\mathbf{R}) \qquad (16)$$

and the transition matrix $\mathbf{T}$ is defined by

$$\mathbf{T} = \mathbf{S} - \mathbf{I} \qquad (17)$$

Note that some authors use slightly different definitions of $\mathbf{T}$.

We can illustrate the use of these matrices by a simple example: the scattering of an electron with initial momentum $\hbar k_p$ off a target with spherically symmetric eigenstates $n = 1, 2, \ldots$. For the partial wave with relative orbital angular momentum $\hbar l$, the scattering amplitude for scattering angle θ and a $p \to n$ transition is

$$f_{np}^l = \frac{-1}{2i(k_n k_p)}(2l + 1)T_{np}^l P_l(\cos \theta) \qquad (18)$$

where P_l is a Legendre polynomial. The total scattering amplitude is a sum over l which converges because the scattering amplitude becomes small for high l

*The work of Smith et al.[34] contains an error which was later corrected.[36,37]

(the classical argument for this is that the impact parameter is l/k_p and scattering is small or zero if this exceeds the range of the potential). For more general collision processes the scattering amplitude may still be expressed in terms of T matrix elements but it involves a multiple-index sum over T matrix elements corresponding to degenerate processes. Further, complicated angular momentum coupling coefficients, which are not needed in this chapter, occur instead of the simple factor $(2l+1)$. In general the integral cross section Q_{ij} for the electron–impact excitation of the $i \to j$ transition (or for elastic scattering when $i = j$) is related to the differential cross section $I_{ij}(\Omega)$ and the scattering amplitude $f_{ji}(\Omega)$ corresponding to scattering solid angle Ω by

$$Q_{ij} = \int I_{ij}(\Omega)\, d\Omega \qquad (19a)$$

$$I_{ij}(\Omega) = (k_j/k_i)|f_{ji}(\Omega)|^2 \qquad (19b)$$

Performing the absolute square in Eq. (19b), we can express the integral and differential cross sections in terms of weighted sums of products of transition matrix elements. The weights include the angular momentum coupling coefficients. The complete formulas are given elsewhere[10,39–42] but they are not given here because we shall not need to define all the angular momentum coupling coefficients which these expressions contain. It is important to use consistent phase conventions for $Y_{lm'}$ and the angular momentum coupling coefficients in calculating these quantities.[42–44] We shall generally deal with the reactance matrix, from which $\mathbf{T}$ and $f_{ji}(\Omega)$ may be calculated, rather than with $\mathbf{T}$ and $f_{ji}(\Omega)$, because $\mathbf{R}$ is real while the latter two quantities are complex.

When the exact reactance matrix cannot be obtained, one must resort to variational methods, to perturbation theory, or to other approximation schemes. In variational methods a variationally correct expression* for the reactance matrix can be written in terms of the reactance matrix obtained from the asymptotic form of a trial function plus an integral over the trial function. For example, we introduce a matrix $\mathbf{L}$ with elements

$$L_{ij} = \langle \psi^i | H - E | \psi^j \rangle \qquad (20)$$

and consider variations of the functions ψ^i and ψ^j with constant τ and constant $\boldsymbol{\alpha}_0 = \mathbf{I}$. In other words, ψ^i and ψ^j are represented by Eq. (1) with the second sum and X_{ip} variable (except for τ and $\boldsymbol{\alpha}_0$). The Kohn variational principle[45–47] is†

$$\delta(\boldsymbol{\alpha}_1 - 2a_0\hbar^{-1}\mathbf{V}^{-1}\mathbf{L}) = 0 \qquad (21)$$

for small variations of the trial function about the exact wave function. Since

---

*A variationally correct expression for a scattering parameter in terms of trial scattering wave functions yields an approximate value for the scattering parameter which depends quadratically (not linearly) on the errors in the trial functions.

†Strictly speaking, this is the Kohn variational principle only when τ is zero. We have considered a generalization to simplify the following discussion.

α_0, τ, and the velocity matrix are known constants in a given application of Eq. (21), Eq. (13) can be used and often is used to rewrite Eq. (21) in terms of **R** instead of α_1. Our treatment in terms of α_1 at fixed α_0 will simplify our subsequent discussion. Thus, given trial functions with asymptotic forms specified by $\alpha_0 = \mathbf{I}$ and $\alpha_1 = \alpha_1^0$, an expression for the $N_{op} \times N_{op}$ coefficient matrix α_1 which is stationary for small variations (with constant τ and $\alpha_0 = \mathbf{I}$) of the trial function about the exact wave function is

$$\alpha_1^K = \alpha_1^0 - 2a_0\hbar^{-1}\mathbf{V}^{-1}\mathbf{L} \tag{22}$$

Use of this variationally correct expression and Eq. (13) provides a prescription for calculating cross sections from any appropriate scattering wave functions.

The Kohn variational principle is even more valuable in providing a means for optimizing trial wave functions of given functional form with variable coefficients. The standard variational derivation shows that if the condition (21) is satisfied for all possible small variations of the set of functions $\{\psi^P\}_{p=1}^{N_{op}}$ consistent with constant τ and $\alpha_0 = \mathbf{I}$, then these functions must be exact solutions of the Schrödinger equation. In practice we impose certain restrictions on the set of functions, such as finite P and M in Eq. (1) and particular choices of the $\{W_m\}_{m=1}^M$. Nevertheless, using the usual variational procedure, the "optimum" set of functions $\{\psi^P\}_{p=1}^{N_{op}}$ consistent with these restrictions may be determined by requiring (21) to be satisfied for all small variations consistent with these restrictions. As usual, the resulting equations for the unknown functions and coefficients in the trial function are the same as would be obtained by requiring that the projection of $(H-E)\psi^P$ have no overlap with the function space spanned by the known or predetermined functions in the trial function. In this respect the derivation and resulting equations are very similar to the derivation of multiconfiguration Hartree–Fock equations for bound–state problems. For example, one may require the variational functional α_1^K to be stationary with respect to point-by-point variation of the functions X_{ip}. If $\hat{O} = 1$, this yields a set of coupled differential equations for the X_{ip} which are called the close coupling equations without exchange.* If $\hat{O}$ is an operator which makes the total wave function antisymmetric to permutation of any two electrons, point-by-point variation of the X_{ip} yields a set of coupled integrodifferential equations for the X_{ip} which are called the close coupling equations.† The actual form of these equations is displayed in the next subsection [see Eq. (50)]. When $M > 0$ and the coefficients c_{mp} are also varied,

---

*This refers to exchange of the scattering electron with the bound electrons. Of course it is still possible to include exchange effects in the Φ_i functions in this formalism.

†An approach halfway between the close coupling methods with and without exchange is the post-symmetrization method.[48] In this case the trial function is optimized with $\hat{O} = 1$ but then a variational correction is calculated with $\hat{O}$ as an antisymmetrizer. This approach is similar in spirit to the Musher–Silbey[49] method for bound-state calculations.

these equations are coupled to algebraic equations for the c_{mp} and the whole set of coupled equations are generally called the correlation method equations if the W_m are thought to represent electron correlation[33,50,51] but they are still called the close coupling equations if the W_m are merely added to allow the X_{ip} to be constrained orthogonal to the bound orbitals.[34,35] The close coupling method and the correlation method are scattering analogs of the multiconfiguration Hartree–Fock equations[52] with numerical radial functions for bound-state problems. In fact they have sometimes been called the continuum Hartree–Fock method.

Numerical radial functions are often used for bound states of atoms and for electron–atom scattering but are less useful for molecular problems because the expansion about the molecular center of mass strongly couples terms with widely different orbital angular momenta. This problem can be alleviated for diatomic molecules by using an expansion in prolate spheroidal coordinates and such an approach has been used both for bound states[53] and for electron scattering[54] for diatomic molecules. This approach is more complicated, however, and is not useful for general polyatomic molecules. The alternative usually adopted for bound-state calculations on molecules is an expansion in nuclear-centered exponential-type or Gaussian-type functions,[55] in Gaussian lobe functions, or in floating Gaussian-type orbitals.[56] This leads to an algebraic problem, i.e., a matrix equation for the coefficients. The resulting methods are the matrix Hartree–Fock method[57]* and the matrix multiconfiguration Hartree–Fock method.[58]† The matrix elements may be evaluated *ab initio*[59] or approximated by techniques such as neglect-of-differential-overlap (NDO)[60] of the $X\alpha$[61] method. These approximation techniques may be parametrized against *ab initio* calculations or against experiments. The scattering analogs of these matrix methods are the matrix variational methods of scattering theory.[46,47] These are also called algebraic variational methods.

In matrix variational methods the X_{ip} are expanded in terms of square-integrable functions with variable coefficients c^i_{ap} and non-square-integrable functions with constant coefficients α_{1ip}, i.e.,

$$X_{ip}(r_{N+1}) = Y_{ip}(r_{N+1}) + \sum_{a=1}^{n_i} c^i_{ap}\eta^i_a(r_{N+1}) \qquad (23)$$

where Y_{ip} is defined by Eqs. (9)–(11) and the variational functional α^K_1 is required to be stationary with respect to variations of the c^i_{ap}, α_{1ip}, and c_{mp} at fixed τ and α_{0ip}. This yields a matrix equation which must be satisfied by the vectors of varied coefficients. Examples of the types of equations which must

---

*This is also called the Hartree–Fock–Roothaan method.
†This is also called the multiconfiguration self-consistent-field method.

actually be solved are given later in this section [see, e.g., Eq. (38) and the discussion following Eq. (41)]. Before working out these details, it will be valuable to consider a few general features of such matrix variational methods. The matrix elements in the matrix equation are integrals involving $(H - E)$ and pairs of terms in the expansion of the wave function. They may be reduced to one-electron and two-electron integrals using standard methods.[46] But, in addition to the usual integrals involving square-integrable functions which occur in bound-state problems, there are integrals involving up to two non-square-integrable one-electron functions.

Matrix variational methods have been successfully applied to *ab initio* electron–atom scattering calculations in the last few years using not only the Kohn variational method described above but also more refined algebraic techniques (see below) which have important advantages in actual calculations. Extensions to electron–molecule scattering are just beginning. Semiempirical methods of evaluating most of the integrals occurring in these methods have not yet been developed or extensively tested but it is possible that such semiempirical methods will eventually prove to be just as successful (or unsuccessful) for scattering problems as for bound-state problems. Some approximation methods for calculating the static potentials and exchange potentials occurring in these integrals have been developed and tested, however, and are discussed in Section 2.1.2.

In using the variational principle (21) it sometimes happens that the variational correction (22) is very large even for a trial function which appears to be fairly accurate at nearby energies or with one less or one more term. In such cases the results are often inaccurate. This problem is not due to an inaccurate representation of the physics in the trial function but is most easily understood in terms of accidental near-singularities of some of the matrices involved in the calculation. This is a defect of the Kohn variational method first discussed in detail by Schwartz.[62] Schwartz was nevertheless able to obtain accurate results by varying a nonlinear parameter in his trial function so that the accidental near-singularities were removed. This is an undesirable computational inconvenience. Saraph *et al.*[63] suggested the problem could be alleviated by choosing the phase angle τ based on the results of a calculation with $P = 1$ and $M = 0$. Nesbet[64] made a more complete analysis of this defect of the variational method and suggested it could be alleviated by choosing τ as either zero or $\pi/2$ based on certain ratios of matrix elements. This is called the anomaly-free method. This analysis has been reviewed elsewhere[46,47] (along with other methods designed to eliminate anomalies and other aspects of algebraic variational methods). The anomaly-free method is not completely free of anomalies.[47,65] Subsequent attempts to improve the method led to the optimized anomaly-free and optimized minimum-norm methods.[65] Experience has shown that none of these methods is completely satisfactory.[66–68] Nevertheless the problem is well enough understood that spurious results due to the singularities of the matrices involved should never be mistaken for accurate ones.*

Malik and Rudge have suggested another procedure for eliminating anomalies by varying nonlinear parameters in the trial function.[70] This is not very efficient for computations. It has also been suggested that in each case several of these methods plus the closely related minimum–norm method[46,47] be applied and the results not be accepted as more accurate than the bounds which can be placed around the results of three or more variational methods.[47,67] This is not too difficult for cases where most of the computer time is spent on the evaluation of elements of $(H-E)$ and relatively little is spent on the algebraic problem. For large molecules and intermediate energies, however, where a large number of basis functions are required and approximation methods may be used for the integrals, the algebraic problem may become very time-consuming. Then the size of the algebraic problem may be profitably reduced using contracted basis functions.[71] However, additional computational simplifications are still desirable. Recently, we have shown[68] that the variational least squares (VLS) method[72] is very successful at avoiding anomalies, it provides results which are continuous functions of energy and any nonlinear parameters in the trial functions, and does not require comparing the results of alternative computational schemes. We recommend this method or the least squares variational method discussed below it for routine production runs and it is described here as an example of the basis function approach to scattering. It should be noted that the same basic one-electron and two-electron integrals occur in all the variationally corrected methods since they involve the calculation of L in the last step.

The VLS method may be motivated by noting that the anomalies in the Kohn method are due to a certain matrix of $(H-E)$ becoming singular. The minimum-norm method and the VLS methods deal with matrices of $(H-E)^2$ in an attempt to avoid such singularities. In fact the VLS method is analogous to the Kohn method but with $(H-E)^2$ replacing $(H-E)$. However, to avoid having to calculate the complicated integrals involving H^2, the minimum-norm and VLS methods effectively use an approximate resolution of the identity, equivalent to applying the closure property to an incomplete basis set.

It is most convenient to set all $n_i = 0$ on the right-hand side of Eq. (23) and include these square-integrable terms in the second sum of Eq. (1). Recalling the meaning of P and M from the two paragraphs following Eq. (12), we see that P in this case is equal to N_{op} and is the number of target eigenstates, all open, included in the first sum in Eq. (1), and M is the number of square-integrable terms in the trial wave function. The VLS method is derived by

---

*Shimamura,[69] however, has shown that the problem of anomalies is far more general than an analysis in terms of singular matrices would suggest and anomalous resonances occur even in trial functions which are not of the linear-combination type. The author is grateful to Dr. Shimamura for valuable correspondence concerning Refs. 47 and 69.

considering the matrix

$$\mathbf{U} = \begin{bmatrix} \mathbf{U}^{00} & \mathbf{U}^{01} & \mathbf{U}^{0W} \\ \mathbf{U}^{10} & \mathbf{U}^{11} & \mathbf{U}^{1W} \\ \mathbf{U}^{W0} & \mathbf{U}^{W1} & \mathbf{U}^{WW} \end{bmatrix} \tag{24}$$

where

$$U_{ij}^{\alpha\beta} = \langle A_{i\alpha}f_i | H - E | \hat{O}A_{j\beta}f_j \rangle, \quad \alpha = 0,1; \beta = 0,1; i = 1,2,\ldots,P; j = 1,2,\ldots,P \tag{25}$$

$$U_{mj}^{W\beta} = U_{jm}^{\beta W} = \langle W_m | H - E | \hat{O}A_{j\beta}f_j \rangle, \quad \beta = 0,1; j = 1,2,\ldots,P; m = 1,2,\ldots,M \tag{26}$$

$$U_{nm}^{WW} = \langle W_n | H - E | \hat{O}W_m \rangle, \quad n = 1,2,\ldots,M; m = 1,2,\ldots,M \tag{27}$$

Therefore $\mathbf{U}$ is a square matrix of order $2P+M$. If we define

$$\mathbf{Z} = \begin{bmatrix} \boldsymbol{\alpha}_0 \\ \boldsymbol{\alpha}_1 \\ \mathbf{c} \end{bmatrix} \tag{28}$$

where $(\mathbf{c})_{mp} = c_{mp}$ and $\mathbf{Z}$ is a rectangular $(2P+M) \times P$ matrix, then if ψ^p of Eq. (1) were exact, it would satisfy

$$(H-E)\psi^p = 0 \tag{29}$$

and therefore it would satisfy the set of equations

$$\int dx_1 \, dx_2 \ldots dx_{N+1} A_{i\alpha}(r_{N+1}) f_i(x_1, x_2, \ldots, x_N, \hat{\mathbf{r}}_{N+1}, \sigma)(H-E)$$
$$\times \psi^p(x_1, x_2, \ldots, x_{N+1}) = 0, \quad i = 1,2,\ldots,P; \alpha = 0,1 \tag{30}$$

$$\int dx_1 \, dx_2 \ldots dx_{N+1} W_n(x_1, x_2, \ldots, x_{N+1})(H-E)\psi_p(x_1, x_2, \ldots, x_{N+1}) = 0$$
$$n = 1,2,\ldots,M \tag{31}$$

which, together for all $p = 1,2,\ldots,P$, are equivalent to the matrix equation

$$\mathbf{UZ} = 0 \tag{32}$$

In general for approximate trial functions, (32) does not hold and it is not possible to make it hold by varying the parameters α_{1ip} and c_{mp} with constant τ and

$$\boldsymbol{\alpha}_0 = \mathbf{I} \tag{33}$$

It is possible to make this equation hold by considering point-by-point variation of the radial functions, but as discussed above, this is not always practical. In the variational least squares method we therefore define a matrix $\mathbf{D}$ of dimension $(2P+M) \times P$ as follows:

$$\mathbf{D} = \mathbf{UZ} \tag{34}$$

and attempt to minimize its norm at constant τ and $\alpha_0 = \mathbf{I}$. To do this we require the $P \times P$ matrix $\mathbf{D}^\dagger\mathbf{D}$ be stationary through first order in small variations, which yields

$$\mathbf{U}^\dagger\mathbf{U}\mathbf{Z} = 0 \tag{35}$$

Since $\alpha_0 = \mathbf{I}$, Eq. (35) represents P sets of $(2P+M)$ inhomogeneous equations, which may be written in partitioned form as

$$\begin{bmatrix} (\mathbf{U}^\dagger\mathbf{U})^{01} & (\mathbf{U}^\dagger\mathbf{U})^{0W} \\ (\mathbf{U}^\dagger\mathbf{U})^{11} & (\mathbf{U}^\dagger\mathbf{U})^{10} \\ (\mathbf{U}^\dagger\mathbf{U})^{W1} & (\mathbf{U}^\dagger\mathbf{U})^{WW} \end{bmatrix} \mathbf{a} = - \begin{bmatrix} (\mathbf{U}^\dagger\mathbf{U})^{00} \\ (\mathbf{U}^\dagger\mathbf{U})^{10} \\ (\mathbf{U}^\dagger\mathbf{U})^{W0} \end{bmatrix} \tag{36}$$

where

$$\mathbf{a} = \begin{pmatrix} \alpha_1 \\ \mathbf{c} \end{pmatrix} \tag{37}$$

Since the unknown vector $\mathbf{a}$ contains only $(P+M)$ linear variational coefficients for each of the P initial states, only a subset $(P+M)$ of these equations can be satisfied simultaneously for each initial state. Wladawsky suggested that the following P sets of $(P+M)$ equations be satisfied exactly:

$$\begin{bmatrix} (\mathbf{U}^\dagger\mathbf{U})^{11} & (\mathbf{U}^\dagger\mathbf{U})^{1W} \\ (\mathbf{U}^\dagger\mathbf{U})^{W1} & (\mathbf{U}^\dagger\mathbf{U})^{WW} \end{bmatrix} \mathbf{a} = \begin{bmatrix} (\mathbf{U}^\dagger\mathbf{U})^{10} \\ (\mathbf{U}^\dagger\mathbf{U})^{W0} \end{bmatrix} \tag{38}$$

Equation (38) may be solved; it yields for α_1

$$\alpha_1 = -(\mathbf{Q}^{11})^{-1}\mathbf{Q}^{10} \tag{39}$$

where for convenience in representing the mathematical solution we have defined

$$\mathbf{Q}^{1\beta} = (\mathbf{U}^\dagger\mathbf{U})^{1\beta} - (\mathbf{U}^\dagger\mathbf{U})^{1W}[(\mathbf{U}^\dagger\mathbf{U})^{WW}]^{-1}(\mathbf{U}^\dagger\mathbf{U})^{W\beta}, \qquad \beta = 0, 1 \tag{40}$$

[The only justification for selecting a subset of equations as Wladawsky did is that it yields useful results. This aspect of the derivation is discussed again in the paragraph before Eq. (43) and in the section containing Eqs. (43)–(49).]

Then the asymptotic form of the optimized trial function yields the following approximation to the reactance matrix [by using Eq. (13) with $\alpha_0 = \mathbf{I}$]:

$$\mathbf{R}_{\mathrm{VLS}} = \mathbf{V}^{1/2}[(\sin \tau) - (\cos \tau)(\mathbf{Q}^{11})^{-1}\mathbf{Q}^{10}][(\cos \tau) + (\sin \tau)(\mathbf{Q}^{11})^{-1}\mathbf{Q}^{10}]\mathbf{V}^{-1/2} \tag{41}$$

This is the zeroth-order variational least squares reactance matrix. Equations (39)–(41) are identical in form to the Kohn variational method except that in the Kohn method $\mathbf{U}^\dagger\mathbf{U}$ is replaced by $\mathbf{U}$ and $\mathbf{R}_{\mathrm{VLS}}$ is replaced by the Kohn zeroth-order reactance matrix $\mathbf{R}_{\mathrm{K}}$. In either method, once the trial function has been optimized one should substitute it into (22) to obtain an improved estimate of the asymptotic form of the wave function since $\mathbf{L}$ is not zero. In fact

$$\mathbf{L} = \mathbf{Z}^{\dagger}\mathbf{U}\mathbf{Z} \qquad (42)$$

From the improved $\boldsymbol{\alpha}_1$ one can compute an improved reactance matrix using (13).

Once the asymptotic form of the wave function is improved variationally, the coefficients may be redetermined by solving a subset of M of the equations (38) for the vector $\mathbf{c}$ in terms of the improved vector $\boldsymbol{\alpha}_1$. This was recommended by Matese and Oberoi,[23] although it has not been justified variationally. Unless one does this, one does not have an approximate wave function whose asymptotic form corresponds to the variationally improved reactance matrix. The asymptotic form of the optimized trial function corresponds to the zeroth-order reactance matrix. For applications which use the whole wave function, e.g., photoionization, it may be preferable to use a wave function corrected by the procedure of Matese and Oberoi.

It has been demonstrated that the wave function in the VLS method generally corresponds to a smaller value of $\mathbf{L}$ than that obtained by several other variational methods.[68] Recalling the definition of $\mathbf{L}$ [Eq. (42)] and the discussion preceding Eq. (32); we see that this means that the wave function in the VLS method satisfies the Schrödinger equation "better" than the wave function obtained in several other variational methods. Further, the zeroth-order reactance matrix is generally more accurate than the zeroth-order reactance matrix of other methods to which it has been compared.[68] Finally, the zeroth-order VLS method is free, for all values of τ, of the anomalies which plague most of the other variational methods discussed above.[72]

It has been suggested that rather than satisfying the last two rows of Eq. (36), as in Eq. (38), one might satisfy, for example, the first and third rows.[68] Schmid and co-workers have developed a similar but even more general method called the least-squares variational method.[73–75] They point out that if ψ^p of Eq. (1) were exact, it would satisfy (29) and therefore, in particular, it would satisfy Eq. (30) with $\alpha = 1$, Eq. (31), and

$$\int dx_1 \, dx_2 \cdots dx_{N+1} \omega_\rho(x_1, x_1, \ldots, x_{N+1})$$
$$\times (H - E)\psi^p(x_1, x_2, \ldots, x_{N+1}) = 0 \qquad \rho = 1, 2, \ldots, C \quad (43)$$

where the ω_ρ are square-integrable or non-square-integrable functions of longer range than the set $\{W_m\}$ and each ω_ρ contains a weight factor whose value can be freely specified. Equation (30) with $\alpha = 1$ and Eqs. (31) and (43) can be written

$$\mathbf{A}\mathbf{a} = \mathbf{b} \qquad (44)$$

where

$$
\mathbf{A} = \begin{bmatrix} \mathbf{U}^{\omega 1} & \mathbf{U}^{\omega W} \\ \mathbf{U}^{11} & \mathbf{U}^{1W} \\ \mathbf{U}^{W1} & \mathbf{U}^{WW} \end{bmatrix}
\tag{45}
$$

$$
\mathbf{b} = \begin{bmatrix} \mathbf{U}^{\omega\omega} \\ \mathbf{U}^{1\omega} \\ \mathbf{U}^{W\omega} \end{bmatrix}
\tag{46}
$$

and $\mathbf{U}^{\omega 1}$, $\mathbf{U}^{\omega W}$, $\mathbf{U}^{\omega\omega}$, $\mathbf{U}^{1\omega}$, and $\mathbf{U}^{W\omega}$ are defined analogously to Eqs. (25)–(27); for example,

$$
U_{\rho i}^{\omega 1} = \langle \omega_\rho | H - E | \hat{O} A_{i1} f_i \rangle
\tag{47}
$$

so that $\mathbf{A}$ and $\mathbf{b}$ are rectangular matrices of dimensions $(C+P+M)\times(P+M)$ and $(C+P+M)\times P$, respectively. For a given trial function with constant τ and $\boldsymbol{\alpha}_0 = \mathbf{I}$, the $(C+P+M)$ equations (44) are incompatible. So we impose the condition that the norm of $(\mathbf{Aa}-\mathbf{b})$ be minimum, which leads to

$$
\mathbf{A}^\dagger \mathbf{A}\mathbf{a} = \mathbf{A}^\dagger \mathbf{b}
\tag{48}
$$

This is the least-squares variational method for optimizing a trial function. The optimized trial function may then be substituted into the Kohn variational expression (22) to obtain a variationally improved reactance matrix.

It is clear that if

$$
\omega_i(x_1, x_2, \ldots, x_{N+1}) = w_i A_{i0}(r_{N+1}) f_i(x_1, x_2, \ldots, x_N, \hat{\mathbf{r}}_{N+1}, \sigma_{N+1})
$$
$$
i = 1, 2, \ldots, P \tag{49}
$$

where w_i are the weight factors and if the weight factors are all set equal to unity and $C = P$, that the least-squares variational method of Schmid and co-workers reduces to the VLS method of Wladawsky.

Schmid and co-workers[73,74] have used the least-squares variational method for potential scattering and nuclear scattering problems. They obtained accurate results which are insensitive to the choice of weight factors and are free of anomalies.

When all the open-channel radial functions are allowed to vary point-by-point, and the exact target wave functions for open channels are known, the variational principles may become minimum principles for certain scattering parameters.[76] In the matrix variational methods, these minimum principles do not hold, but a quasiminimum principle can be very useful.[77] If the integrals are approximated, however, the quasiminimum principle does not hold. When the exact target wave functions are not known, a subsidiary minimum principle may be useful.[78]

Unfortunately space does not allow a fuller discussion of these points here, but the reader should be aware that these aspects of the calculations are a current research area.

The best illustrations of accurate calculations using the matrix variational methods are the electron–atom calculations of Nesbet and co-workers, who used the optimized anomaly-free variational method.[50] In some cases they also used a semiempirical net correlation energy parameter.

When the many-electron Hamiltonian is substituted into Eqs. (25)–(27), (30), (31), (43), and (47), these integrals reduce to products of one-electron and two-electron integrals.[46] Methods have been developed for performing these integrals using nuclear-centered exponential-type basis functions and various non-square-integrable functions for electron–atom scattering.[79] For molecules it is easier to evaluate the integrals if Gaussian-type basis functions are used for the square-integrable basis functions[80] or if integral approximations such as mentioned before Eq. (23) are used. The most difficult integrals to evaluate are those containing free waves and another approach to simplifying the integrals is to extract scattering information from an expansion of the wave function which involves only square-integrable (L^2) basis functions. These methods are considered in Section 2.2. Chung and Ajmera[81] developed a procedure for carrying out calculations by the Kohn variational method in which free waves are used in the expansion but the most difficult integrals (those involving two free waves) are not evaluated. This will simplify the calculation, but in large calculations the computer time is often dominated by the much larger numbers of easier integrals.

2.1.2 Approximations to Potential Terms

The close coupling equations for the radial functions, in the case where the second term in (1) is omitted, may be written

$$\left[\frac{-d^2}{dr^2}+\frac{l_i(l_i+1)}{r^2}-k_i^2\right]F_{ip}(r)+2\sum_j\left[V_{ij}(r)F_{jp}(r)+\int dr'\,K_{ij}(r,r')F_{jp}(r')\right]=0 \quad (50)$$

where the direct radial potentials are given by

$$V_{ij}(r)=\int d\hat{\mathbf{r}}\,\hat{V}_{ij}(\mathbf{r}) \quad (51)$$

in terms of the direct potentials defined in the following way:

$$\hat{V}_{ij}(\mathbf{r}_{N+1})=\int dx_1\,dx_2\ldots dx_N\,d\sigma_{N+1}f_i^*(x_1,x_2,\ldots,x_N,\hat{\mathbf{r}}_{N+1},\sigma_{N+1})V_{\text{int}}$$
$$\times f_j(x_1,x_2,\ldots,x_N,\hat{\mathbf{r}}_{N+1},\sigma_{N+1}) \quad (52)$$
$$V_{\text{int}}=H(x_1,x_2,\ldots,x_{N+1})-H_i(x_1,x_2,\ldots,x_N)-T(\mathbf{r}_{N+1}) \quad (53)$$

and the exchange potential terms are

$$\int dr' K_{ij}(r, r') F_{jp}(r') = \int d\hat{\mathbf{r}}\, d\mathbf{r}'\, \hat{K}_{ij}(\mathbf{r}, \mathbf{r}') F_{jp}(r') Y_{l_j m_j'}(\hat{\mathbf{r}}') \tag{54}$$

with kernels

$$\hat{K}_{ij}(\mathbf{r}_{N+1}, \mathbf{r}_N) = -N \int dx_1\, dx_2 \cdots dx_{N-1}\, d\sigma_N\, d\sigma_{N-1}\, f_i^*(x_1, x_2, \ldots, x_N, \hat{\mathbf{r}}_{N+1}, \sigma_{N+1})$$

$$\times [H(x_1, x_2, \ldots, x_{N+1}) - E] \Phi_j(x_1, x_2, \ldots, x_{N-1}, X_{N+1}) \chi_{m_j}(\sigma_N) \tag{55}$$

The physical meaning of the omission of the second term in (1) is that the scattering electron is represented by a single orbital in every term in Eq. (1); some workers would say that "the scattering electron is not explicitly correlated." Then the equations for the radial function parts F_{op} of the scattering electron orbitals assume the form (50). The interpretation of Eq. (50) is familiar because it has the same form as the multiconfiguration Hartree–Fock equation for bound-state radial functions. The difference is that the scattering electron is unbound so at least one and maybe all of the radial functions F_{ip} are non-square-integrable. Then the diagonal potential terms contain the nuclear attraction, electronic screening, and exchange effects, while the nondiagonal terms are responsible for coupling of channels, i.e., inelastic effects, by both nonexchange and exchange mechanisms. Because the equations in the form (50) are closely analogous to the equations for bound states and because the approximations we will consider are closely analogous to those used for bound states, Eq. (50) provides a good basis for discussion of the potential terms. However, the same potentials occur, albeit in a more complicated algebraic framework, even if the second term in (1) is included and they occur as parts of the matrix elements in **U** when the radial functions are expanded in a basis set in the matrix variational methods. Note that Eqs. (51) and (54) are functions of r. The r-independent integrals discussed in the previous subsection [see, e.g., Eqs. (24)–(27)] involve matrix elements of these potentials with the basis functions of r. Next we consider the approximate evaluation of these potentials by methods similar to those used for approximate calculations on bound states.

For most atomic calculations Φ_i and Φ_j are built up using a basis set of orthonormal one-electron orbitals $P_{k l_k}(r) Y_{l_k m_k'}(\hat{r})$. Then it is convenient for simplifying the exchange potential to constrain the free orbitals $F_{ip}(r) Y_{l_i m_i'}(\hat{r})$ to be orthogonal to the members of this set.* This constraint is enforced by the use of Lagrange multipliers μ_{ik} and adds terms like $\delta_{l_i l_k} \mu_{ik} P_{k l_k}(r)$ to Eq. (50).[35]† Similar orthogonality constraints are useful for electron–molecule scattering.

---

*This need not be a constraint on the total wave function if appropriate square-integrable terms are added in the second sum of Eq. (1). Inclusion of such terms adds additional terms to Eq. (50).

†An alternative to this orthogonalization procedure is discussed by Smith *et al.*[36] Orthogonalization is usually not employed for electron scattering by the hydrogen atom.[50,68,71]

When $i = j$, $V_{ij}(\mathbf{r}_{N+1})$ with f_i replaced by Φ_i is called the static potential (for electron scattering). Its negative, the static potential for positron or proton scattering, has been called the electrostatic potential. One can envisage a whole series of approximations to the static potential corresponding to a given electronic state of a given atom or molecule. However, since it is a one-electron property of the target charge distribution, one expects the Hartree–Fock approximation to be fairly accurate[82] and we need only consider a sequence of approximations to the *ab initio* Hartree–Fock static potential. Accurate static potentials for atoms may be easily determined.[83] To obtain accurate *ab initio* one-electron properties for molecules, one must use an extended basis set including polarization functions. In a few cases[84–87] such accurate calculations have been carried out for ground states and have been used as standards for testing more approximate calculations.[85,86,88–90] At small r, the static potential is dominated by the attractive interactions of the scattering electron with the screened nuclei. At large r the static potential is dominated by the interaction of the electron with the multipole moments of the target. Thus it is necessary to use approximations which yield accurate charge densities near the nuclei and accurate multipole moments, respectively, to obtain these two features of the potential correctly.

There has been much interest in the electrostatic potential as an approximation of the interaction energy of certain electrophilic reagents with organic molecules. In this context the general shape of the electrostatic potential energy surface has been mapped for many molecules and the errors introduced in such surfaces by the approximation methods of quantum chemistry have been studied. This work has been reviewed by Scrocco and Tomasi.[91]

A particularly interesting study of CNDO/2 and INDO approximation methods for evaluating such potentials is included in two articles by Giessner-Prettre and Pullman.[92] They distinguish four levels of approximation, which are most easily discussed in reverse order from their numbering:

(iv) The coefficients of the NDO wave function are interpreted as coefficients of orthogonalized atomic orbitals (OAOs) obtained from the NDO method's minimum basis set of exponential-type functions by Löwdin's method.[93] From these coefficients one may obtain by a transformation the deorthogonalized coefficients, i.e., the coefficients of the original nonorthogonal exponential-type functions. From these the static potential is calculated correctly. In this case differential overlap cannot be neglected during the calculation of the static potential because it is the OAOs rather than the AOs which are assumed to have zero differential overlap. This calculation differs from an *ab initio* calculation only in that the core electrons are frozen into the nuclei and the density matrix of the target is obtained by a calculation using approximate integrals. The latter is a simplification of Φ_i rather than of the step by which V_{ii} is obtained from Φ_i.

(iii) In this case the coefficients of the NDO wave function are interpreted as coefficients of the nonorthogonal atomic orbitals. Then the static wave function is computed using the NDO method's density matrix with the basic assumption of neglect of differential overlap of basis functions on different centers.

(ii) Approximation (ii) is like approximation (iii) except for two differences: (a) intraatomic differential overlap is also neglected (this is called the XI approximation) and (b) the entire valence charge distribution on a given center is treated as if the charge were in the valence s orbital (this is called the V_{ss} approximation).

(i) Approximation (i) is like approximation (ii) but the interactions of the target electron with the external charge are approximated using the formula used in the NDO method for repulsion integrals between s orbitals (the sign must be switched if the external charge is positive).

They did not study the XI approximation independently of the V_{ss} approximation. They found that deorthogonalization [approximation (iv)] yields the best results but that the V_{ss} method [approximation (ii)] yields results of useful accuracy very inexpensively. However, some caution is warranted in using these conclusions because the *ab initio* results to which they make comparisons are computed with limited basis sets which do not include polarization functions.

We have compared calculations of the static potential for the ground state of N_2 at its equilibrium internuclear separation using three *ab initio* calculations to INDO [approximation (iii)] and INDO XI calculations[86] and to a calculation using a sum of modified atomic densities.[89] The latter calculation is called Massey's method; for this calculation the atomic density is divided into a contribution from core electrons and a contribution from valence electrons and the latter is contracted uniformly in an attempt to account for one of the effects of bond formation. Van-Catledge[94] has also computed the INDO-V_{ss} result [approximation (ii)] for comparison with these results. Finally, we have performed calculations (INDO/1s and INDO XI/1s) in which the NDO calculation is modified by pulling the core electrons out of the nucleus and adding the sum of atomic core densities to the NDO valence electronic charge distributions.[86] These calculations are compared in Tables 1–3. In these tables the [432] and [53] calculations are extended-basis *ab initio* calculations, where the former includes polarization functions. The [21] calculation is a minimum-basis-set calculation using a basis set of accurate Hartree–Fock atomic orbitals.

Some of the methods are also compared in Fig. 1.[94]*

Massey's method is fairly accurate at small r. It is easy to incorporate accurate atomic densities in this method, whereas the basis set of exponential-type functions used in the NDO methods does not yield an accurate atomic

---

*The author is grateful to Prof. F. A. Van-Catledge for supplying these isopotential maps.

Table 1. *Static Potential in Hartrees for Electron–N_2 Scattering for Collinear Approach as a Function of Distance in Bohrs from Electron to Center of N_2*

$V,^a$ h

r, a_0	Ab initio			IA	INDO	INDO/1s	INDO XI	INDO XI/1s	INDO-V_{ss}
	[432]	[53]	[21]						
0	-1.882	-2.806	-2.439	-1.943	-1.761	-1.761	-1.761	-1.761	-1.919
0.6	-7.076	-7.195	-7.510	-6.974	-6.804	-6.858	-6.845	-6.899	-3.161
1.2	-2.813(1)	-2.813(1)	-2.831(1)	-2.793(1)	-2.521(1)	-2.795(1)	-2.519(1)	-2.793(1)	-4.236
1.8	-2.195	-2.139	-2.191	-2.197	-1.885	-1.885	-2.110	-2.110	-1.667
2.4	-3.207(-1)	-2.747(-1)	-2.773(-1)	-3.769(-1)	-9.235(-2)	-9.235(-2)	-2.816(-1)	-2.816(-1)	-4.478(-1)
3.0	-2.648(-1)	1.758(-3)	6.250(-3)	-7.450(-2)	9.309(-2)	9.309(-2)	-2.191(-2)	-2.191(-2)	-9.428(-2)
3.6	1.510(-2)	^{b}	3.662(-2)	-1.603(-2)	7.512(-2)	7.512(-2)	7.630(-3)	7.630(-3)	-1.699(-2)
4.2	1.566(-2)	2.592(-2)	2.990(-2)	-3.639(-3)	4.891(-2)	4.891(-2)	7.379(-3)	7.379(-3)	-2.756(-3)

^{a} Numbers in parentheses are powers of ten by which the entries should be multiplied.
^{b} Not calculated.

Table 2. Static Potential in Hartrees for Electron–N_2 Scattering for Approach at 45° Angle to Bond Axis as Function of Distance in Bohrs from Electron to Center of N_2

	$V,$[a] h						
	Ab initio						
r, a_0	[432]	[53]	[21]	IA	INDO, INDO/1s	INDO XI, INDO XI/1s	INDO-V_{ss}
0	−1.882	−2.086	−2.439	−1.943	−1.761	−1.761	−1.919
0.6	−2.661	−2.803	−3.084	−2.600	−2.945[b]	−2.483[b]	−2.040
1.2	−1.805	−1.850	−1.993	−1.651	−1.800	−1.603	−1.428
1.8	−5.523(−1)	−5.489(−1)	−6.092(−1)	−4.712(−1)	−4.603(−1)	−4.257(−1)	−5.434(−1)
2.4	−1.362(−1)	−1.240(−1)	−1.494(−1)	−1.115(−1)	−7.180(−2)	−8.120(−2)	−1.414(−1)
3.0	−3.099(−2)	−2.179(−2)	−3.234(−2)	−2.600(−2)	5.236(−4)	−1.286(−2)	−2.929(−2)
3.6	−6.017(−3)	[c]	−4.608(−3)	−6.166(−3)	8.927(−3)	−1.380(−3)	−5.235(−3)
4.2	−3.06(−4)	2.963(−3)	1.432(−3)	−1.491(−3)	7.643(−3)	3.265(−4)	−8.452(−4)

[a]Numbers in parentheses are powers of ten by which the entries should be multiplied.
[b]The INDO/1s and INDO XI/1s static potentials are the same (to four significant figures) as the INDO and INDO XI ones, respectively, for all entries in this table except $r = 0.6a_0$, where the results including 1s orbitals are more negative by 0.001.
[c]Not calculated.

Table 3. *Static Potential in Hartrees for Electron–N_2 Scattering for Perpendicular Approach as Function of Distance in Bohrs from Electron to Center of N_2*

	V^a, h						
	Ab initio						
r, a_0	[432]	[53]	[21]	IA	INDO[b]	INDO XI[b]	INDO-V_{ss}[b]
0.0	−1.882	−2.087	−2.439	−1.943	−1.761	−1.761	−1.919
0.6	−1.210	−1.374	−1.616	−1.215	−1.534	−1.063	−1.337
1.2	−4.417(−)	−5.391(−1)	−6.338(−1)	−4.108(−1)	−5.924(−1)	−3.185(−1)	−5.187(−)
1.8	−1.352(−1)	−1.880(−1)	−2.155(−1)	−1.112(−1)	−2.143(−1)	−7.211(−2)	−1.389(−1)
2.4	−4.388(−2)	−7.187(−2)	−7.848(−2)	−2.828(−2)	−9.257(−2)	−1.739(−2)	−2.950(−2)
3.0	−1.724(−2)	−3.260(−2)	−3.374(−2)	−7.098(−3)	−4.886(−2)	−6.260(−3)	−5.383(−3)
3.6	−8.526(−3)	c	−1.755(−2)	−1.785(−3)	−2.940(−2)	−3.411(−3)	−8.839(−4)
4.2	−5.160(−3)	−1.082(−2)	−1.078(−2)	−4.516(−4)	−1.914(−2)	−2.251(−3)	−1.344(−4)

[a] Numbers in parentheses are powers of ten by which the entries should be multiplied.
[b] The contribution from the $1s$ orbitals does not affect the first four significant figures of the static potential at 90°.
[c] Not calculated.

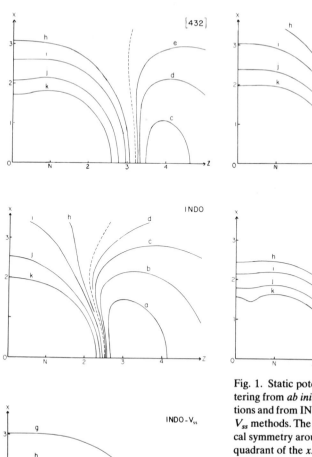

Fig. 1. Static potential for electron–N_2 scattering from *ab initio* ([432] and [21]) calculations and from INDO, INDO XI, and INDO-V_{ss} methods. The static potential has cylindrical symmetry around the z axis and only one quadrant of the xz plane is shown. The N on the z axis indicates the position ($z = R_e/2$) of a nucleus and the distances are in bohrs. The dashed line is the zero contour and the other contours have the following value in hartrees: (a) 0.05100, (b) 0.02550, (c) 0.01275, (d) 0.00637, (e) 0.00319, (f) 0.00159, (g) −0.00159, (h) −0.01594, (i) −0.03187, (j) −0.07968, (k) −0.15936.

density, although it is accurate enough for many purposes. (The NDO methods were not originally parametrized with the goal of obtaining accurate charge distributions through all space. A version using different basis functions and reparametrized with this goal would be valuable.)

The [21] and INDO static potentials show positive regions behind the nitrogen nucleus which are closer to the nucleus and more repulsive at the maximum than the positive region in the [432] potential. This is consistent with density difference maps for N_2, which show that minimum-basis set[95,96] and INDO[96] charge densities show too much buildup of electron charge in the lone-pair region and not enough between the nuclei. Wherever the NDO calculations agree with the [432] results better than the [21] results do, it is due to a cancellation of effects which cannot be expected in general.

Massey's method and the V_{ss} approximation predict that all the multipole moments of the target are zero and the static potential is everywhere negative. Thus while these methods have some applicability at small r, they are bad at large r.

Overall the INDO XI/1s approximation yields a static potential in best overall agreement with the *ab initio* results at small r. The contribution from the 1s orbitals is important only near the nuclei but provides significant improvement there.[86] The INDO XI/1s calculation is economical enough that the small-r part of the static potential can be calculated inexpensively at the many geometries necessary for a complete scattering calculation.

At large r the static potential is dominated by the electron's interaction with the lowest order nonnegligible electric multipole moment of the target. Dipole moments are predicted fairly accurately by NDO wave functions (CO, partly because the dipole moment is small, is a well-known exception[90]).

Some progress has been made in the semiempirical calculation of quadrupole moments.[97] However, quadrupole moments arise from large cancellation of electronic and nuclear contributions and are very sensitive to approximations. For scattering calculations the quadrupole moment is generally required as a function of internuclear distance and it seems unlikely that semiempirical methods can yield the necessary accuracy[98] as a function of internuclear distance. Information on multipole moments available from any source is often incorporated directly into electron scattering calculations without requiring consistency with the quadrupole moment which would be obtained from the wave function or static potential used for the rest of the calculation.*

Green and co-workers have developed an "independent-particle model" for electron–atom potentials and have used this to generate potentials for electron–N_2 scattering.[100] However, they made an arbitrary change in the potential in the regions of the nuclear singularities to effect a numerical simplification in their calculations.

An even simpler approximation scheme has been used for electron scattering by Itikawa.[101] He has represented molecules as a system of point charges placed on the atoms in such a way as to give the correct value of the dipole moment. Of course such a model does not give an accurate potential at small r.

It has sometimes been the practice to further approximate the static potential by expanding it in spherical harmonics about the molecular center of mass,

$$V_{ii}(\mathbf{r}) = \sum_{L} \sum_{M^L} V_{iiLM^L}(r) Y_{LM^L}(\hat{\mathbf{r}})$$

(56)

*See, for example, the work of Itikawa and Takayanagi.[99]

This expansion is rapidly convergent at large r, where it is a multipole-moment expansion, but it is very slowly convergent near the nuclei due to the strong singularities there.[102,103]

Semiempirical molecular orbital (SEMO) calculations can be used to economically map the electronic excitation energies as functions of molecular geometries.[104] While these calculations may often be inaccurate, they are still useful for interpreting electron–impact spectra. But there are indications that the excited-state charge distributions obtained by SEMO calculations are sometimes less accurate than the excited-state energetics or the ground-state charge distributions.[105] Thus it may be more difficult to approximate excited-state static potentials than ground-state ones in some cases.

When $i \neq j$, the integral of Eq. (52) is a transition potential. If states i and j may be connected by an electric dipole-allowed transition, then $V_{ij}(\mathbf{r}_{N+1})$ is proportional to the transition dipole moment divided by r_{N+1}^2 in the large-r_{N+1} limit. Because of its long range, this asymptotic part of the transition potential is very important. The difficulties of calculating transition dipole moments, or equivalently optical oscillator strengths, using approximate wave functions are well known for both atoms[106] and molecules.[107]

When $i = j$, the nonlocal potential of Eq. (54) is the diagonal radial exchange potential. There has been much work on the approximation of this exchange potential for bound-state calculations. The use of the Slater exchange potential and its modified version, the Xα method, has been particularly successful.[61,108] In this case the exchange potential is approximated by the local potential

$$W_X(\mathbf{r}) = \alpha_X W_S(\mathbf{r}) \tag{57}$$

where α_X is a parameter and $W_S(\mathbf{r})$ is the Slater exchange potential,

$$W_S(\mathbf{r}) = -3e^2 K_F/(2\pi) \tag{58}$$

where $\hbar K_F$ is the Fermi momentum,

$$K_F = (3\pi^2)^{1/3}/r_S \tag{59}$$

and r_S is the electron gas parameter

$$r_S = \rho^{-1/3} \tag{60}$$

where ρ is the electron density. Kohn and Sham[109a] and Gaspar[109b] have derived the value 2/3 for α_X, although values a little larger than this are often used. Although such $\rho^{1/3}$-type potentials have been applied to electron scattering problems several times, the derivation from the free-electron-gas model is not applicable to the scattering problem. The Slater potential is averaged over the electrons of an atom and the Kohn–Sham potential is for the exchange interaction of an electron at the Fermi level. But the scattering electron has an energy above the Fermi level. This was originally pointed out by Hara[110] and

Mittleman and Watson.[111] Hara suggested that correct application of the free-electron-gas model to scattering problems leads to the following approximation to the exchange potential when $i = j$ and $\Phi_i(x_1, x_2, \ldots, x_N)$ is a closed-shell wave function with doubly occupied spatial orbitals. The exchange potential is to be given by the free-electron-gas approximation

$$W_{\text{Hara}}(\mathbf{r}) = -2e^2 K_F(\mathbf{r}) F(\eta)/\pi \tag{61}$$

where

$$F(\eta) = \frac{1}{2} + \frac{1-\eta^2}{4\eta} \ln \left| \frac{1+\eta}{1-\eta} \right| \tag{62}$$

$$\eta = K(\mathbf{r})/K_F(\mathbf{r}) \tag{63}$$

and $\hbar^2 K^2(\mathbf{r})/2m$ is an approximation to the local kinetic energy of the scattering electron. Further, Hara suggested the latter be approximated using

$$K^2(\mathbf{r}) = [2m(T_i + U_i)/\hbar^2] + K_F^2(\mathbf{r}) \tag{64}$$

which has an obvious physical justification at small r. But Eq. (64) clearly has the wrong limit at large r since it implies the local kinetic energy there is $T_i + U_i$ instead of T_i. Mittleman and Watson derived Eqs. (61)–(63) with $K^2(\mathbf{r})$ given by $2mT_i$ as a high-energy approximation to the exchange potential in the Thomas–Fermi approximation. Riley and Truhlar[112] have suggested that the deficiency of Eq. (64) be corrected by setting $U_i = 0$ and have called the resulting approximation the asymptotically adjusted free-electron-gas exchange potential (AAFEGE potential). As T_i tends to zero, the AAFEGE potential tends to the Kohn–Sham potential as expected since the AAFEGE model treats a zero-energy scattering electron as if it is at the Fermi level. But at high energy

$$W_{\text{AAFEGE}}(\mathbf{r}) \underset{\text{high } T_i}{\cong} -\pi e^2 \hbar^2 \rho(\mathbf{r})/mT_i \tag{65}$$

which is proportional to the first rather than the one-third power of the density. The right-hand side of Eq. (65) was also derived as the high-energy limit by Mittleman and Watson and clearly illustrates the well-known fact that exchange is less important at higher T_i.

Riley and Truhlar[112] have also derived an approximation to the exchange potential which treats the scattering electron semiclassically. For the case when $i = j$ and $\Phi_i(x_1, x_2, \ldots, x_N)$ is a closed-shell wave function with doubly occupied spatial orbitals, the resulting potential is

$$W_{\text{SCE}}(\mathbf{r}) = \tfrac{1}{2}[T_i - V_{ii}(\mathbf{r})] - \tfrac{1}{2}\{[T_i - V_{ii}(\mathbf{r})]^2 + \alpha^2\}^{1/2} \tag{66}$$

$$\alpha^2 = (4\pi e^2 \hbar^2/m)\rho(\mathbf{r}) \tag{67}$$

This potential is called the semiclassical exchange (SCE) potential. It has the same high-energy limit as the AAFEGE and Hara exchange potentials.

Table 4. Phase Shifts in Radians for Electron Scattering in the Static-Exchange Approximation with and without the Nonlocal Exchange Potential and with Three Other Exchange Approximations[a]

Target	l	No exchange	Nonlocal exchange	Hara exchange	AAFEGE	SCE
He	0	2.65	2.99	2.96	3.03	3.01
	1	0.00009	0.0004	0.0002	0.002	0.002
Ar	0	9.15	9.3	9.25	9.68	9.37
	1	3.15	6.3	6.28	6.29	6.28

[a]The initial translational energy T_i is equal to 0.14 eV; l is the orbital angular momentum of the scattering electron.

Table 5. Phase Shifts in Radians for Electron Scattering in the Static-Exchange Approximation with and without the Nonlocal Exchange Potential and with Three Other Exchange Approximations[a]

Target	l	No exchange	Nonlocal exchange	Hara exchange	AAFEGE	SCE
He	0	1.761	2.436	2.304	2.442	2.422
	1	0.010	0.042	0.023	0.068	0.076
	2	0.0003	0.001	0.001	0.002	0.003
Ar	0	8.200	8.647	8.561	8.746	8.658
	1	3.882	6.001	5.909	6.042	5.938
	2	0.010	0.045	0.024	0.061	0.061

[a]The initial translational energy T_i is equal to 3.40 eV; l is the orbital angular momentum of the scattering electron.

Although these potentials (Hara, AAFEGE, and SCE) are local in coordinate space, they are not, strictly speaking, local potentials since they depend parametrically on the asymptotic translational energy of the scattering electron. However, they are as easy to use as local potentials. These approximations have been tested against the Hartree–Fock nonlocal exchange potential for electron scattering from ground-state helium and argon.[112] The resulting phase shifts are given in Tables 4–6.

The tables show that these approximate exchange potentials are accurate enough to be useful for many applications. They are more accurate for Ar than for He, as expected theoretically, since they are more appropriate for high-density than for low-density regions. Since they are as easy to calculate and use as direct potentials, they should be very useful for electron–molecule scattering where exchange has previously proved to be more difficult to include.

An interesting feature of the results of Table 4 is that the approximate exchange potentials are good enough that the zero-energy limit of the phase shift* satisfies Swann's generalization[113] of Levinson's theorem,[114] *viz.,* the

---

*The phase shift is ordinarily defined only modulo π. For the present discussion it must be put on an absolute basis by requiring it to be a continuous function of T_i which tends to zero as $T_i \to \infty$.

phase shift for orbital angular momentum l tends to $(n_{bl} + n_{pl})\pi$, where n_{pl} is the number of orbitals of symmetry l from which the scattering electron is excluded by the Pauli exclusion principle. Thus for He, $n_{p0} = 1$ and $n_{p1} = 0$, while for Ar, $n_{p0} = 3$ and $n_{p1} = 2$. In all cases the number of bound states n_{bl} is zero.

An alternative method of including exchange in electron scattering from closed-shell systems has been suggested and applied by Burke, Chandra, and Gianturco,[87,115–118] based on a treatment of electron–hydrogen atom scattering in the triplet state by Lippmann and Schey.[119] These workers force the scattering electron's wave function for appropriate symmetries to be orthogonal to the n_p orbitals from which the scattering electron is excluded by the Pauli principle. This ensures at least n_p nodes in the inner region of the wave function so that Swann's theorem may be satisfied. They found good agreement for N_2 with an earlier calculation[120] using the nonlocal exchange potential. It is clear, however, that the orthogonality requirement cannot give the whole effect of exchange; e.g., it gives no exchange effect for $l = 1$ for He. Also, Burke *et al.* encountered convergence difficulties using the orthogonalization procedure for CO.[103]

The semiclassical exchange approximation is not restricted to closed-shell targets. It has also been pointed out that the semiclassical exchange approximation can also be applied to terms with $i \neq j$ and that its high-energy limit should be a particularly easy-to-use approximation for such terms.[112] Yet for $i = j$ the high-energy limit has been shown to be remarkably accurate in the intermediate- and high-energy regions.

In the discussion of the use of INDO approximation we distinguished approximations in the bound-state wave functions Φ_i from approximations to the integrals involving such functions which occur in scattering theory. The general effect of inaccuracies in Φ_i functions on the results of scattering calculations requires more study. The reader should carefully distinguish in this regard the Φ_i functions corresponding to open channels, which are input to the

Table 6. *Phase Shifts in Radians for Electron Scattering in the Static-Exchange Approximation with and without the Nonlocal Exchange Potential and with Three Other Exchange Approximations[a]*

Target	l	No exchange	Nonlocal exchange	Hara exchange	AAFEGE	SCE
He	0	1.076	1.279	1.236	1.262	1.257
	1	0.196	0.327	0.274	0.297	0.315
	2	0.044	0.074	0.064	0.072	0.076
Ar	0	6.417	6.722	6.692	6.716	6.664
	1	4.215	4.548	4.544	4.570	4.500
	2	1.498	1.837	1.873	1.916	1.840

[a] The initial translational energy T_i is equal to 54.40 eV; l is the orbital angular momentum of the scattering electron.

scattering calculations and cannot be reoptimized during it,[47] from the Φ_i corresponding to closed channels and the W_n. These latter functions are essentially arbitrary and are chosen specifically to enhance convergence of the scattering wave function (see Section 4).

Only for electron–hydrogen atom scattering can the exact Φ_i functions be used. The problem of using approximate Φ_i functions in multichannel calculations has recently been studied by several workers.[13,74,78,121] An earlier study was carried out by Delves.[122]

2.2. L^2 Expansions

Many methods have now been developed for extracting scattering information from a calculation using square-integrable (L^2) basis functions. These methods are most closely related to methods used for bound-state calculations on molecules and thus it can be expected that bound-state *computer programs* and *approximation methods* will be adapted for such scattering calculations.

Square-integrable approximations to resonances have been used for both atoms and molecules for many years.[123] Many *ab initio* calculations have been carried out for resonances in electron scattering by atoms and H_2. Semiempirical calculations have been performed for resonances in electron scattering from N_2,[124] CO_2,[125] and H_2O[126] and used to discuss vibrational excitation and dissociative attachment.

Recently the use of square-integrable approximations has been extended to also include the treatment of nonresonant scattering. Such methods hold great promise for electron–molecule scattering calculations; so far, however, most of the calculations which have been performed have been *ab initio* electron–atom scattering calculations. A few *ab initio* electron–molecule calculations have been performed and references are given in Section 3.1.

Several different methods have been used for L^2 scattering calculations. In one approach (for example, that used in derivative matrix theory*) the wave function is obtained in an L^2 basis in a finite volume enclosing the target. The size of the volume is arbitrary, except that it should be large enough so that the solution of the Schrödinger equation in the external region is simple. The scattering information is obtained by matching the wave functions of the internal and external regions on the surface of the volume. A reexpansion technique may be used so that all the integrals except overlap integrals may still be evaluated over all space (not just the interior region) as is done in bound-state calculations and computer programs. In another approach[129] (Jacobi matrix theory) a particular complete L^2 basis set $\{\phi_n\}_{n=0}^{\infty}$ is chosen for

---

*This is also called R matrix theory or, to avoid confusion with the reactance matrix, NR matrix theory. It has recently been reviewed by Burke and Robb.[127] Some alternative techniques have been described by Schneider.[128]

which a zeroth-order Hamiltonian (for neutral targets, $H - V_{int}$) can be solved exactly. Then the remaining part of H (for neutrals, V_{int}) is approximated by an $N \times N$ matrix representation in the set $\{\phi_n\}_{n=0}^{N-1}$ and the Schrödinger equation with this approximation is solved in this basis. In another approach[130] the transition matrix or Fredholm determinant is calculated for complex energies or coordinates, where the wave function is square-integrable and only L^2 basis functions are needed, and continued (extrapolated) to the real axis to obtain physical scattering information. Other approaches to L^2 scattering, the T matrix projection methods, are discussed elsewhere.[21,131] In these methods, overlap integrals involving regular free waves are required.

3. Neglect of Electronic Excitation Except for Final State

3.1. Strong-Coupling, Static-Exchange, and Distorted-Wave Approximations

At high energies, virtually excited states do not play an important role, inelastic cross sections become small, and the close coupling equations may be terminated at two states for inelastic collisions and at one state for elastic collisions. The former is called the strong-coupling approximation and the latter is called the static-exchange approximation.

The strong-coupling approximation is useful over a wider energy range for cases where two states are much more strongly coupled to each other than to all other states, e.g., for the resonance transition in alkali scattering.[132] But the strong-coupling approximation may be useful at intermediate energies for other optically allowed transitions, too.[133]

The static-exchange approximation is not expected to be valid at low energies, but it has recently been used for electron–molecule scattering at low energies for testing electron–molecule scattering formulations involving single-center numerical radial functions,[120] the Kohn variational method with nonsquare-integrable basis functions and nuclear-centered Gaussian basis functions,[80] and L^2 methods using Gaussian basis functions.[134,135]

Various versions of the distorted-wave approximation for excitation collisions have been derived from the strong-coupling approximation,[6,136] from the two-potential formalism,[137,138] from the exact formula for the transition amplitude,[16] and from many-body perturbation theory.[21,139] They have recently been tested for several problems.[16,136,137,140]

3.2. High-Energy Approximations

At energies high enough that electronic states other than the initial and final ones may be neglected, many other simplifications of the static-exchange and strong-coupling approximations may often be made. These approximations generally treat the scattering electron on quite a different basis than the bound electrons and these approximation methods may bear little resemblance to the approximation schemes used for bound states and discussed elsewhere in this volume. Some of them have recently been reviewed elsewhere.[141–144]

From the point of view of the target (rather than the electron–target system), however, these high-energy approximation schemes often take the form of a transition matrix element involving the initial and final states of the target and a one-electron operator.[27] For example, the first Born approximation[141,143,145,146] for the $i \to f$ transition involves the operator $\exp[i(\mathbf{k}_f - \mathbf{k}_i)\mathbf{r}_1]$. Thus these high-energy theories require accurate charge densities for elastic scattering and accurate transition densities for inelastic scattering and may be evaluated using any approximations for the target states which give accurate values of these quantities.[147,148]

4. Inclusion of Effect of Omitted Electronic States by Approximate Polarization Potentials

At intermediate and low energies many electronic states must be included in the expansion of the electron scattering wave function. Static-exchange and strong-coupling calculations are generally inadequate and expansions in target eigenstates converge slowly. For this reason it has become popular to use target pseudostates[51,67,149] or perturbed first-order functions[150]* for the expansion of wave functions. For elastic scattering a calculation involving one polarized orbital is often surprisingly accurate.[151,152] Perturbed first-order functions have also been used in a modified distorted-wave formalism.[153]

Alternatively, the omitted states may be included implicitly by using modified potentials. The dipole polarization of the target makes a long-range contribution to such potentials which is very important and for this reason the difference between the modified potential matrix and the potential matrix occurring in the truncated close coupling equations is often called the polarization potential matrix.

Following Volkin,[154] "by a generalized potential we refer to a simplified Hamiltonian which will yield a scattering solution having the same asymptotic behavior in a certain group of channels as that which the scattering state of the full Hamiltonian, with the corresponding boundary conditions, has in these channels."† Such a generalized potential must be complex if some of the

---

*The perturbation is the interaction of the target with the incident electron or a multipole component of this interaction. Such a perturbed first-order function may be called a polarized orbital,[151] although this term has also been used to describe the free wave associated with the perturbed first-order function.

†These potentials are also called equivalent potentials[155,156] or generalized optical potentials.[157]

channels that are not explicitly included are open; the imaginary part accounts for loss of flux to these channels. Exact formal theories for such generalized potentials are well known[154–156,158] and are most easily written using the Feshbach projection operators.

The diagonal elements of such potentials have been well studied and approximations to them have been heavily used.* In practice, diagonal elements of the polarization potential are usually represented as real potentials of the form (letting $\mathbf{r}_{N+1} = \mathbf{r}$)

$$V_{\text{pol}}(\mathbf{r}) = -\frac{e^2 \alpha(\hat{\mathbf{r}})}{2r^4} f(\mathbf{r}) \tag{68}$$

where $\alpha(\hat{\mathbf{r}})$ is the static dipole polarizability of the target as a function of its orientation in space and $f(\mathbf{r})$ is function which at small r cuts off the strong, unphysical singularity of the r^{-4} factor. The dipole term dominates because at large r the leading term in the perturbation of the target there is

$$H_1 = r^{-2} \sum_{i=1}^{N} r_i P_l(\cos \delta_i) \tag{69}$$

where δ_i is the angle between $\mathbf{r}$ and $\mathbf{r}_i$. The static polarizability is used in (68) because it may be shown that the adiabatic polarization approximation, in which the target response is calculated for a fixed position of the scattering electron and the polarization potential is the interaction of the scattering electron with the perturbed target, is correct in the large-r limit for energies below the first inelastic threshold.[160] But the adiabatic polarization approximation overestimates the response near the nucleus at small r where the scattering electron's velocity is large and exchange effects become inseparable from polarization effects. For atomic targets, the polarized-orbital polarization potential, in which the perturbation is taken to be equal to that given in Eq. (69) for all $r > r_i$ and to be zero for $r < r_i$ and the perturbed wave function is also zero for $r < r_i$, appears to give reasonable, and sometimes even remarkably accurate, results for electron–atom scattering.[151,159] For molecules, however, this approximation may be less valid since the nuclei are not at the origin.[161] The correct form of $f(\mathbf{r})$ for molecules is not known, although a polarized-orbital-type polarization potential has been obtained by Lane and Henry[162] for H_2.†

For molecules, $\alpha(\hat{\mathbf{r}})$, $V_{\text{pol}}(\hat{\mathbf{r}})$, and $f(\hat{\mathbf{r}})$ should also depend on the nuclear coordinates. But very little is known quantitatively about the dependence of $\alpha(\hat{\mathbf{r}})$ on internuclear distances and the dependence of $f(\hat{\mathbf{r}})$ on internuclear distances has always been neglected.

There have been only a few attempts to include the imaginary part of the diagonal polarization potential, but both empirical and *ab initio* approaches have been taken.[144,164]

---

*A review and extensive references are given elsewhere.[151,159]
†See also the calculation by Hara.[163]

Nondiagonal elements of the polarization potential have likewise received very little attention,[144,157,159,165] although there are many studies of the effect of intermediate electronic states on electron scattering processes and these studies (some of which even use the Feshbach formalism[33]) could provide a basis for parametrizing generalized potentials.

Huo has developed a closely related conceptual potential called the effective potential.[166] While a generalized potential is determined so that the scattering calculated for it in a truncated close coupling approximation is exact, the effective potential of Huo is defined such that the nonexchange scattering calculated for it in the first Born approximation is exact.

Generalized potentials have been considered in a variety of approximations to the scattering; for example, the Born approximation,[159] the eikonal approximation,[167] the static-exchange approximation,[168] and coupled channels approaches.

The Feshbach projection-operator approach has recently been applied to the analysis of semiempirical approximation schemes for molecular bound states,[105] and this work might provide some ideas about how to extend such schemes to electron scattering problems.

ACKNOWLEDGMENTS

This work was supported in part by the National Science Foundation under NSF grant No. GP-28684.

The author is an Alfred P. Sloan Research Fellow.

References

1. H. S. W. Massey and E. H. S. Burhop, *Electronic and Ionic Impact Phenomena*, 2nd ed., Oxford, London (1969), Vol. I.
2. H. S. W. Massey, E. H. S. Burhop, and H. B. Gilbody, *Electronic and Ionic Impact Phenomena*, 2nd ed., Oxford, London (1969), Vol. II.
3. L. G. Christophorou, *Atomic and Molecular Radiation Physics*, Wiley–Interscience, London (1971).
4. T.-Y. Wu and T. Ohmura, *Quantum Theory of Scattering*, Prentice-Hall, Englewood Cliffs, N.J. (1962).
5. M. L. Goldberger and K. M. Watson, *Collision Theory*, Wiley, New York (1964).
6. N. F. Mott and H. S. W. Massey, *Theory of Atomic Collisions*, 3rd ed., Oxford, London (1965).
7. G. F. Drukarev, *The Theory of Electron–Atom Collisions*, Academic Press, London (1965).
8. S. Geltman, *Topics in Atomic Collision Theory*, Academic Press, New York (1969).
9. B. H. Bransden, *Atomic Collision Theory*, Benjamin, New York (1970).
10. K. Smith, *The Calculation of Atomic Collision Processes*, Wiley–Interscience, New York (1971).
11. R. A. Bonham and M. Fink, *High-Energy Electron Scattering*, Van Nostrand–Reinhold, New York (1974).

12. E. S. Chang, *Phys. Rev. A* **9**, 1644–1655 (1974) and references therein.
13. R. Blau, L. Rosenberg, and L. Spruch, *Phys. Rev.* **11**, 200–209 (1975).
14. R. S. Berry, *Ann. Rev. Phys. Chem.* **20**, 257–406 (1969); S. Trajmar, J. K. Rice, and A Kuppermann, *Adv. Chem. Phys.* **18**, 15–90 (1970).
15. E. N. Lassettre, *Can. J. Phys.* **47**, 1733–1744 (1969).
16. S. Geltman and M. B. Hidalgo, *J. Phys. B.* **4**, 1299–1307 (1971).
17. J. Geiger and K. Wittmaack, *Z. Phys.* **187**, 433–443 (1965).
18. L. R. B. Elton, *Proc. Phys. Soc. (Lond.) A* **63**, 1115–1124 (1950) and references therein.
19. B. Hahn, D. G. Ravenhall, and R. Hofstadter, *Phys. Rev.* **101**, 1131–1142 (1956); R. Hofstadter, *Rev. Mod. Phys.* **28**, 214–258 (1956).
20. P. W. Langhoff, *Chem. Phys. Lett.* **22**, 60–64 (1973); P. W. Langhoff and C. T. Corcoran, *J. Chem. Phys.* **61**, 146–159 (1974); P. W. Langhoff, J. Sims, and C. T. Corcoran, *Phys. Rev. A* **10**, 829–841 (1974); U. Fano and C. M. Lee, *Phys. Rev. Lett.* **31**, 1573–1576 (1973); A. Dalgarno, H. Doyle, and M. Oppenheimer, *Phys. Rev. Lett.* **29**, 1051–1052 (1972); H. Doyle, M. Oppenheimer, and A. Dalgarno, *Phys. Rev. A* **11**, 909–915 (1975); J. T. Broad and W. P. Reinhardt, *J. Chem. Phys.* **60**, 2182–2183 (1974).
21. C. W. McCurdy, Jr., T. N. Rescigno, D. L. Yeager, and V. McKoy, in *Modern Theoretical Chemistry, Vol. 3: Methods of Electronic Structure Theory* (H. Schaefer, ed.), Plenum, New York (1977), Chap. 9.
22. J. W. Cooper, *Phys. Rev.* **128**, 681–693 (1962); W. R. Garrett and H. T. Jackson, *Phys. Rev.* **153**, 28–35 (1967); R. J. W. Henry and L. Lipsky, *Phys. Rev.* **153**, 51–56 (1967); R. J. W. Henry, *Phys. Rev.* **162**, 56–63 (1967); E. J. McGuire, *Phys. Rev.* **175**, 20–30 (1968); J. J. Matese and R. W. LaBahn, *Phys. Rev.* **188**, 17–23 (1969).
23. J. J. Matese and R. S. Oberoi, *Phys. Rev. A* **4**, 569–579 (1971).
24. S. Geltman, *J. Quant. Spectrosc. Radiat. Transfer* **13**, 601–613 (1973).
25. C. J. Ballhausen and A. E. Hansen, *Ann. Rev. Phys. Chem.* **23**, 15–38 (1972).
26. A. Temkin and E. C. Sullivan, *Phys. Rev. Lett.* **33**, 1057–1060 (1974), and references therein.
27. D. G. Truhlar, D. C. Cartwright, and A. Kuppermann, *Phys. Rev.* **175**, 113–133 (1968) and references therein.
28. E. S. Chang and U. Fano, *Phys. Rev. A* **6**, 173–185 (1972).
29. W. R. Garrett, *Mol. Phys.* **24**, 465–487 (1972).
30. N. Chandra and F. A. Gianturco, *Chem. Phys. Lett.* **24**, 326–330 (1974).
31. M. Lieber, L. Rosenberg, and L. Spruch, *Phys. Rev. D* **5**, 1330–1346 (1972) and references therein.
32. M. R. H. Rudge, *Rev. Mod. Phys.* **40**, 564–590 (1968) and references therein; I. E. McCarthy, *J. Phys. B* **6**, 2358–2367 (1973); W. D. Robb, S. P. Rountree, and T. Burnett, *Phys. Rev. A* **11**, 1193–1199 (1975).
33. P. G. Burke and A. J. Taylor, *Proc. Phys. Soc. (Lond.)* **88**, 549–562 (1966); A. J. Taylor and P. G. Burke, *Proc. Phys. Soc. (Lond.)* **92**, 336–344 (1967); P. G. Burke and A. J. Taylor, *J. Phys. B* **2**, 44–51 (1969).
34. K. Smith, R. J. W. Henry, and P. G. Burke, *Phys. Rev.* **147**, 21–28 (1966).
35. H. E. Seraph, M. J. Seaton, and J. Shemming, *Phil. Trans. R. Soc. (Lond.) A* **246**, 77–105 (1969).
36. K. Smith, M. J. Conneely, and L. A. Morgan, *Phys. Rev.* **177**, 196–203 (1969).
37. R. J. W. Henry, P. G. Burke, and A.-L. Sinfailam, *Phys. Rev.* **178**, 218–225 (1969).
38. K. Smith and L. A. Morgan, *Phys. Rev.* **165**, 110–122 (1968).
39. J. M. Blatt and L. C. Biedenharn, *Rev. Mod. Phys.* **24**, 258–272 (1952).
40. A. M. Lane and R. G. Thomas, *Rev. Mod. Phys.* **30**, 257–353 (1958).
41. P. G. Burke, in *Atomic Collision Processes* (S. Geltman, K. T. Mahanthappa, and W. E. Brittin, eds.), Gordon and Breach, New York (1969), pp. 1–55.
42. M. A. Brandt, D. G. Truhlar, and R. L. Smith, *Comput. Phys. Commun.* **5**, 456–477 (1973); Errata and addenda **7**, 177 (1974).
43. M. LeDourneuf, H. E. Seraph, M. J. Seaton, and Vo Ky Lan, *J. Phys. B* **5**, L87–L90 (1972).
44. D. G. Truhlar, C. A. Mead, and M. A. Brandt, *Adv. Chem. Phys.* **33**, 296–344 (1975).

45. W. Kohn, *Phys. Rev.* **74**, 1763–1772 (1948).
46. F. E. Harris and H. H. Michels, *Meth. Comput. Phys* **10**, 143–210 (1971).
47. D. G. Truhlar, J. Abdallah, Jr., and R. L. Smith, *Adv. Chem. Phys.* **25**, 211–293 (1974).
48. D. G. Truhlar and J. Abdallah, Jr., *Phys. Rev. A* **9**, 297–300 (1974).
49. J. I. Musher, *MTP (Med. Tech. Publ. Co.) Int. Rev. Sci.: Theor. Chem., Ser. One* **1**, 1–40 (1972).
50. A.-L. Sinfailam and R. K. Nesbet, *Phys. Rev. A* **6**, 2118–2125 (1972); R. S. Oberoi and R. K. Nesbet, *Phys. Rev. A* **8**, 2969–2979 (1973); L. D. Thomas, R. S. Oberoi, and R. K. Nesbet, *Phys. Rev. A.* **10**, 1605–1611 (1974); L. D. Thomas and R. K. Nesbet, *Phys. Rev. A* **11**, 170–173 (1975); L. D. Thomas and R. K. Nesbet, *Phys. Rev. A* **12**,, 2369–2377 (1975); L. D. Thomas and R. K. Nesbet, *Phys. Rev. A* **12**, 2378–2382 (1975); R. K. Nesbet, *Phys. Rev. A* **12**, 444–450 (1975).
51. S. P. Rountree, E. R. Smith, and R. J. W. Henry, *J. Phys. B* **7**, L167–L170 (1974).
52. A. P. Jucys, *Adv. Chem. Phys.* **14**, 191–206 (1969).
53. E. A. McCullough, Jr., *Chem. Phys. Lett.* **24**, 55–58 (1974); W. R. Fimple and S. P. White, *Int. J. Quantum Chem.* **9**, 301–324 (1975).
54. K. Takayanagi, *Prog. Theor. Phys. Suppl.* **40**, 216–248 (1967); S. Hara, *J. Phys. Soc. Japan* **27**, 1009–1019 (1969).
55. T. H. Dunning, Jr., and P. J. Hay, in *Modern Theoretical Chemistry, Vol. 3: Methods of Electronic Structure Theory* (H. Schaefer, ed.), Plenum, New York (1977), Chap. 1.
56. A. A. Frost, in *Modern Theoretical Chemistry, Vol. 3: Methods of Electronic Structure Theory* (H. Schaefer, ed.), Plenum, New York (1977), Chap. 2.
57. C. C. J. Roothaan, *Rev. Mod. Phys.* **23**, 69–89 (1951); R. K. Nesbet, *Rev. Mod. Phys.* **35**, 552–557 (1963).
58. A. C. Wahl and G. Das, in *Modern Theoretical Chemistry, Vol. 3: Methods of Electronic Structure Theory* (H. Schaefer, ed.), Plenum, New York (1977), Chap. 3.
59. F. E. Harris and H. H. Michels, *Adv. Chem. Phys.* **13**, 205–266 (1967); S. Huzinaga, *Prog. Theor. Phys. Suppl.* **40**, 52–77 (1967).
60. G. Klopman and R. C. Evans, in *Modern Theoretical Chemistry, Vol. 7: Semiempirical Methods of Electronic Structure Calculation—Part A, Techniques* (G. A. Segal, ed.), Plenum, New York (1977), Chap. 2.
61. J. W. D. Connolly, in *Modern Theoretical Chemistry, Vol. 7: Semiempirical Methods of Electronic Structure Calculation—Part A, Techniques* (G. A. Segal, ed.), Plenum, New York (1977), Chap. 4.
62. C. Schwartz, *Ann. Phys. (N.Y.)* **16**, 36–50 (1961); C. Schwartz, *Phys. Rev.* **124**, 1468–1471 (1961).
63. H. E. Seraph, M. J. Seaton, and J. Shemming, *Proc. Phys. Soc. (Lond.)* **89**, 27–34 (1966).
64. R. K. Nesbet, *Phys. Rev.* **175**, 134–142 (1968); R. K. Nesbet, *Phys. Rev.* **179**, 60–70 (1969).
65. R. K. Nesbet and R. S. Oberoi, *Phys. Rev. A* **6**, 1855–1862 (1972).
66. A.-L. Sinfailam and R. K. Nesbet, *Phys. Rev. A* **7**, 1987–1994 (1973).
67. J. Callaway and J. W. Wooten, *Phys. Rev. A* **9**, 1924–1931 (1974).
68. J. Abdallah, Jr., and D. G. Truhlar, *J. Chem. Phys.* **60**, 4670–4675 (1974).
69. I. Shimamura, *Proc. Phys. Soc. Japan* **31**, 852–870 (1971).
70. F. B. Malik, *Ann. Phys. (N.Y.)* **20**, 464–478 (1962); F. B. Malik, in *Atomic Collision Processes* (M. R. C. McDowell, ed.), North-Holland, Amsterdam (1964), pp. 80–85; M. R. H. Rudge, *J. Phys. B* **6**, 1788–1796 (1973); R. J. Drachman, *J. Phys. B* **7**, L76–L78 (1974).
71. J. Abdallah, Jr., and D. G. Truhlar, *J. Chem. Phys.* **61**, 30–36 (1974).
72. I. Wladawsky, *J. Chem. Phys.* **58**, 1826–1832 (1973).
73. E. W. Schmid and K. H. Hoffmann, *Nucl. Phys. A* **175**, 443–448 (1971); E. W. Schmid and J. Schwager, *Nucl. Phys. A* **180**, 434–448 (1972).
74. J. Schwager and E. W. Schmid, *Nucl. Phys. A* **205**, 168–176 (1973).
75. E. W. Schmid, *Nuovo Cimento* **18A**, 771–786 (1973).
76. I. C. Percival, *Phys. Rev.* **119**, 159–164 (1960); Y. Hahn, T. F. O'Malley, and L. Spruch, *Phys. Rev.* **134**, B397–B404 (1964); Y. Hahn, T. F. O'Malley, and L. Spruch, *Phys. Rev.* **134**, B911–B917 (1964); R. Sugar and R. Blankenbeckler, *Phys. Rev.* **136**, B472–B491 (1964);

W. A. McKinley and J. Macek, *Phys. Lett.* **10**, 210–212 (1964); M. Gailitis, *Zh. Eksp. Teor. Fiz.* **47**, 160–166 (1964); M. Gailitis, *Soviet Phys.—JETP* **20**, 107–111 (1965); Y. Hahn and L. Spruch, *Phys. Rev.* **153**, 1159–1164 (1967).

77. Y. Hahn, *Phys. Rev. A* **4**, 1881–1886 (1971); D. G. Truhlar and R. L. Smith, *Phys. Rev. A* **6**, 233–239 (1972); K. T. Chung and J. C. Y. Chen, *Phys. Rev. Lett.* **27**, 1112–1114 (1971); K. T. Chung and J. C. Y. Chen, *Phys. Rev. A* **6**, 686–693 (1972).

78. L. Rosenberg and L. Spruch, *Phys. Rev. A* **10**, 2002–2015 (1974).

79. J. Abdallah, Jr. and D. G. Truhlar, *Comput. Phys. Commun.* **9**, 327–336 (1975) and references therein.

80. K. Onda, *J. Phys. Soc. Japan* **36**, 826–838 (1974).

81. K. T. Chung and M. P. Ajmera, *Phys. Rev. A* **8**, 2895–2897 (1973); M. P. Ajmera and K. T. Chung, *Phys. Rev. A* **10**, 1013–1018 (1974).

82. C. Møller and M. S. Plesset, *Phys. Rev.* **46**, 618–622 (1934).

83. T. G. Strand and R. A. Bonham, *J. Chem. Phys.* **40**, 1686–1691 (1964).

84. R. W. B. Ardill and W. D. Davidson, *Proc. R. Soc. (Lond.) A* **304**, 465–477 (1968); F. H. M. Faisal, *J. Phys. B* **3**, 636–640 (1970).

85. F. A. Gianturco and J. H. Tait, *Chem. Phys. Lett.* **12**, 589–595 (1972); J. Almlöf and Å. Støgård, *Chem. Phys. Lett.* **29**, 1260–1268 (1974).

86. D. G. Truhlar, F. A. Van-Catledge, and T. H. Dunning, Jr., *J. Chem. Phys.* **57**, 4788–4799 (1972).

87. F. A. Gianturco and N. Chandra, *Proc. Sect. Sci. Isr. Acad. Sci. Humanit.* **1974**, 219–241.

88. D. G. Truhlar and J. K. Rice, *J. Chem. Phys.* **52**, 4480–4501 (1970); Erratum **55**, 2005 (1971).

89. D. G. Truhlar, *Chem. Phys. Lett.* **15**, 486–489 (1972).

90. D. G. Truhlar and F. A. Van-Catledge, *J. Chem. Phys.* **59**, 3207–3213 (1973).

91. E. Scrocco and J. Tomasi, *Top. Curr. Chem.* **42**, 95–170 (1973).

92. C. Giessner-Prettre and A. Pullman, *Theor. Chim. Acta* **25**, 83–88 (1972); C. Giessner-Prettre and A. Pullman, *Theor. Chim. Acta* **33**, 91–94 (1974).

93. P.-O. Löwdin, *J. Chem. Phys.* **18**, 365–375 (1970).

94. F. A. Van-Catledge, On INDO wave functions as sources of molecular electrostatic potentials, unpublished.

95. P. R. Smith and J. W. Richardson, *J. Phys. Chem.* **71**, 924–930 (1967).

96. F. A. Van-Catledge, *J. Phys. Chem.* **75**, 763–769 (1974).

97. H. Meyer and A. Schweig, *Chem. Phys. Lett.* **6**, 975–988 (1972).

98. D. G. Truhlar, *Int. J. Quant. Chem.* **6**, 975–988 (1972); D. C. Cartwright and T. H. Dunning, Jr., *J. Phys. B* **7**, 1776–1781 (1974).

99. Y. Itikawa and K. Takayanagi, *J. Phys. Soc. Japan* **26**, 1254–1264 (1969).

100. T. Sawada, P. S. Ganas, and A. E. S. Green, *Phys. Rev. A* **9**, 1130–1135 (1974).

101. Y. Itikawa, *J. Phys. Soc. Japan* **30**, 835–842 (1971).

102. D. E. Golden, N. F. Lane, A. Temkin, and E. Gerjuoy, *Rev. Mod. Phys.* **43**, 642–678 (1971).

103. P. G. Burke, N. Chandra, and F. A. Gianturco, *Mol. Phys.* **27**, 1121–1137 (1974).

104. A. Chutjian and G. Segal, *J. Chem. Phys.* **57**, 3069–3082 (1972).

105. K. F. Freed, in *Modern Theoretical Chemistry, Vol. 7: Semiempirical Methods of Electronic Structure Calculations, Part A, Techniques* (G. A. Segal, ed.), Plenum, New York (1977), Chap. 7.

106. H. Freidrich and E. Trefftz, *J. Quant. Spectrosc. Radiat. Transfer* **9**, 333–359 (1969); R. H. Garstang, *Astron. J.* **79**, 418–420 (1974); O. Sinanoglu, *Nucl. Instr. Methods* **110**, 193 (1973).

107. R. L. Ellis and H. H. Jaffé, this volume, Chap. 2.

108. J. C. Slater, *Phys. Rev.* **81**, 385–390 (1951); J. C. Slater and J. H. Wood, *Int. J. Quant. Chem. Symp.* **4**, 3–34 (1971); J. C. Slater, *Adv. Quant. Chem.* **6**, 1–92 (1972); J. C. Slater and K. H. Johnson, *Phys. Rev. B* **5**, 844–853 (1972); K. H. Johnson, *Adv. Quant. Chem.* **7**, 143–185 (1973); I. Lindgren and A. Rosen, *Case Stud. At. Phys.* **4**, 93–196 (1974).

109. (a) W. Kohn and L. J. Sham, *Phys. Rev.* **140**, A1133–A1138 (1965); (b) R. Gaspar, *Acta Phys. Acad. Sci. Hung.* **3**, 263–286 (1954).

110. S. Hara, *J. Phys. Soc. Japan* **22**, 710–718 (1967).
111. M. H. Mittleman and K. M. Watson, *Ann. Phys.* **10**, 268–279 (1960).
112. M. E. Riley and D. G. Truhlar, *J. Chem. Phys.* **63**, 2182–2191 (1975).
113. P. Swann, *Proc. R. Soc. (Lond.) A* **228**, 10–33 (1955).
114. N. Levinson, *K. Dan. Vidensk. Selsk. Mat. Fys. Medd.* **25**, no. 9 (1949).
115. P. G. Burke and N. Chandra, *J. Phys. B* **5**, 1696–1711 (1972).
116. N. Chandra and P. G. Burke, *J. Phys. B* **6**, 2355–2357 (1973).
117. N. Chandra, *Comput. Phys. Commun.* **5**, 417–429 (1973).
118. N. Chandra and F. A. Gianturco, *Chem. Phys. Lett.* **24**, 326–330 (1974).
119. B. A. Lippmann and H. M. Schey, *Phys. Rev.* **121**, 1112–1118 (1968).
120. P. G. Burke and A.-L. Sinfailam, *J. Phys. B* **3**, 641–659 (1970).
121. R. Peterkop and L. Rabik, *J. Phys. B* **4**, 1440–1449 (1971); M. J. Lavan and G. L. Payne, *Phys. Rev. C* **3**, 2156–2167 (1971); R. J. Drachman, *J. Phys. B* **5**, L30–L32 (1972); Y. Hahn, *Phys. Rev. A* **5**, 309–317 (1972); T. S. Murtaugh and W. P. Reinhardt, *J. Chem. Phys.* **59**, 4900–4906 (1973); H. E. Seraph, *J. Phys. B* **6**, L243–L246 (1973); E. Wichman and P. Heiss, *J. Phys. B* **7**, 1042–1051 (1974); R. Blau, L. Rosenberg, and L. Spruch, *Phys. Rev. A* **10**, 2246–2256 (1974); W. D. Robb, *J. Phys. B* **8**, L46–L49 (1975).
122. L. M. Delves, *Nucl. Phys.* **29**, 326–334 (1962).
123. H. S. Taylor, *Adv. Chem. Phys.* **18**, 91–147 (1970) and references therein; M. Fels and A. U. Hazi, *Phys. Rev. A* **5**, 1236–1249 (1972) and references therein; G. D. Doolen, *J. Phys. B* **8**, 525–528 (1975).
124. A. Herzenberg, in *Fundamental Interactions in Physics* (B. Kursonoglu and A. Perlmutter, eds.), Plenum Press, New York (1973), Vol. 2, pp. 261–284, and references therein.
125. C. R. Claydon, G. A. Segal, and H. S. Taylor, *J. Chem. Phys.* **52**, 3387–3398 (1970).
126. C. R. Claydon, G. A. Segal, and H. S. Taylor, *J. Chem. Phys.* **54**, 3799–3816 (1971).
127. P. G. Burke, *Comments At. Mol. Phys.* **4**, 157–162 (1973); P. G. Burke, *Comput. Phys. Commun.* **6**, 288–302 (1974); P. G. Burke and W. D. Robb, *Adv. At. Mol. Phys.* **11**, 143–214 (1975).
128. B. I. Schneider, *Chem. Phys. Lett.* **31**, 237–241 (1975).
129. E. J. Heller and H. A. Yamani, *Phys. Rev. A* **9**, 1201–1208 (1974); E. J. Heller and H. A. Yamani, *Phys. Rev. A* **9**, 1209–1214 (1974); H. A. Yamani and L. Fishman, *J. Math. Phys.* **16**, 410–420 (1975); E. J. Heller, *Phys. Rev. A* **12**, 1222–1231 (1975).
130. W. P. Reinhardt, *Comput. Phys. Commun.* **6**, 303–315 (1974) and references therein; J. Nuttall, *Comput. Phys. Commun.* **6**, 331–345 (1974) and references therein; P. W. Langhoff and W. P. Reinhardt, *Chem. Phys. Lett.* **24**, 495–500 (1974).
131. B. I. Schneider, *Chem. Phys. Lett.* **25**, 140–142 (1974).
132. G. D. Carse, *J. Phys. B* **5**, 1928–1937 (1972); P. G. Burke and J. F. B. Mitchell, *J. Phys. B* **6**, L161–L164 (1973); D. F. Korff, S. Chung, and C. C. Lin, *Phys. Rev. A* **7**, 545–556 (1973).
133. D. G. Truhlar, S. Trajmar, W. Williams, S. Ormonde, and B. Torres, *Phys. Rev. A* **8**, 2475–2482 (1973).
134. T. N. Rescigno, C. W. McCurdy, Jr., and V. McKoy, *Phys. Rev. A* **11**, 825–829 (1975).
135. T. G. Winter and N. F. Lane, *Chem. Phys. Lett.* **30**, 363–366 (1975); B. I. Schneider, *Phys. Rev. A* **11**, 1957–1962 (1975).
136. M. S. Pindzola and H. P. Kelly, *Phys. Rev. A* **11**, 221–229 (1975).
137. W. N. Shelton, E. S. Leherissey, and D. H. Madison, *Phys. Rev. A* **3**, 242–250 (1971); D. H. Madison and W. N. Shelton, *Phys. Rev. A* **7**, 449–513 (1973).
138. W. N. Shelton and E. S. Leherissey, *J. Chem. Phys.* **54**, 1130–1136 (1971).
139. G. Csanak, H. S. Taylor, and R. Yaris, *Phys. Rev. A* **3**, 1322–1328 (1971); G. Csanak, H. S. Taylor, and D. N. Tripathy, *J. Phys. B* **6**, 2040–2054 (1973).
140. L. D. Thomas, B. S. Yarlagadda, G. S. Csanak, and H. S. Taylor, *Comput. Phys. Commun.* **6**, 316–330 (1974); L. D. Thomas, G. Csanak, H. S. Taylor, and B. S. Yarlaggadda, *J. Phys. B* **7**, 1719–1733 (1974); D. R. Flower, *J. Phys. B* **4**, 697–705 (1971); D. R. Flower and J. M. Launay, *J. Phys. B* **5**, L207–L210 (1972); M. B. Hidalgo and S. Geltman, *J. Phys. B* **5**, 617–626 (1972); M. D. Hershkowitz and M. J. Seaton, *J. Phys. B* **6**, 1176–1187 (1973); D. H. Madison and W. N. Shelton, *Phys. Rev. A* **7**, 514–523 (1973); L. D. Thomas, G. Csanak, H. S. Taylor, and B. S. Yarlagadda, *J. Phys. B* **7**, 1719–1733 (1974).

141. M. Inokuti, *Rev. Mod. Phys.* **43**, 297–347 (1971).
142. M. R. H. Rudge, *Adv. At. Mol. Phys.* **9**, 47–126 (1973); C. J. Joachain and C. Quigg, *Rev. Mod. Phys.* **46**, 279–324 (1974); B. L. Moiseiwitsch, *Comput. Phys. Commun.* **6**, 372–376 (1974); C. J. Joachain, *Comput. Phys. Commun.* **6**, 358–371 (1974); E. Gerjuoy and B. K. Thomas, *Rep. Prog. Phys.* **37**, 1345–1431 (1974).
143. K. L. Bell and A. E. Kingston, *Adv. At. Mol. Phys.* **10**, 53–130 (1974).
144. B. H. Bransden, in *Phys. Electron. At. Collisions, Invited Lect. Prog. Rep. Int. Conf., 8th* (1973), pp. 400–16, and references therein.
145. E. N. Lassettre, *J. Chem. Phys.* **43**, 4479–4486 (1965); E. N. Lassettre, A. Skerbele, and M. A. Dillon, *J. Chem. Phys.* **50**, 1829–1839 (1969).
146. M. Inokuti, in *Phys. Electron. At. Collisions, Invited Pap. Prog. Rep. Int. Conf., 7th* (1971), pp. 327–40.
147. M. Inokuti and M. R. C. McDowell, *J. Phys. B* **7**, 2382–2395 (1974).
148. S. Chung and C. C. Lin, *Appl. Opt.* **10**, 1790–1794 (1971); S. Chung and C. C. Lin, *Phys. Rev. A* **6**, 988–1002 (1972); S. Chung and C. C. Lin, *Phys. Rev. A* **9**, 1954–1964 (1974); J. W. Liu, *J. Chem. Phys.* **59**, 1988–1998 (1973); J. W. Liu and V. H. Smith, Jr., *J. Phys. B* **6**, L275–L279 (1973); J. W. Liu and V. H. Smith, Jr., *Chem. Phys. Lett.* **30**, 63–68 (1975); S. Ishi and K. Tanaka, *J. Phys. Soc. Japan* **37**, 1073–1076 (1974); K. J. Miller and M. Krauss, *J. Chem. Phys.* **47**, 3754–3762 (1967); K. J. Miller, S. R. Mielczarek, and M. Krauss, *J. Chem. Phys* **51**, 26–32 (1969); K. J. Miller, *J. Chem. Phys.* **51**, 5235–5240 (1969); K. J. Miller, *J. Chem. Phys.* **62**, 1759–1768 (1975).
149. P. G. Burke, D. F. Gallaher, and S. Geltman, *J. Phys. B* **2**, 1142–1154 (1969) and references therein; R. J. Damburg, in *Phys. Electron. At. Collisions, Invited Pap. Prog. Rep. Int. Conf., 7th* (1971), pp. 200–215; S. Geltman, in *Phys. Electron. At. Collisions, Invited Pap. Prog. Rep. Int. Conf., 7th* (1971), pp. 216–231; P. G. Burke and J. F. B. Mitchell, *J. Phys. B* **6**, 320–328 (1973); P. G. Burke and J. F. B. Mitchell, *Phys. B* **7**, 665–673 (1974); P. G. Burke, K. A. Berrington, M. LeDourneuf, and Vo Ky Lan, *J. Phys. B* **7**, L531–L534 (1974); M. R. Flannery and K. J. McCann, *J. Phys. B* **7**, L223–L227 (1974); J. Callaway and J. W. Wooten, *Phys. Rev. A* **11**, 1118–1120 (1975).
150. N. Feautrier, H. van Regemorter, and Vo Ky Lan, *J. Phys. B* **4**, 670–680 (1971); Vo Ky Lan, *J. Phys. B* **5**, 242–249 (1972); Vo Ky Lan, N. Feautrier, M. LeDourneuf, and H. van Regemorter, *J. Phys. B* **5**, 1506–1516 (1972).
151. R. J. Drachman and A. Temkin, *Case Stud. At. Coll. Phys.* **2**, 399–481 (1972).
152. J. Callaway, *Comput. Phys. Commun.* **6**, 265–274 (1974).
153. M. D. Lloyd and M. R. C. McDowell, *J. Phys. B* **2**, 1313–1322 (1969); M. R. C. McDowell, L. A. Morgan, and V. P. Myerscough, *J. Phys. B.* **6**, 1435–1451 (1973); M. R. C. McDowell, V. P. Myerscough, and V. Narain, *J. Phys. B* **7**, L195–L197 (1974).
154. H. C. Volkin, *Phys. Rev.* **155**, 1177–1190 (1967).
155. M. H. Mittleman and R. Pu, *Phys. Rev.* **126**, 370–372 (1962).
156. M. H. Mittleman, *Adv. Theor. Phys.* **1**, 283–315 (1965).
157. R. T. Pu, University of California Lawrence Radiation Laboratory Technical Report 10878 (June 1963).
158. H. Feshbach, *Ann. Rev. Nucl. Sci.* **8**, 49–105 (1958); H. Feshbach, Ann. Phys. (N. Y.) **5**, 357–390 (1958); H. Feshbach, *Ann. Phys. (N.Y.)* **19**, 287–313 (1962).
159. D. G. Truhlar, J. K. Rice, S. Trajmar, and D. C. Cartwright, *Chem. Phys. Lett.* **9**, 299–305 (1971) and references therein.
160. M. H. Mittleman and K. M. Watson, *Phys. Rev.* **113**, 198–211 (1959); M. H. Mittleman, *Ann. Phys. (N.Y.)* **14**, 94–106 (1961); C. J. Kleinman, Y. Hahn, and L. Spruch, *Phys. Rev.* **165**, 53–61 (1968).
161. N. F. Lane, in *Fundamental Interactions in Physics* (B. Kursunoglu and A. Perlmutter, ed.), Plenum Press, New York (1973), Vol. 2, pp. 297–298.
162. N. F. Lane and R. J. W. Henry, *Phys. Rev.* 173, 183–180 (1968).
163. S. Hara, *J. Phys. Soc. Japan* **27**, 1262–1267 (1969).
164. V. D. Ob'edkov, in *Int. Conf. Phys. Electron. At. Collisions, 5th, Abstr. Pap.* (1967), pp. 456–57; J. E. Purcell, R. A. Berg, and A. E. S. Green, *Phys. Rev. A* **2**, 107–111 (1970); J. E. Purcell, R. A. Berg, and A. E. S. Green, *Phys. Rev. A* **3**, 508–510 (1970); C. J. Joachain and

M. H. Mittleman, *Phys. Rev. A* **4**, 1492–1499 (1971); C. J. Joachain and R. Vanderpoorten, *J. Phys. B* **6**, 622–641 (1973); F. W. Byron and C. J. Joachain, *Phys. Lett.* **49A**, 306–308 (1974); J. B. Furness and I. E. McCarthy, *J. Phys. B* **6**, 2280–2291 (1973).

165. M. H. Mittleman, *Ann. Phys. (N.Y.)* **14**, 94–106 (1961); R. J. Damburg and S. Geltman, *Phys. Rev. Lett.* **20**, 485–487 (1968); J. K. Rice, D. G. Truhlar, D. C. Cartwright, and S. Trajmar, *Phys. Rev. A* **5**, 762–782 (1972); G. Y. Csanak and H. S. Taylor, *J. Phys. B* **6**, 2055–2071 (1973); C. B. O. Mohr, *J. Phys. B* **8**, 388–392 (1975).

166. W. Huo, *J. Chem. Phys.* **56**, 3468–3481 (1972); W. Huo, *J. Chem. Phys.* **57**, 4800–4813 (1972).

167. L. Sakar, B. C. Saha, and A. K. Ghosh, *Phys. Rev. A* **8**, 236–243 (1973); T. N. Chang, R. T. Poe, and P. Ray, *Phys. Rev. Lett.* **31**, 1097–1099 (1973).

168. V. M. Martin, M. J. Seaton, and J. G. B. Wallace, *Proc. Phys. Soc. (Lond.)* **72**, 701–710 (1958); R. W. LaBahn and J. Callaway, *Phys. Rev.* **147**, 28–40 (1966); J. C. Callaway, R. W. La Bahn, R. T. Pu, and W. M. Duxler, *Phys. Rev.* **168**, 12–21 (1968).

Author Index

The suffixes A and B on the page numbers indicate the volume (Part A or Part B, respectively) in which the citation appears. Boldface page numbers indicate a chapter in one of these two volumes.

Molecule Index

The suffixes A and B on the page numbers indicate the volume (Part A or Part B, respectively) in which the citation appears.

Acetaldehyde, 47A, 25B, 31B

Acetone, 47A, 57A, 61A

Allene, 48A, 51A, 60A, 63A, 23B, 30B, 31B, 186B, 187B, 195B

Allyl radical, 54A, 206B

Aniline, 57A, 61A

Anthracene, 55A

Anthracene$^-$, 207B

9-10-Anthrasemiquinone, 208B

ArH$_2$, 174A

Azulene, 207A

Azulene$^-$, 206B

BF$_3$, 46A, 194B

BH$_2$, 191A

BH$_3$, 46A, 52A

BeH$_2$, 184A

Benzaldehyde, 47A

Benzene, 51A, 55A, 57A, 60A, 63A, 64A, 152A, 216A, 231A, 8B, 23B, 31B, 34B, 36B, 67B, 84B, 92B, 186B, 195B

Benzene$^-$, 207B

Benzonitrile$^-$, 206B, 208B

Benzosemiquinone$^-$, 207B, 208B

Benzyl radical, 52A, 206B

Bicyclobutane, 30B

Biphenyl, 89B

Biphenylene, 22A

Biphenylene$^-$, 206B

Butadiene, 57A, 58A, 60A, 96A, 152A, 216A, 23B, 30B, 31B, 42B, 106B, 118B

Butadiene$^-$, 207B

Butanol, 57A

2-Butyne, 30B, 31B, 186B

CF$_3$, 201B

CF$_4$, 194B

CHF$_3$, 194B

CH$_2$, 52A, 55A, 188A, 191A, 195A

CH$_2$F$_2$, 61A, 194B

CH$_3$, 52A, 55A, 206B

CH$_3$CN, 25B

CH$_3$CO, 199B

CH$_3$F, 47A, 194B, 195B

CH$_3$NC, 25B

CH$_3$NH$_2$, 57A, 25B, 31B, 34B

CH$_3$N$_2$CH$_3$, 26B

CH$_3$OH, 47A, 61A, 25B, 31B

CH$_4$, 188A, 8B, 25B, 31B, 34B, 39B, 186B, 187B, 195B

CO, 44A, 45A

CO$_2$, 46A, 52A, 25B, 31B, 34B, 68B

CO$_2^-$, 199B

C$_2$H$_2$, 45A, 51A, 60A, 63A, 64A, 188A, 8B, 23B, 31B, 34B, 36B, 44B,

C$_2$H$_2$ (cont'd) 185B, 186B, 187B, 194B, 195B

C$_2$H$_2$F$_2$, 195B

C$_2$H$_3$, 199B

C$_2$H$_3$F, 195B

C$_2$H$_4$, 45A, 51A, 57A, 63A, 64A, 152A, 188A, 235A, 8B, 23B, 31B, 34B, 36B, 39B, 43B, 105B, 185B, 186B, 187B, 194B, 195B, 241B

C$_2$H$_5$, 52A, 55A, 206B

C$_2$H$_6$, 45A, 51A, 57A, 58A, 60A, 63A, 64A, 97A, 150A, 8B, 23B, 31B, 34B, 36B, 39B, 186B, 187B, 194B, 195B

C$_3$H$_3$, 8B, 30B

C$_3$H$_4$, 8B, 30B, 31B, 34B, 186B, 187B

C$_3$H$_6$, 47A, 64A, 152A, 8B, 23B, 30B, 31B, 34B, 44B, 88B, 187B

C$_3$H$_7$, 187B

C$_3$H$_8$, 51A, 64A, 8B, 23B, 31B, 34B, 187B

C$_4$H$_6$, 34B

C$_4$H$_8$, 58A, 60A, 63A, 167A, 8B, 23B

C$_4$H$_9$, 187B

C$_4$H$_{10}$, 150A, 8B

C$_5$H$_6$, 187B

C$_5$H$_{10}$, 151A

C$_5$H$_{12}$, 60A, 150A, 8B

C$_6$H$_{14}$, 151A, 8B

Subject Index

The suffixes A and B on the page numbers indicate the volume (Part A or Part B, respectively) in which the citation appears.